Lecture Notes
in Control and Information Sciences   208

Editor: M. Thoma

Springer-Verlag London Ltd.

Hassan Khalil, Joe Chow and Petros Ioannou (Eds)

Proceedings of Workshop on Advances in Control and its Applications

 Springer

Series Advisory Board

Editors

Hassan K. Khalil, Professor
Department of Electrical Engineering, Michigan State University, East Lansing, MI 48824-1226, USA

Joe H. Chow, Professor
Electrical, Computer, and Systems Engineering Department, Rensselaer Polytechnic Institute, Troy, NY 12180-3590, USA

Petros A. Ioannou, Professor
Department of Electrical Engineering-Systems, University of Southern California, Los Angeles, CA 90089-2563, USA

ISBN 978-3-540-19993-9

British Library Cataloguing in Publication Data
Workshop on Advances in Control and Its Applications
   Proceedings of Workshop on Advances in
   Control and Its Applications. - (Lecture
   Notes in Control & Information Sciences; Vol.208)
   I. Title II. Khalil, Hassan K.
   III. Series
   629.8312
ISBN 978-3-540-19993-9   ISBN 978-3-540-39384-9 (eBook)
DOI 10.1007/978-3-540-39384-9

Library of Congress Cataloging-in-Publication Data
A catalog record for this book is available from the Library of Congress

© Springer-Verlag London 1996
Originally published by Springer-Verlag London Limited in 1996

Typesetting: Camera ready by authors

69/3830-543210 Printed on acid-free paper

**TO PROFESSOR PETAR V. KOKOTOVIC
ON THE OCCASION OF HIS 60TH BIRTHDAY**

Preface

On Saturday, May 21, 1994, colleagues and former students of Professor Petar V. Kokotovic gathered at the University of Illinois, Urbana-Champaign, to celebrate his 60th birthday. The "Workshop on Advances in Control and its Applications" was organized by Professor Tamer Basar of the University of Illinois together with the three coeditors of this Proceedings to honor Professor Kokotovic. The efforts of the organizers to keep the workshop a surprise for Petar was apparently successful. The day-long presentations and the evening banquet in the new Grainger library on the Urbana-Champaign campus were enjoyable events. They brought back a lot of fond memories to Petar and his friends. To Petar's students, the day was a special occasion to show their gratitude to the man they affectionately call PK. They respect his expert knowledge in both theory and industrial applications, his dedication to his profession, and his personal mettle.

All the speakers at the workshop were former students of PK. Their papers, arranged in the order of presentation at the workshop, cover a wide range of topics in both control theory and applications. The paper by Liubakka, Rhode, and Winkelman (presented by Rhode) uses sensitivity methods and slow adaptation to design an adaptive controller for automotive speed control. The paper by Datta and Ioannou (presented by Ioannou) gives a modified model reference adaptive controller with improved performance and examines the tradeoff between nominal performance versus robustness to unmodeled dynamics. The paper by Kanellakopoulos gives an account of recent advances in adaptive nonlinear control. The paper by Khalil presents a methodology for nonlinear output feedback control using high-gain observers. It illustrates the use of saturation nonlinearities to overcome the peaking phenomenon. The paper by Jamshidi describes a number of hardware implementation schemes for fuzzy control systems. The paper by Niinomi and Krogh (presented by Krogh) gives algorithms for modeling and analysis of a hybrid system, that is, a system with both discrete and continuous state models. The paper by Lin, Saberi, Sannuti, and Shamash (presented by Sannuti) gives an explicit eigenstructure assignment procedure by low and high gain state feedback that achieves perfect regulation. The paper by Young examines the role of manifolds in system reduction and feedback designs which exploit time-scale separation. The paper by Zaad and Khorasani (presented by Khorasani) utilizes the concept of integral manifolds to design a dynamical composite control strategy for nonminimum phase singularly perturbed systems and applies it to flexible link manipulators. The paper by Taylor applies sampled-data techniques to nonlinear singularly perturbed systems and illustrates the results by a permanent magnet synchronous motor example. The paper by Chow presents a new nonlinear model reduction formulation for large power systems, based on slow coherency and aggregation ideas. The paper by Ahmed-Zaid, Jang, Awed-Badeeb, and Taleb (presented by Ahmed-Zaid) investigates the aggregation of synchronous generators with small and large induction motors in a hybrid multimachine representation of power systems.

Many of the topics covered in these papers started in collaboration with

Petar, and are still his favorite topics. Liubakka's paper builds on the sensitivity points method developed by Petar in the sixties. The papers by Khalil, Lin, Young, Zaad, Taylor, and Chow use singular perturbation theory, the major theme of Petar's work from the late sixties till the early eighties. the influence of his pioneering work in the seventies on the connection between high-gain feedback and singular perturbation is evident in the papers by Khalil, Lin, and Young. The concept of composite control of two-time-scale systems that was developed by Petar in the late seventies is used in the papers by Zaad and Taylor. His work on the use of integral manifolds in the analysis and design of two-time-scale systems in the early eighties is used in the papers by Young, and Zaad. His breakthrough discovery in the early eighties of the relationship between the time scales of slow and fast dynamics on one hand and strong and week, or dense and sparse connections, on the other hand, is the impetus for the aggregation methods of power systems presented in the papers by Chow and Ahmed-Zaid. His work on nonlinear control in the late eighties which revealed the destabilizing effect of the peaking phenomenon associated with high-gain feedback has a direct impact on the paper by Khalil. His work on robustness of adaptive control in the eighties is the starting point of a large body of work that leads into the paper by Datta. Finally, the confluence of nonlinear and adaptive control ideas, the thrust of Petar's work since the late eighties, represents the bulk of development in nonlinear adaptive control theory, surveyed by Kanellakopoulos.

In addition to the speakers, whose names appeared in the preceding paragraphs, the workshop was attended by several colleagues who collaborated with Petar over the years, including Les Fink, Zoran Gajic, Abe Haddad, Harold Javid, Alan Laub, Jim Winkelman, and the University of Illinois friends: Tamer Basar, Ken Jenkins, Jure Medanic, M.A. Pai, Bill Perkins, Peter Sauer, and Mark Spong. Some of them were joined in the evening banquet by their wives, including Petar's wife Anna Bergman. We thank all of them for attending the presentations and speaking at the evening banquet. We also thank Rose Harris (Petar's secretary for many years) and Sue Jenkins for their speeches at the banquet. We are grateful to Tamer Basar for his efforts in organizing this workshop. We also thank his secretary Becky Lonberger for her excellent work before and during the workshop. Finally, we extend our thanks to Imke Mowbray and Christopher Greenwell of Springer-Verlag for their cooperation in publishing this proceedings.

Hassan Khalil
Joe Chow
Petros Ioannou

This article was processed using the LaTeX macro package with LLNCS style

Contents

Petar V. Kokotovic

Petar V. Kokotovic was born on March 18, 1934, in Belgrade, Yugoslavia. He received his Dipl. Ing. and Magistar degrees from the University of Belgrade in 1958 and 1962, respectively, and his Ph.D. degree in 1965 from the Institute of Control Sciences, Moscow, U.S.S.R. His interest in feedback control started while working at a power station in France in 1956, and at a magnetic laboratory in Germany in 1957. In 1958 he joined the Pupin Institute in Belgrade, first as a control engineer and then as head of the process control division. In 1965 he was invited to the United States to present a series of seminars on sensitivity analysis to more than 20 universities and research institutions. Soon after, he joined the ECE Department and Coordinated Science Laboratory, University of Illinois, Champaign-Urbana, where he remained for 25 years. While in Illinois, he supervised 25 Ph.D. students, received the Eminent Faculty Award and held the endowed Grainger Chair. In 1991 Professor Kokotovic joined the University of California, Santa Barbara, as Co-director of the newly formed Center for Control Engineering and Computation.

In the early 1960's Kokotovic developed the sensitivity points method, a precursor to adaptive control that is still in industrial use for automatic tuning of controller parameters. In the late 1960's and 1970's, Kokotovic pioneered singular perturbation techniques for multi-time-scale analysis and composite design of control systems and flight trajectories. This methodology has spurred many extensions and applications, numerous books and hundreds of papers. In applying it to large-scale systems, Kokotovic and coworkers discovered a fundamental relationship between slow and fast dynamic phenomena and strong and weak connections of subsystems. This result became a cornerstone of an efficient program for dynamic equivalencing of power systems.

In the early 1980's, Kokotovic and coworkers identified the main forms of adaptive systems instability caused by unmodeled dynamics. This led to robust redesigns of adaptive controllers which made them more suitable for applications. Kokotovic's most recent research is in nonlinear control, both robust and adaptive, where he and his coworkers have developed a novel recursive design procedure - *backstepping*, and solved several open adaptive stabilization and tracking problems.

As a long-term industrial consultant, Professor Kokotovic contributed to the design of first control computers for car engines at Ford, and to power system stability analysis at General Electric.

Professor Kokotovic is a Fellow of the IEEE, received two outstanding IEEE Transactions Paper Awards (1983 and 1993) and delivered the 1991 IEEE Control Systems Society Bode Prize Lecture. In 1990 he was awarded the Triennial Quazza Medal by the International Federation of Automatic Control, and in 1995 he was awarded the IEEE Control Systems Field Award.

This article was processed using the LaTeX macro package with LLNCS style

PETAR V. KOKOTOVIĆ

SELECTED PUBLICATIONS 1964 - 1994
ARRANGED BY TOPICS

1 SENSITIVITY ANALYSIS

Survey

1.1 P. V. Kokotović and R. Rutman. Sensitivity of automatic control systems (survey). *Automation and Remote Control*, 26:727–748, 1965.

Algebraic Sensitivity

1.2 P. V. Kokotović and D. Siljak. The sensitivity problem in continuous and sampled-data linear systems by generalized Mitrovic method. *IEEE Trans. Part II*, 83:321–324, 1964.

1.3 P. V. Kokotović and D. Siljak. Automatic analog solution of algebraic equations and plotting of root loci by generalized Mitrovic method. *IEEE Trans. Part II*, 83:324–328, 1964.

Sensitivity Points

1.4 P. V. Kokotović. Structural method for simultaneous obtaining of sensitivity functions in linear feedback systems. *Automatika*, 2:96–99, 1964. (Also: The structure of the parameter influence analyzer. *Proc. 2nd IFAC Congress*, 517–518, Basle, 1963).

1.5 P. V. Kokotović and D. Djokovic. A homogeneity property of the solution of linear differential equations with constant coefficients. *Mathematical Review*, 16:44–48, Belgrade 1964.

1.6 P. V. Kokotović. Method of sensitivity points. In R. Tomovic, *Sensitivity Analysis of Dynamic Systems*, 57–78, McGraw-Hill, New York, NY, 1964.

1.7 P. V. Kokotović. Method of sensitivity points in the investigation and optimization of linear control systems. *Automation and Remote Control*, 25:1670–1676, 1964. (Also: *Sensitivity Points*, Soviet Encyclopedia of Modern Engineering, pp. 1512–1517).

1.8 P. V. Kokotović and R. Rutman. Sensitivity matrices and their modeling. *Automation and Remote Control*, 27:1067–1079, 1966. (Also: On the determination of sensitivity functions with respect to the change of system order. *Proc. IFAC Symp. Sensitivity Analysis*, 131–142, Pergamon Press, London, 1964).

Parameter Optimization and Adaptation

1.9 S. Bingulac and P. V. Kokotović. Automatic optimization of linear feedback control systems on an analog computer. *Proc. Int. Assoc. Analog Computation*, 8:12–17, Brussels, 1965.

1.10 P. V. Kokotović, J. Medanić, M. Vušković, and S. Bingulac. Sensitivity method in the experimental design of adaptive control systems. *Proc. 3rd IFAC Congress*, 45B.1–45B.12, London, 1966.

1.11 P. V. Kokotović and J. Heller. Direct and adjoint sensitivitiy equations for parameter optimization. *IEEE Trans. Automatic Control*, AC-12:609–610, 1967.

Optimally Sensitive Systems

1.12 P. V. Kokotović, J. B. Cruz, Jr., J. Heller, and P. Sannuti. Synthesis of optimally sensitive systems. *Proc. IEEE*, 56:1318–1324, 1968.

1.13 P. V. Kokotović and A. Haddad. Estimators for optimally sensitive systems. *Proc. IEEE*, 56:1399–1400, 1968.

1.14 P. V. Kokotović and D. Salmon. Design of linear feedback controllers for nonlinear plants. *IEEE Trans. Automatic Control*, AC-14:289–292, 1969.

1.15 P. V. Kokotović, J. Heller, and P. Sannuti. Sensitivity comparison of optimal controls. *Int. J. Control*, 9:111–115, 1969.

1.16 P. V. Kokotović and J. B. Cruz, Jr. An approximation theorem for linear optimal regulators. *J. Math. Analysis Appl.*, 27:249–252, 1969.

1.17 M. Jamshidi, G. D'Ans, and P. V. Kokotović. Application of a parameter-imbedded Riccati equation. *IEEE Trans. Automatic Control*, AC-15:682–683, 1970.

1.18 P. V. Kokotović, A. H. Haddad, and J. B. Cruz, Jr. Design of control systems with random parameters. *Int. J. Control*, 13:981–992, 1971.

1.19 P. V. Kokotović and W. R. Perkins. Deterministic parameter estimation for near-optimum feedback control. *Automatica*, 7:439–444, 1971.

2 SINGULAR PERTURBATIONS AND OPTIMAL CONTROL

Books

B.1 P. V. Kokotović, H. K. Khalil, and J. O'Reilly. *Singular Perturbation Methods in Control: Analysis and Design*. Academic Press, New York, NY, 1986. (Review by M. Mangel, SIAM Review, p. 674, 1988).

B.2 P. V. Kokotović and H. K. Khalil, editors. *Singular Perturbations in Systems and Control*. IEEE Press, 1986.

B.3 P. V. Kokotović, A. Bensoussan, and G. Blankenship, editors. *Singular perturbations and asymptotic analysis in control systems. Lecture Notes in Control and Information Sciences*, 90, Springer-Verlag, 1986.

B.4 P. V. Kokotović and W. Perkins, editors. *Singular perturbations: Order reduction in control systems design*. ASME, 1972.

Surveys

2.1 P. V. Kokotović. Applications of singular perturbation techniques to control problems. *SIAM Review*, 26:501–550, 1984. (In French: Perturbations singulieres en controle optimal, in *Outils et Modeles Mathematiques Pour l'Automatique*, I. D. Landau, Ed., C.N.R.S., Paris, 135–175, 1983).

2.2 P. V. Kokotović, R. E. O'Malley, Jr., and P. Sannuti. Singular perturbations and order reduction in control theory – an overview. *Automatica*, 12:123–132, 1976.

2.3 V. R. Saksena, J. O'Reilly, and P. V. Kokotović. Singular perturbations and time-scale methods in control theory: Survey 1976–1983. *Automatica*, 20:273–293, 1984.

Linear Regulator Design

2.4 P. Sannuti and P. V. Kokotović. Near-optimum design of linear systems by a singular perturbation method. *IEEE Trans. Automatic Control*, AC-14:15–22, 1969.

2.5 A. H. Haddad and P. V. Kokotović. Note on singular perturbation of linear state regulators. *IEEE Trans. Automatic Control*, AC-16:279–281, 1971.

2.6 P. V. Kokotović and R. A. Yackel. Singular perturbation of linear regulators: basic theorems. *IEEE Trans. Automatic Control*, AC-17:29–37, 1972.

2.7 R. A. Yackel and P. V. Kokotović. A boundary layer method for the matrix Riccati equation. *IEEE Trans. Automatic Control*, AC-18:17–24, 1973.

2.8 R. R. Wilde and P. V. Kokotović. Stability of singularly perturbed systems and networks with parasitics. *IEEE Trans. Automatic Control*, AC-17:245–246, 1972.

2.9 P. V. Kokotović. A Riccati equation for block-diagonalization of ill-conditioned systems. *IEEE Trans. Automatic Control*, AC-20:812–814, 1975.

2.10 J. H. Chow and P. V. Kokotović. A decomposition of near-optimum regulators for systems with slow and fast modes. *IEEE Trans. Automatic Control*, AC-21:701–705, 1976.

2.11 A. H. Haddad and P. V. Kokotović. Stochastic control of linear singularly perturbed systems. *IEEE Trans. Automatic Control*, AC-22:815–821, 1977.

2.12 K. D. Young, P. V. Kokotović, and V. I. Utkin. A singular perturbation analysis of high gain feedback systems. *IEEE Trans. Automatic Control*, AC-22:931–938, 1977.

2.13 J. H. Chow, J. J. Allemong, and P. V. Kokotović. Singular perturbation analysis of systems with sustained high frequency oscillations. *Automatica*, 14:271–279, 1978.

2.14 J. J. Allemong and P. V. Kokotović. Eigensensitivities in reduced order modeling. *IEEE Trans. Automatic Control*, AC-25:821–822, 1980.

2.15 V. R. Saksena and P. V. Kokotović. Singular perturbation of the Popov-Kalman-Yakubovich lemma. *Systems and Control Letters*, 1:65–68, 1981.

2.16 K. K. D. Young and P. V. Kokotović. Analysis of feedback-loop interactions with actuator and sensor parasitics. *Automatica*, 18:577–582, 1982.

Optimal Control

2.17 P. V. Kokotović and P. Sannuti. Singular perturbation method for reducing the model order in optimal control design. *IEEE Trans. Automatic Control*, AC-13:377–384, 1968.

2.18 P. Sannuti and P. V. Kokotović. Singular perturbation method for near-optimum design of high-order non-linear systems. *Automatica*, 5:773–779, 1969.

2.19 R. R. Wilde and P. V. Kokotović. A dichotomy in linear control theory. *IEEE Trans. Automatic Control*, AC-17:382–383, 1972.

2.20 R. R. Wilde and P. V. Kokotović. Optimal open- and closed-loop control of singularly perturbed linear systems. *IEEE Trans. Automatic Control*, AC-18:616–625, 1973.

2.21 P. V. Kokotović and A. H. Haddad. Controllability and time-optimal control of systems with slow and fast modes. *IEEE Trans. Automatic Control*, AC-20:111–113, 1975.

2.22 P. V. Kokotović and A. H. Haddad. Singular perturbations of a class of time-optimal controls. *IEEE Trans. Automatic Control*, AC-20:163–164, 1975.

2.23 S. H. Javid and P. V. Kokotović. A decomposition of time scales for iterative computation of time-optimal controls. *J. Optim. Theory Appl.*, 21:459–468, 1977.

2.24 B. D. O. Anderson and P. V. Kokotović. Optimal control problems over large time intervals. *Automatica*, 23:355–363, 1987.

Applications of Optimal Control

2.25 G. D'Ans, P. V. Kokotović, and D. Gottlieb. A nonlinear regulator problem for a model of biological waste treatment. *IEEE Trans. Automatic Control*, AC-16:341–347, 1971.

2.26 G. D'Ans, P. V. Kokotović, and D. Gottlieb. Time-optimal control for a model of bacterial growth. *J. Optim. Theory and Appl.*, 7:61–69, 1971.

2.27 G. D'Ans, D. Gottlieb, and P. V. Kokotović. Optimal control of bacterial growth. *Automatica*, 8:729–736, 1972.

2.28 P. V. Kokotović and G. Singh. Minimum-energy control of a traction motor. *IEEE Trans. Automatic Control*, AC-17:92–95, 1972.

3   LARGE SCALE SYSTEMS

Books

B.5 J. H. Chow, editor, B. Avramovic, P. V. Kokotović, G. Peponides, and J. Winkelman. *Time-Scale Modeling of Dynamic Networks. Lecture Notes in Control and Information Sciences*, 46, Springer-Verlag, 1982.

B.6 P. V. Kokotović, J. B. Cruz, Jr., J. V. Medanić and W. R. Perkins, editors. Systems engineering for power: Organizational forms for large scale systems-Volume II. US Department of Energy, Conf-79090-P3, 1979.

Tutorial

3.1 P. V. Kokotović. Subsystems, time scales and multimodeling. *Automatica*, 17:789–795, 1981.

Multimodeling

3.2 H. K. Khalil and P. V. Kokotović. Control strategies for decision makers using different models of the same system. *IEEE Trans. Automatic Control*, AC-23:289–298, 1978.

3.3 H. K. Khalil and P. V. Kokotović. Control of linear systems with multiparameter perturbations. *Automatica*, 15:197–207, 1979.

3.4 H. K. Khalil and P. V. Kokotović. D-stability and multi-parameter singular perturbation. *SIAM J. Control and Optim.*, 17:56–65, 1979.

3.5 H. K. Khalil and P. V. Kokotović. Feedback and well-posedness of singularly perturbed Nash games. *IEEE Trans. Automatic Control*, AC-24:699–708, 1979.

Coherency and Aggregation

3.6 P. V. Kokotović, B. Avramovic, J. H. Chow, and J. R. Winkelman. Coherency based decomposition and aggregation. *Automatica*, 18:47–56, 1982.

3.7 R. G. Phillips and P. V. Kokotović. A singular perturbation approach to modeling and control of Markov chains. *IEEE Trans. Automatic Control*, AC-26:1087–1094, 1981.

3.8 F. Delebecque, J. P. Quadrat, and P. V. Kokotović. A unified view of aggregation and coherency in networks and Markov chains. *Int. J. Control*, 40:939–952, 1984.

3.9 J. H. Chow and P. V. Kokotović. Time scale modeling of sparse dynamic networks. *IEEE Trans. Automatic Control*, AC-30:714–722, 1985.

Power Systems and Networks

3.10 B. Avramovic, P. V. Kokotović, J. R. Winkelman, and J. H. Chow. Area decomposition for electromechanical models of power systems. *Automatica*, 16:637–648, 1980.

3.11 G. Peponides, P. V. Kokotović, and J. H. Chow. Singular perturbations and time scales in nonlinear models of power systems. *IEEE Trans. Circuits and Systems*, CAS-29:758–767, 1982.

3.12 P. V. Kokotović, J. J. Allemong, J. R. Winkelman, and J. H. Chow. Singular perturbation and iterative separation of time scales. *Automatica*, 16:23–33, 1980.

3.13 J. R. Winkelman, J. H. Chow, J. J. Allemong and P. V. Kokotović. Multi-time-scale analysis of a power system. *Automatica*, 16:35–43, 1980.

3.14 J. R. Winkelman, J. H. Chow, B. C. Bowler, B. Avramovic, and P. V. Kokotović. An analysis of interarea dynamics of multi-machine systems. *IEEE Trans. Power App. Systems*, PAS-100:754–763, 1981.

3.15 B. H. Krogh and P. V. Kokotović. Feedback control of overloaded networks. *IEEE Trans. Automatic Control*, AC-29:704–711, 1984.

Weakly Coupled Systems

3.16 P. V. Kokotović. Feedback design of large linear systems, in *Feedback Systems*, J. B. Cruz, Jr., editor, 99–137. McGraw-Hill, New York, 1972.

3.17 P. V. Kokotović, W. R. Perkins, J. B. Cruz, Jr., and G. D'Ans. ϵ-coupling method for near-optimum design of large-scale linear systems. *Proc. IEE*, 116:889–892, London, 1969.

3.18 P. V. Kokotović and G. Singh. Optimization of coupled non-linear systems. *Int. J. Control*, 14:51–64, 1971.

4   ADAPTIVE CONTROL

Books

B.7 P. A. Ioannou and P. V. Kokotović. *Adaptive Systems with Reduced Models, Lecture Notes in Control and Information Sciences, 47.* Springer-Verlag, New York, 1983.

B.8 B. D. O. Anderson, R. R. Bitmead, C. R. Johnson, Jr., P. V. Kokotović, R. L. Kosut, I. Mareels, L. Praly, and B. D. Riedle. *Stability of Adaptive Systems: Passivity and Averaging Analysis.* MIT Press, Cambridge, MA, 1986. (Russian translation published by Mir, 1989).

Survey

P. V. Kokotović. Recent trends in feedback design: an overview. *Automatica*, 21:225–236, 1985.

Robust Design

4.1 P. A. Ioannou and P. V. Kokotović. An asymptotic error analysis of identifiers and adaptive observers in the presence of parasitics. *IEEE Trans. Automatic Control*, AC-27:921–927, 1982. (Best IEEE Trans. Paper Award).

4.2 P. A. Ioannou and P. V. Kokotović. Robust redesign of adaptive control. *IEEE Trans. Automatic Control*, AC-29:202–211, 1984.

4.3 P. A. Ioannou and P. V. Kokotović. Instability analysis and improvement of robustness of adaptive control. *Automatica*, 20:583–594, 1984.

4.4 P. A. Ioannou and P. V. Kokotović. Decentralized adaptive control of interconnected systems with reduced-order models. *Automatica*, 21:401–412, 1985.

Stability-Instability Criteria

4.5 B. D. Riedle and P. V. Kokotović. A stability-instability boundary for disturbance free slow adaptation with unmodeled dynamics. *IEEE Trans. Automatic Control*, AC-30:1027–1030, 1985. (also 23rd CDC, 1984).

4.6 B. D. Riedle and P. V. Kokotović. Stability analysis of an adaptive system with unmodeled dynamics. *Intl. J. Control*, 41:389–402, 1985.

4.7 P. V. Kokotović, B. D. Riedle, and L. Praly. On a stability criterion for continuous slow adaptation. *Systems and Control Letters*, 6:7–14, 1985.

4.8 B. Riedle, L. Praly, and P. Kokotović. Examination of the SPR condition in output error parameter estimation. *Automatica*, 22:495–498, 1986.

Integral Manifold Method

4.9 B. D. Riedle and P. V. Kokotović. Integral manifolds of slow adaptation. *IEEE Trans. Automatic Control*, AC-31:316–323, 1986. (Also: Integral manifold approach to slow adaptation, 24th IEEE CDC, Ft. Lauderdale, FL, 1985.)

4.10 B. D. Riedle and P. V. Kokotović. Stability of slow adaptation for non-SPR systems with disturbances. *IEEE Trans. Automatic Control*, AC-32:451–455, 1987.

4.11 S. T. Hung, P. V. Kokotović, and J. R. Winkelman. Self tuning of conflicting oscillatory modes: a case study. *IEEE Trans. Automatic Control*, AC-34:250–253, 1989.

Bifurcations and Drift Instabilities

4.12 B. Riedle, B. Cyr, and P. Kokotović. Disturbance instabilities in an adaptive system. *IEEE Trans. Automatic Control*, AC-29:822–824, 1984. (Also: Bifurcating equilibria of adaptive control systems. *Proc. 1984 American Control Conference*, 238–240, 1984).

4.13 D. A. Schoenwald and P. V. Kokotović. Global and local stability properties of a non-SPR output error estimator. *Int. J. Control*, 50:937–953, 1989.

4.14 D. A. Schoenwald and P. V. Kokotović. Boundedness conjecture for an output error adaptive algorithm. *Int. J. Adaptive Control and Signal Processing*, 4:27–47, 1990.

4.15 R. H. Middleton and P. V. Kokotović. Boundedness properties of simple, indirect adaptive-control systems. *IEEE Trans. Automatic Control*, AC-37:1989–1994, 1992.

Applications of Adaptive Control

4.16 D. E. Henderson, P. V. Kokotović, J. L. Schiano, and D. S. Rhode. Adaptive control of an arc welding process. *IEEE Control Systems Magazine*, 13:49-53, 1993.

4.17 M. K. Liubakka, D. S. Rhode, J. R. Winkelman and P. V. Kokotović, Adaptive automotive speed control. *IEEE Trans. Automatic Control*, 38:1011-1020, 1993.

5 NONLINEAR CONTROL

Survey

5.1 P. V. Kokotović. Recent trends in feedback design: an overview. *Automatica*, 21:225–236, 1985.

Nonlinear Composite Control

5.2 J. H. Chow and P. V. Kokotović. Near-optimal feedback stabilization design of a class of nonlinear singularly perturbed systems. *SIAM J. Control and Optim.*, 16:756–770, 1978.

5.3 J. H. Chow and P. V. Kokotović. Two-time-scale feedback design of a class of nonlinear systems. *IEEE Trans. Automatic Control*, AC-23:438–443, 1978.

5.4 J. H. Chow and P. V. Kokotović. A two-stage Lyapunov-Bellman feedback design of a class of nonlinear systems. *IEEE Trans. Automatic Control*, AC-26:656–663, 1981.

Slow Manifold Method

5.5 G. Peponides and P. V. Kokotović. Weak connections, time scales, and aggregation of nonlinear systems. *IEEE Trans. Circuits and Systems*, CAS-30:416–421, 1983.

5.6 K. Khorasani and P. V. Kokotović. Feedback linearization of a flexible manipulator near its rigid body manifold. *Systems and Control Letters*, 6:187–192, 1985.

5.7 K. Khorasani and P. V. Kokotović. A corrective feedback design for nonlinear systems with fast actuators. *IEEE Trans. Automatic Control*, AC-31:67–69, 1986.

5.8 M. W. Spong, K. Khorasani, and P. V. Kokotović. An integral manifold approach to the feedback control of flexible joint robots. *IEEE J. of Robotics and Automation*,RA-3:291–300, 1987.

5.9 R. Marino and P. V. Kokotović. A geometric approach to nonlinear singularly perturbed control systems. *Automatica*, 24:31–41, 1988.

5.10 P. W. Sauer, S. Ahmed-Zaid, and P. V. Kokotović. An integral manifold approach to reduced order dynamic modeling of synchronous machines. *IEEE Trans. Power Systems*, PS-3:17–23, 1988.

5.11 P. V. Kokotović and P. W. Sauer. Integral manifold as a tool for reduced-order modeling of nonlinear systems: A synchronous machine case study. *IEEE Trans. Circuits and Systems*, 36:403–410, 1989.

5.12 H. K. Khalil and P. V. Kokotović. On stability properties of nonlinear systems with slowly-varying inputs. *IEEE Trans. Automatic Control*, 36:229, 1991.

Feedback Linearization and Stabilization

5.13 P. V. Kokotović and R. Marino. On vanishing stability regions in nonlinear systems with high-gain feedback. *IEEE Trans. Automatic Control*, AC-31:967–970, 1986.

5.14 J. W. Grizzle and P. V. Kokotović. Feedback linearization of sampled-data systems. *IEEE Trans. Automatic Control*, AC-33:857–859, 1988.

5.15 J. Hauser, S. Sastry, and P. V. Kokotović. Zero dynamics of regularly perturbed systems may be singularly perturbed. *Systems and Control Letters*, 13:299–314, 1989.

5.16 H. J. Sussman and P. V. Kokotović. The peaking phenomenon and the global stabilization of nonlinear systems. *IEEE Trans. Automatic Control*, AC-36:424-439, 1991.

5.17 J. Hauser, S. S. Sastry, and P. V. Kokotović. Nonlinear control via approximate input-output linearization: the ball and beam example. *IEEE Trans. Automatic Control*, AC-37:392-398, 1992.

5.18 A. Isidori, S. S. Sastry, P. V. Kokotović and C. I. Byrnes. Singularly perturbed zero dynamics of nonlinear systems. *IEEE Trans. Automatic Control*, AC-37:1625-1631, 1992.

Backstepping Design

5.19 P. V. Kokotović and H. Sussmann. A positive real condition for global stabilization of nonlinear systems. *Systems and Control Letters*, 13:125-133, 1989.

5.20 A. Saberi, P. V. Kokotović and H. J. Sussmann. Global stabilization of partially linear composite systems. *SIAM J. Control and Optim.*, 28:1491–1503, 1990.

5.21 I. Kanellakopoulos, P. V. Kokotović and A. S. Morse. A toolkit for nonliner feedback design. *Systems and Control Letters*, 18:83-92, 1992.

5.22 R. A. Freeman and P. V. Kokotović. Design and comparison of globally stabilizing controllers for an uncertain nonlinear system. *Systems, Models and Feedback: Theory and Applications*, A. Isidori and T. J. Tarn, (Eds.), 249-264, Birkhauser, June 1992.

5.23 R. A. Freeman and P. V. Kokotović. Design of "softer" robust nonlinear control laws. *Automatica*, 29:1425–1437, 1993.

6   ADAPTIVE NONLINEAR CONTROL

Book

B9 P. V. Kokotović, (ed). Foundations of adaptive control. *Lecture Notes in Control and Information Sciences*, Vol. 160. Srpinger-Verlag, Berlin, 1991.

Tutorial

6.1 P. V. Kokotović, (ed). The joy of feedback: nonlinear and adaptive. *Control Systems Magazine*, 12:7-17, 1992. (Revised text of the 1991 IEEE Bode Prize Lecture).

Design with Matching Conditions

6.2 D. G. Taylor, P. V. Kokotović, R. Marino, and I. Kanellakopoulos. Adaptive regulation of nonlinear systems with unmodeled dynamics. *IEEE Trans. Automatic Control*, AC-34:405-412, 1989.

6.3 S. S. Sastry and P. V. Kokotović. Feedback linearization in the presence of uncertainties. *Int. J. Adaptive Control and Signal Processing*, 2:327-346, 1988.

6.4 I. Kanellakopoulos, P. V. Kokotović, and R. Marino. An extended direct scheme for robust adaptive nonlinear control. *Automatica*, 27:247-255, 1991.

6.5 R. Marino, P. Tomei, I. Kanellakopoulos, and P. V. Kokotović. Adaptive tracking for a class of feedback linearizable systems. *IEEE Trans. Automatic Control*, 1314-1319, 1994.

Adaptive Backstepping Design

6.6 I. Kanellakopoulos, P. V. Kokotović, and A. S. Morse. Systematic design of adaptive controllers for feedback linearizable systems. *IEEE Trans. Automatic Control*, AC-36:1241-1253, 1991. (Outstanding IEEE Trans. Paper Award).

6.7 I. Kanellakopoulos, P. V. Kokotović, and A. S. Morse. Adaptve nonlinear control with incomplete state information. *Int. J. Adaptive Control and Signal Processing*, 6:367-394, 1992.

6.8 M. Krstic, I. Kanellakopoulos, and P. V. Kokotović. Adaptive nonlinear control without overparametrization. *Systems and Control Letters*, 19:177-185, 1992.

6.9 M. Krstic, P. V. Kokotović, and I. Kanellakopoulos. Transient-performance improvement with a new class of adaptive controllers. *Systems and Control Letters*, 21:451-461, 1993.

Estimation-Based Design

6.10 A. Teel, R. Kadiyala, P. V. Kokotović, and S. Sastry. Indirect techniques for adaptive input-output linearization of non-linear systems. *Int. J. Control*, 53:193-222, 1991.

6.11 I. Kanellakopoulos, P. V. Kokotović, and A. S. Morse. Adaptive output-feedback control of systems with output nonlinearities. *IEEE Trans. Automatic Control*, AC-37:1666-1682, 1992.

6.12 M. Krstic, I. Kanellakopoulos, and P. V. Kokotović. Nonlinear design of adaptive controllers for linear systems. *IEEE Trans. Automatic Control*, 39:738-752, 1994.

6.13 M. Krstic and P. V. Kokotović. Observer-based schemes for adaptive nonlinear state-feedback control. *Int. J. Control*, 59:1373-1381, 1994.

Compensation of Dead-Zone and Backlash

6.14 G. Tao and P. V. Kokotović. Adaptive control of systems with backlash. *Automatica*, 29:323-335, 1993.

6.15 G. Tao and P. V. Kokotović. Adaptive control of plants with unknown dead-zones. *IEEE Trans. Automatic Control*, 39:59-68, 1994.

This article was processed using the LaTeX macro package with LLNCS style

List of Former Ph.D. Students ‡

1. P. Sannuti (1968), Rutgers University.
2. C. Hadlock (1970), A. DeLittle.
3. M. Jamshidi (1971), University of New Mexico.
4. G. Singh (1971), India.
5. R. Yackel (1971), Martin Marietta.
6. G. D'Ans (1972), Belgium.
7. R. Wilde (1972), U.S. Air Force.
8. J. Chow (1977), Rensselaer.
9. H. Javid (1977), Acrowood.
10. K.D. Young (1977), Lawrence Livermore.
11. J. Allemong (1978, with M. E. Van Valkenburg), American Electric.
12. H. Khalil (1978), Michigan State University.
13. B. Avramovic (1980), Energy and Control Consultants
14. R. Philips (1980), General Electric.
15. P. Ioannou (1982), University of Southern California.
16. B. Krogh (1982), Carnegie-Mellon University.
17. G. Peponides (1982), Pacific Communications Sciences
18. K. Khorasani (1985 with M. Pai), Concordia University.
19. B. Riedle (1986), Ford.
20. S. Hung (1988), Ford.
21. C. Tseng (1988), University of Santa Clara.
22. D. Rhode (1990), Ford.
23. I. Kanellakopoulos (1991), UCLA.
24. J. Schiano (1991), Penn State.
25. D. Recker (1992), Ford.
26. M. Krstic (1994), University of Maryland.
27. P. C. Yeh (1994), Taiwan.

This article was processed using the LaTeX macro package with LLNCS style

Adaptive Automotive Speed Control

M. K. Liubakka, D.S. Rhode and J. R. Winkelman

Vehicle Controls Engineering
Electronics Division, Ford Motor Company
Dearborn, MI 48121

Abstract. Modern automotive speed control systems are designed to provide smooth throttle movement, zero steady state speed error, good speed tracking over varying road slopes, and robustness to system variations and operating conditions. Additionally, there is a need to minimize the number of controller calibrations for different vehicle applications. All of the above objectives cannot be simultaneously met by conventional fixed gain controllers which need different calibrations for different vehicle lines. With such requirements, an adaptive controller offers benefits over a conventional controller provided its complexity does not significantly exceed that of a conventional controller.

To limit the controller complexity, the adaptive design in this study is based on sensitivity analysis and slow adaptation using gradient methods. This design method allows the use of our a priori knowledge about the plant model in order to determine a stability region for a reduced order adaptive controller, in this case a simple PI controller. The adaptive algorithm, driven by the vehicle response to road load torque disturbances, tunes a PI controller to continuously minimize a single performance based cost functional for each different vehicle over varying road terrain. This results in performance not possible with a fixed gain controller. The adaptive controller has been tested on a number of vehicles with excellent results, some of which are presented here.

1 Introduction

One of the main goals for an automobile speed control (cruise control) system from a manufacturer's point of view is to provide acceptable performance over a wide range of vehicle lines and operating conditions. Ideally, this is to be achieved with one control module, without recalibration for different vehicle lines. For commonly used proportional feedback controllers, no single controller gain is adequate for all vehicles and all operating conditions. Such simple controllers no longer have the level of performance expected by customers.

The complexity of speed control algorithms has increased through the years to meet the more stringent customer performance expectations. The earliest systems simply held the throttle in a fixed position [1]. In the late 1950's speed control systems with feedback appeared [2]. These used proportional feedback of the speed error, with the gain typically chosen so that 6 to 10 mph of error would pull full throttle. The next enhancement was proportional control with an integral preset or bias input [3]. This helped to minimize steady state error as well as speed droop when the system was initialized. Only with the recent availability of inexpensive microprocessors have more sophisticated control strategies been implemented. PID controllers, Optimal LQ regulators, Kalman Filters, Fuzzy Logic, and Adaptive algorithms have all been tried [4]-[10].

Still, it is hard to beat the performance of a well tuned PI controller for speed control. The problem is how to keep the PI controller well tuned, since both the system and operating conditions vary greatly. The optimal speed control gains are dependent on:

- Vehicle Parameters (engine, transmissions, weight, load, etc.)
- Vehicle Speed
- Torque Disturbances (road slope, wind, etc.)

Gain scheduling over vehicle speed is only a partial solution, because the vehicle parameters are not constant and torque disturbances are not measurable. Much testing and calibration work has been done to tune PI gains for a controller which works across more than one car line, but as new vehicles are added, retuning is often necessary. For example, with a PI speed control, low power cars will generally need higher gains than high power cars. This suggests a need for adaptation to vehicle parameters. For an individual car, the best performance on flat roads is achieved with low integral gain, while rolling hill terrain requires high integral gain. This suggests a need for adaptation to disturbances.

Our goal was to build an adaptive controller that outperforms its fixed gain competitors, yet retains their simplicity and robustness. This goal has been achieved with a slow adaptation design using a sensitivity based gradient algorithm. This algorithm, driven by the vehicle response to unmeasured load torque disturbances, adjusts the proportional and integral gains, K_p and K_i, to minimize a quadratic cost functional. Through simulations and experiments a single cost functional was found which, when minimized, resulted in satisfactory speed control performance for each vehicle and all operating conditions. Adaptive minimization of this cost functional improved the performance of every tested vehicle over varying road terrain (flat, rolling hills, steep grades etc...). This is not possible with a fixed gain controller.

Our optimization-type adaptive design has several advantages. The slow adaption of only two adjustable parameters is simple and makes use of knowledge already acquired about the vehicle. The main requirement for slow adaptation is the existence of a fixed gain controller which provides the desired performance when properly tuned. Since the PI control meets this requirement and is well understood, the design and implementation of an adaptive control with good robustness properties become fairly easy tasks. With only two adjustable parameters, all but perfectly flat roads provide sufficient excitation for parameter convergence and local robustness. These properties are strengthened by the sensitivity filter design and speed-dependent initialization.

The main idea of the adaptive algorithm employed here comes from a sensitivity approach proposed in the 1960's [11], but soon abandoned because of its instabilities in fast adaptation. Under the ideal model-matching conditions, such instabilities do not occur in more complex schemes developed in the 1970-1980's. However, the ideal model-matching requires twice as many adjustable parameters as the dynamic order of the plant. If the design is based on a reduced order model, the resulting unmodeled dynamics may cause instability and robust redesigns are required. This difficulty motivated our renewed interest in *a sensitivity-based approach in which both the controller structure and the adjustable parameters are free to be chosen independently of the plant order.* Such an approach would be suitable for adaptive tuning of simple controllers to higher order plants, if a verifiable condition for its stability could be found. For this purpose we employ the "pseudogradient condition," recently derived by Rhode [12]-[13] using the averaging results of [14]-[16]. A brief outline of this derivation is given in the Appendix. From the known bounds on vehicle parameters and torque disturbances, we evaluate, in the frequency domain, a "phase-uncertainty envelope." Then we design a sensitivity filter to guarantee that the pseudogradient condition is satisfied at all points encompassed by the envelope.

2 Design Objectives

The automobile speed control is simpler than many other automotive control problems - engine efficiency and emissions, active suspension, four-wheel steering, - to name only a few. It is, therefore, required that the solution to the speed control problem be simple. However, this simple solution must also satisfy a set of challenging performance and functional requirements. A typical list of these is as follows:

Performance requirements

- Speed tracking ability for low frequency commands.
- Torque disturbance attenuation for low frequencies, with zero steady-state error for large grades (within the capabilities of the vehicle powertrain).
- Smooth and minimal throttle movement.
- Robustness of the above properties over a wide range of operating conditions.

Functional requirements

- Universality: the same control module must meet the performance requirements for different vehicle lines without recalibration.
- Simplicity: design concepts and diagrams should be understandable to automotive engineers with basic control background.

The dynamics which are relevent for this design problem are organized in the form of a generic vehicle model in Fig. 1. To represent vehicles of different lines, (Escort, Mustang, Taurus, etc.) individual blocks will contain different parameters and, possibly, slightly different structures (e.g. manual or automatic transmission.) Although the first order vehicle dynamics are dominant, there are other blocks with significant higher frequency dynamics. Nonlinearities like dead-zone, saturation, multiple gear-ratios and backlash are also present. In

4

conjunction with large variations of incremental gain (e.g. low or high power engine), these nonlinearities may cause limit cycles which should be suppressed, especially if noticeable to the driver. There are two inputs: the speed set-point y_{set} and the road load disturbance torque T_{dis}. While the response to set-point changes should be acceptable, the most important performance requirement is the accomodation of the torque disturbance. The steady-state error caused by a constant torque disturbance (e.g. constant slope road) must be zero. For other types of roads (e.g. rolling hills) a low frequency specification of disturbance accommodation is defined.

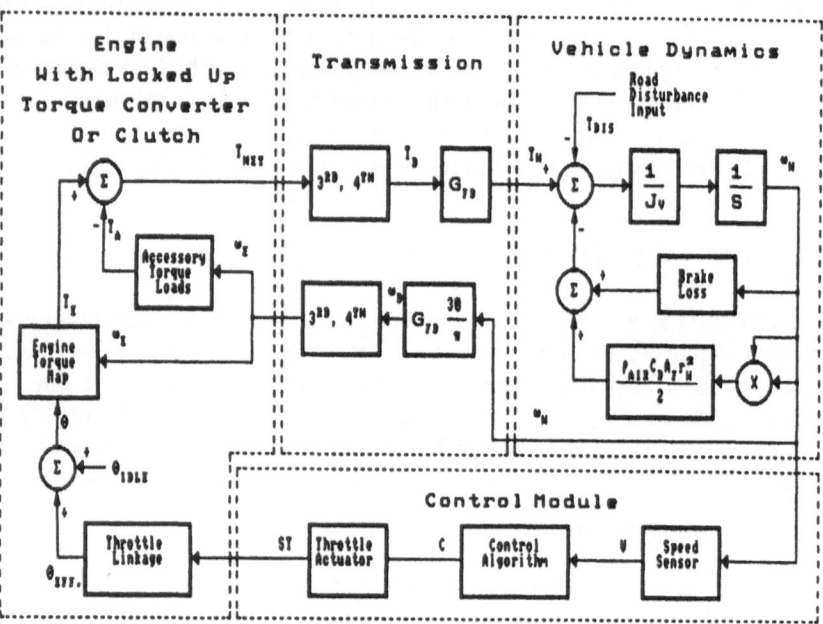

Fig. 1. Vehicle model for speed control. Different vehicle lines are represented by different structures and parameters of individual blocks.

This illustrative review of design objectives, which is by no means complete, suffices to motivate the speed control design presented in the Sections 3 and 4. Typical results with test vehicles presented in Section 5 show how the above requirements have been satisfied.

3 The Design Concept

The choice of a design concept for mass production differs substantially from an academic study of competing theories, in this case numerous ways to design an adaptive scheme. With physical constraints, necessary safety nets and diagnostics, the implementation of an analytically conceived algorithm may appear similar to an "expert," "fuzzy," or "intelligent" system. Innovative terminologies respond to personal tastes and market pressures, but the origins of most successful control designs are often traced to some fundamental concepts. The most enduring among these are PI control and gradient-type optimization algorithms. Recent theoretical results on conditions for stability of such algorithms reduce the necessary ad-hoc fixes required to assure reliable performance. They are an excellent starting point for many practical adaptive designs and can be expanded by additional nonlinear compensators and least-square modifications of the algorithm.

For our design, a PI controller is suggested by the zero steady-state error requirement, as well as by earlier speed control designs. In the final design, a simple signed quadratic feedback was added to the PI controller. This quadratic term responds aggressively to large speed errors. Near zero error, this feedback term is insignificant, and did not affect the design of the adaptive controller. The decision that the PI controller gains, K_p and K_i, be adaptively tuned was reached after it was confirmed that a controller with gain scheduling based on speed cannot satisfy performance requirements for all vehicle lines under all road load conditions. Adaptive control is chosen to eliminate the need for costly recalibration and satisfy the universal functionality requirement.

The remaining choice was that of a parameter adaptation algorithm. Based on the data about the vehicle lines and the fact that the torque disturbance is not available for measurement, the choice of an optimization-based algorithm was made. A reference-model approach was not followed, because no single model could specify the desired performance for the wide range of dynamics and disturbances. On the other hand, an optimization based algorithm requires a single cost functional whose minimization leads to acceptable performance for all vehicles and operating conditions. To construct the cost functional, a priori knowledge of the fixed gain controller was used. In the past, the fixed gain controller was tuned to provide as small a speed error as possible without excessive throttle motion. From this experience, it was apparent that both the tracking error and throttle motion must be minimized to achieve acceptable performance. This lead to the construction of a single quadratic cost function which penalized both speed error and throttle motion. Since this functional mirrors the tuning objectives of the fixed gain controller, the adaptive algorithm was not only robust to changes in vehicle parameters and torque disturbances, its performance was equivalent to that of a properly tuned fixed parameter controller. For a given vehicle subjected to a given torque disturbance, the minimization of the cost functional generates an optimal pair of the PI controller gains, K_p and K_i. In this sense, the choice of a single cost functional represents an implicit map from the set of vehicles and operating conditions to an admissible region in the parameter plane (K_p, K_i). This region was chosen to be a rectangle with preassigned bounds.

The task of parameter adaptation was to minimize the selected quadratic cost functional for each unknown vehicle and each unmeasured disturbance. A possible candidate was an indirect adaptive scheme with an estimator of the unknown vehicle and disturbance model parameters and an on-line LQ optimization algorithm. In the particular system under study, the frequency content in the disturbance was significantly faster than the plant dynamics. This resulted in difficulties in estimating the disturbance. After some experimentation, this scheme was abandoned in favor of a sensitivity-based scheme, which more directly led to adaptive minimization of the cost functional and made better use of the knowledge acquired during its construction.

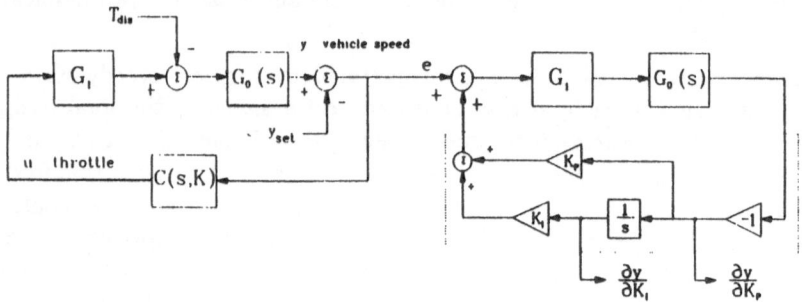

Fig. 2. The system and its copy generate sensitivity functions for optimization of the PI controller parameters K_p and K_i.

The sensitivity-based approach to parameter optimization exploits the remarkable sensitivity property of linear systems: *the sensitivity function (i.e. partial derivative) of any signal in the system with repect to any constant system parameter can be obtained from a particular cascade connection of the system and its copy.* For a linearized version of the vehicle model in Fig. 1, the sensitivities of the vehicle speed error $e = y - y_{\text{set}}$ with respect to the PI controller parameters K_p and K_i are obtained as in Fig. 2, where $G_0(s)$ and $G_1(s)$ represent the vehicle and powertrain dynamics respectively, $C(s, K) = -K_p - \frac{K_i}{s}$, and the control variable u is throttle position. This result can be derived by differentiation of

$$e(s, K) = \frac{1}{1 + C(s, K)G_1(s)G_0(s)} (y_{\text{set}} - G_0(s)T_{\text{dis}}) \qquad (1)$$

with respect to $K = [K_p, K_i]$, namely

$$\frac{\partial e}{\partial K} = \frac{\partial C}{\partial K} \frac{G_1(s)G_0(s)}{1 + C(s, K)G_1(s)G_0(s)} e(s, K) \qquad (2)$$

where $\frac{\partial C}{\partial K} = (-1, -\frac{1}{s})$. Expressions analogous to (2) can be obtained for the control sensitivities $\frac{\partial u}{\partial K}$. Our cost functional also uses a high-pass filter $F(s)$ to penalize higher frequencies in u, that is $\tilde{u}(s, K) = F(s)u(s, K)$. The sensitivities of $\tilde{u}$ are simply obtained as $\frac{\partial \tilde{u}}{\partial K} = F(s)\frac{\partial u}{\partial K}$.

When the sensitivity functions are available, a continuous gradient algorithm for the PI controller parameters is

$$\frac{dK_p}{dt} = -\epsilon \left(\beta \tilde{u} \frac{\partial \tilde{u}}{\partial K_p} + (1 - \beta) e \frac{\partial e}{\partial K_p} \right)$$

$$\frac{dK_i}{dt} = -\epsilon \left(\beta \tilde{u} \frac{\partial \tilde{u}}{\partial K_i} + (1 - \beta) e \frac{\partial e}{\partial K_i} \right)$$

(3)

where the adaptation speed determined by ϵ must be kept sufficiently small so that the averaging assumption (K_p and K_i are constant) is approximately satisfied. With ϵ small, the method of averaging is applicable to (3) and proves that, as $t \to \infty$, the parameters K_p and K_i converge to an ϵ–neighborhood of the values which minimize the quadratic cost functional

$$J = \int_0^\infty \left(\beta \tilde{u}^2 + (1 - \beta) e^2 \right) dt.$$

(4)

With a choice of the weighting coefficient, β, to be discussed later, our cost functional will be (4). Thus, (3) is a convergent algorithm which can be used to minimize this functional when the system is known so that its copy can be employed to generate the sensitivities needed in (3). In fact, our computational procedure for finding a cost functional good for all vehicle lines and operating conditions made use of this algorithm.

Unfortunately, when the vehicle parameters are unknown, the exact gradient algorithm (3) cannot be used, because then the copy of the system is not available. In other words, an algorithm employing exact sensitivities is not suitable for adaptive control. A practical escape from this difficulty is to generate some approximations of the sensitivity functions

$$\psi_1 \approx \frac{\partial \tilde{u}}{\partial K_p}, \quad \psi_2 \approx \frac{\partial \tilde{u}}{\partial K_i}, \quad \psi_3 \approx \frac{\partial e}{\partial K_p}, \quad \psi_4 \approx \frac{\partial e}{\partial K_i}$$

(5)

and employ them in a "pseudogradient" algorithm

$$\frac{dK_p}{dt} = -\epsilon \left(\beta \tilde{u} \psi_1 + (1 - \beta) e \psi_3 \right)$$

$$\frac{dK_i}{dt} = -\epsilon \left(\beta \tilde{u} \psi_2 + (1 - \beta) e \psi_4 \right).$$

(6)

A filter used to generate ψ_1, ψ_2, ψ_3, and ψ_4 is called a *pseudosensitivity filter*. The fundamental problem in the design of the pseudosensitivity filter is not only to guarantee that the algorithm (6) will converge, but also that the values to which it converges are close to those which minimize the chosen cost functional.

4 Adaptive Controller Implementation

The adaptive speed control algorithm presented here is fairly simple and easy to implement, but care must be taken when choosing its free parameters and designing pseudosensitivity filters. This section discusses the procedure used to achieve a robust system and provide the desired speed control performance.

4.1 Pseudosensitivity Filter

While testing the adaptive algorithm it becomes obvious that the gains K_p and K_i and the vehicle parameters vary greatly for operating conditions and vehicles. This makes it impossible to implement the exact sensitivity filters for the gradient algorithm (3). Our approach is to generate a "pseudogradient" approximation of $\partial J/\partial K$, satisfying the stability and convergence conditions summarized in the Appendix.

In the Appendix, the two main requirements for stability and convergence are: a Persistently Exciting (PE) input condition and a "pseudogradient condition", which in our case is a *phase condition on the nominal sensitivity filters*. Since we are using a reduced order controller with only two adjustable parameters, the PE condition is easily met by the changing road loads. Road disturbances have an approximate frequency spectrum centered about zero which drops off with the square of frequency. This meets the PE condition for adapting two gains, K_p and K_i.

To satisfy the pseudogradient condition, the phase of the pseudosensitivity filter must be within $\pm 90°$ of the phase of the actual sensitivity at the dominant frequencies. To help guarantee this for a wide range of vehicles and operating conditions, we varied the system parameters in the detailed vehicle model to generate an envelope of possible exact sensitivities. Then the pseudosensitivity filter was chosen near the center of this envelope. An important element of the phase condition is the fact that it is a condition on the sum of frequencies; that is, the phase condition is most important in the range of frequencies where there are dominant dynamics. If the pseudosensitivity filters do not meet the phase conditions at frequencies where there is little dominant spectral content, the algorithm may still be convergent provided the phase conditions are strongly met in the region of dominant dynamics. Thus, the algorithm possesses an inherent robustness property.

Figure 3 shows the gain and phase plots for the psuedosensitivity filter $\partial y/\partial K_p$ The other three sensitivities are left out for brevity. Figure 4 shows the $\pm 90°$ phase boundary (solid lines) along with exact sensitivity phase angles (dashed lines) as vehicle inertia, engine power, and the speed control gains are varied over their full range. From this plot it is seen that the chosen pseudosensitivity filter meets the pseudogradient condition along with some safety margin to accommodate unmodeled dynamics.

4.2 Choice of ϵ and β

For implementation, the first parameters which must be chosen are the adaptation gain ϵ and the weighting in the cost functional, β. The adaptation gain ϵ determines the speed of adaptation and should be chosen based on the slowest dynamics of the system. For speed control the dominant dynamics result from the vehicle inertia and have a time constant on the order of 30-50 seconds. To avoid interaction between the adaptive controller states and those of the plant, the adaptation should be approximately an order of magnitude slower than the

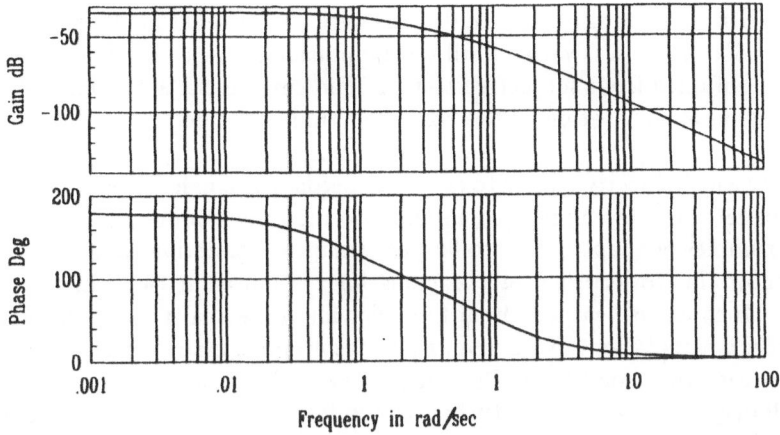

Fig. 3. Gain and Phase Plot of Pseudosensitivity $\partial y / \partial K_p$

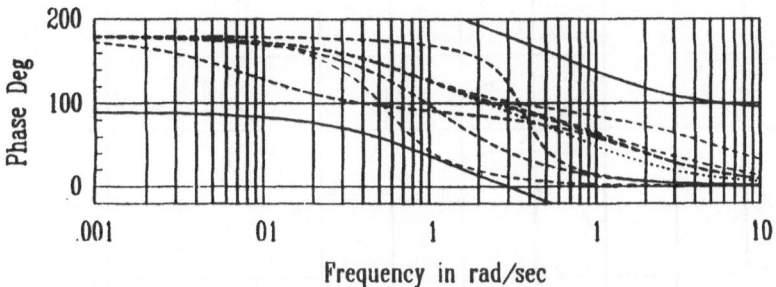

Fig. 4. Envelope of Possible Exact Sensitivity Phase Angles

plant. As shown in [14]-[16], this rule of thumb allows one to use the frozen parameter system and averaging to analyze stability of the adaptive system. The adaptation law takes up to several minutes to converge depending on initial conditions and the road load disturbances.

The two extreme choices of β are $\beta = 0$ and $\beta = 1$. For the first extreme, $\beta = 0$, we are penalizing only speed error and the adaptation will tune to an unacceptable high gain controller. High gain will cause too much throttle movement resulting in nonsmooth behavior as felt by the passengers, and the system will be less robust from a stability point of view. For the second case, $\beta = 1$, the adaptation will try to keep a fixed throttle angle and will allow large speed errors. Obviously some middle values for the weightings are desired. Increasing β reduces unnecessary throttle movement, while to improve tracking and transient speed errors we need to decrease β.

The choice of the weighting was based on experience with tuning standard PI speed controllers. Much subjective testing has been performed on vehicles to obtain the best fixed gains for the PI controller when the vehicle parameters are known. With this information, simulations were run on a detailed vehicle model with a small ϵ, and β was varied until the adaptation converged to approximately the gains obtained from subjective testing. The cost functional weighting, β, is a different parameterization of the controller tuning problem. For development engineers, who may not be control system engineers, β represents a tunable parameter which directly relates to customer needs. Although simpler, this allows for a broader range of engineering inputs into the tuning process.

As examples of adaptive controller performance, simulation results in Figs. 5 and 6 show the trajectories of the gains for a vehicle on two different road terrains. Much can be learned about the behavior of the adaptive algorithm from these types of simulations. In general, K_p varies proportionally with vehicle speed and K_i varies with road load. The integral gain, K_i, tends toward low values for small disturbances or for disturbances too fast for the vehicle to respond to. This can be seen in Fig. 5. Conversely, K_i tends toward high values for large or slowly varying road disturbances, as can be seen in Fig. 6.

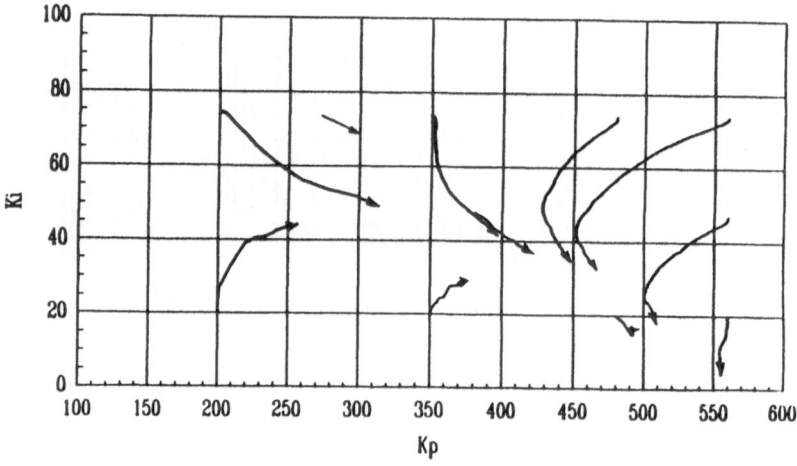

Fig. 5. Gain Trajectories on Flat Ground at 30 mph

4.3 Modifications For Improved Robustness

Additional steps have been taken to insure robust performance in the automotive environment. First, the adaptation is turned off if the vehicle is operating in regions where the modeling assumptions are violated. These include operation at closed throttle or near wide open throttle, during start up transients, and when the set speed is changing. When the adaptation is turned off, the gains are frozen, but the sensitivities are still computed.

Care has been taken to avoid parameter drift due to noise and modeling imperfections. Two common ways to reduce drift are projection or a deadband on

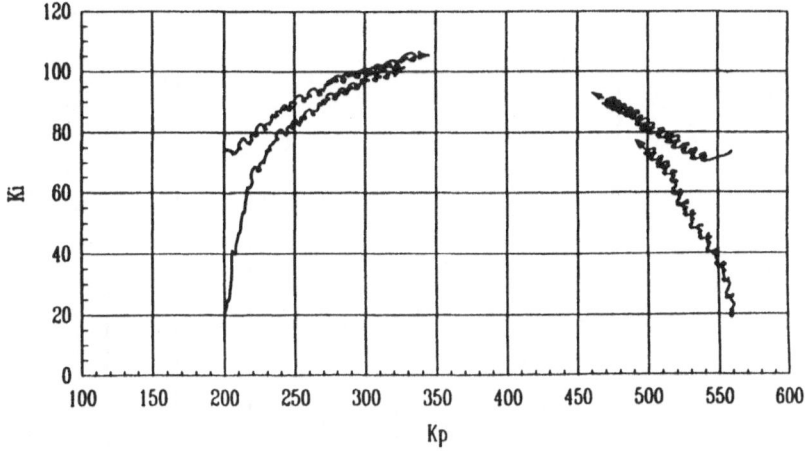

Fig. 6. Gain Trajectories on Low Frequency Hills at 30 mph

error. Projection, which limits the range over which the gains may adapt, is more attractive given the a priori knowledge of reasonable gains for the speed control system. To minimize computation a simple projection is used, constraining the tuned gains to a predetermined set shown in Fig. 7.

There are other unmeasurable disturbances which can affect performance, such as the driver overriding speed control by use of the throttle. This condition which can cause large gain changes cannot be detected immediately because throttle position is not measured. To minimize these unwanted gain changes, the rate at which the gains can adapt is limited. If the adaptation adjusts quicker than a predetermined rate, the adaptation gain ϵ is lowered, limiting the rate of change of the gains K_p and K_i.

The final parameters to choose are the initial estimates for K_p and K_i. Since the adaptation is fairly slow, a poor choice of initial gains can cause poor start-up performance. For quick convergence, the initial controller gains are scheduled with vehicle speed at the points A, B, and C. Figure 7 shows these initial gains along with the range of possible tuning. The proportional relationship between vehicle speed and the optimal controller gains can be seen.

4.4 Cost Reduction

The implementation cost of a speed control algorithm is very important to the acceptance and ultimately the monitary success of the product. Although this cost depends upon many things, mimimizing the RAM, ROM and computation necessary to implement the algorithm is an important factor. Comparing the complexity of the adaptive control algorithm to a conventional PI controller, it is apparent that the sensitivity filters and parameter update equations add significantly to the computational complexity of the speed control system. Al-

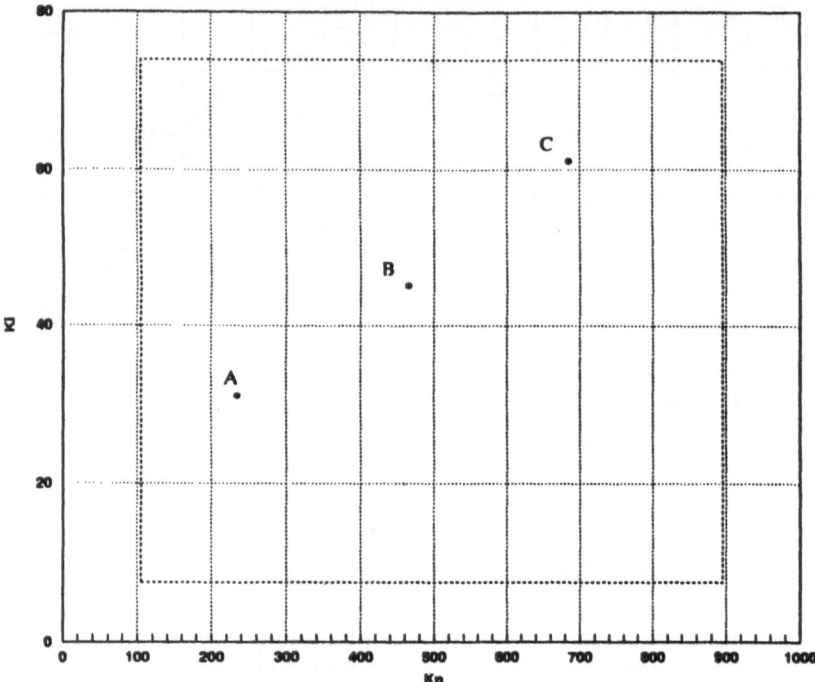

Fig. 7. Limits and Initial Conditions for Control Gains A) $y_{set} < 35$ mph;   B) 35mph $\leq y_{set} < 50$mph;   C) $y_{set} \geq 50$ mph.

though some increase in complexity is expected to produce a much more robust controller, several steps were taken to minimize its impact.

To minimize the ROM requirements of the adaptive algorithm, extensive use of subroutines were used. To implement the pseudosensitivity filters, each filter was decomposed into a series of first order low pass filters. Common elements were combined, resulting in the structure of Figure 8. Then a common low pass filter subroutine was used to implement this structure. This structure reduced the amount of ROM necessary to implement the adaptive algorithm. Since the amount of RAM needed for the algorithm depends primarily upon the number of states used, partitioning the sensitivity filters into a cascade of first order filters did not increase the RAM required.

The computational requirements of the algorithm depend upon the number of numerical operations involved and the sampling rate of the algorithm. One of the fundamental reasons for selecting a sensitivity based approach was to limit the complexity of the controller and adaptation algorithm. To further reduce the computational requirements, the parameter update law and controller sample rates were reduced as much as possible. With care to avoid aliasing, the computational requirements of the controller were reduced significantly. The resulting adaptive system provided improved performance and robustness with an implementation cost that was only marginally greater than the conventional

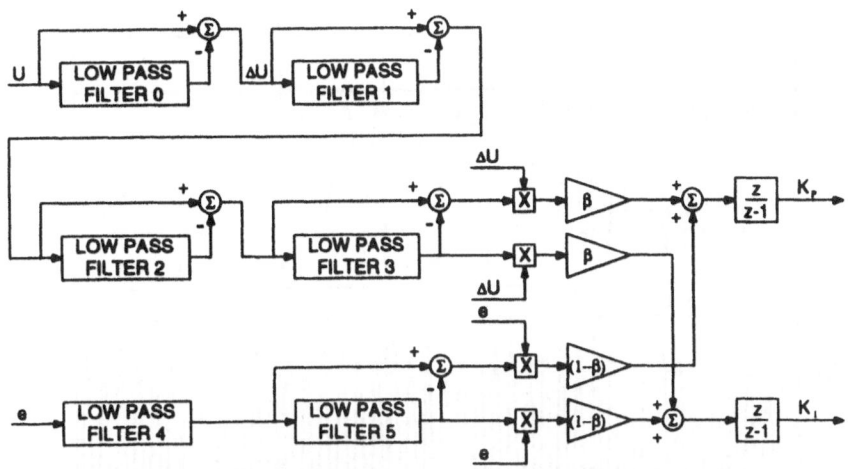

Fig. 8. Simplified Cascade Sensitivity Filter Implementation

fixed gain controller.

5   Performance in Test Vehicles

The adaptive algorithm discussed in this paper has been tested in a number of vehicles with excellent results. In this section, the perfomance of the adaptive algorithm is presented by examining four different test scenarios.

For the first test, a vehicle was selected which exhibited a limit cycle or surge condition at low speeds or on down slopes [7]. The limit cycles are partially due to nonlinearities such as the idle stop which limits throttle movements. The first set of data, Figs. 9 to 11 shows the low speed limit cycle. The high frequency (0.2 Hz) and amplitude make it noticeable to the driver. The adaptive controller greatly improves performance by decreasing the amplitude and frequency to the point where the driver cannot feel the limit cycle. Looking at the control gains during this test it is seen that the gains initially decrease to reduce the limit cycle then continue to adjust in small amounts for the varying road loads as the trade-off between speed tracking and throttle movement is made. The parameter histories repeat themselves as the experiment was performed on a closed course. Only the first cycle is shown. Of course, a fixed parameter controller could be adjusted to minimize this limit cycle. However, the controller gains which would minimize the limit cycle behavior would not produce acceptable perfomance on rolling hill terrain. The adaptive controller yielded acceptable performance under all operating conditions.

The next test demonstrated the ability of the adaptive algorithm to adjust for different vehicle types. The next three Figs., 12 to 14, show the PI gains for three different vehicles traveling over the same road at 40 mph. As expected, even though the initial gains are acceptable, each vehicle tunes to a different set of optimal gains for this road. One thing to notice in Fig. 12 is the initial

14

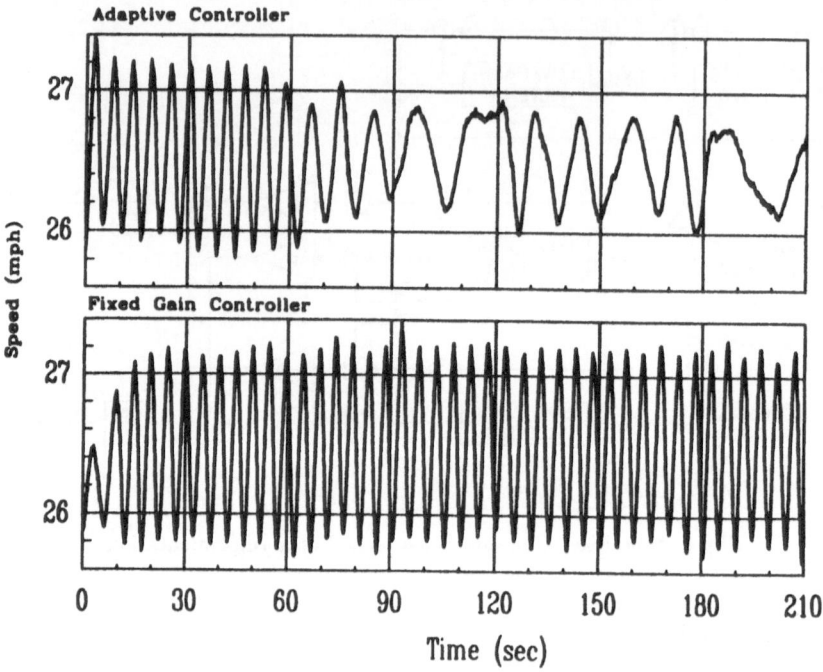

Fig. 9. Vehicle With Low Speed Limit Cycle (Adaptive : Top, Fixed Gain : Bottom)

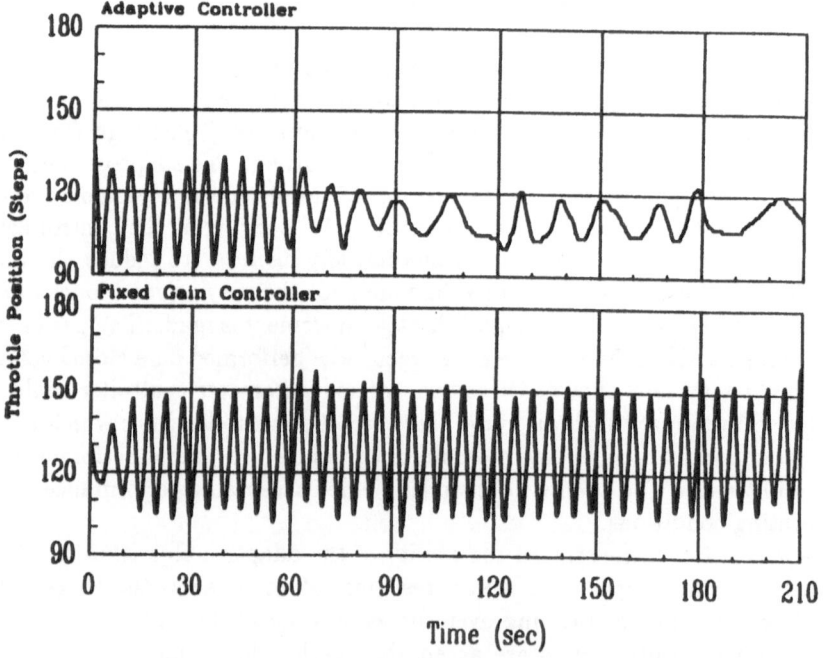

Fig. 10. Vehicle With Low Speed Limit Cycle

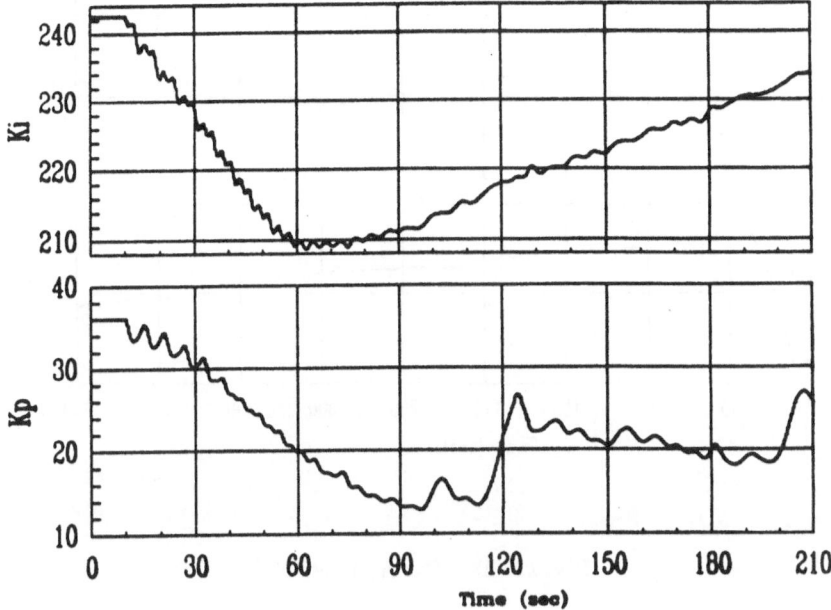

Fig. 11. Controller Gains

increase in K_i. This occurred because the integral preset of the control did not match this vehicle causing an initial overshoot or droop in speed. To the adaptive algorithm, this offset looks like a large road disturbance. Again, since these tests were performed on a closed course and the gains eventually become periodic.

The third test shows the performance of the system on a road which produces a relatively high frequency road disturbance. Figs. 15 to 17 show a vehicle driving on a freeway which passes below the grade level to pass under surface roads. An approximate road profile for a section of this road is shown in Fig. 18. Because these disturbances are relatively brief, the adaptive controller cannot greatly improve the tracking ability of the vehicle, but much of the high frequency behavior has been removed. For this disturbance, the proportional gain tunes up and the integral gain tunes down. The integral gain decreases to reduce limit cycle behavior since the road disturbance is too fast for the vehicle to track.

In the final test, a vehicle is tested with varying loads to demonstrate the robustness of the algorithm to vehicle parameter changes. The response of the system is compared for a vehicle unloaded and loaded with 3000 lbs. For this vehicle the load represented 60% of the unloaded weight. In Figs. 19 and 20, the parameters histories are shown. In both cases, the integral gain is reduced, while the proportional gain adjusts appropriately for the weight of the vehicle. As can be seen in Figs. 21 and 22, after adaptation, the speed errors are equivalent, although more throttle is necessary for the loaded vehicle. A fixed gain controller can be tuned to yield good performace in either loading condition, but a 60% change in vehicle weight will produce a significant change in the performace. The

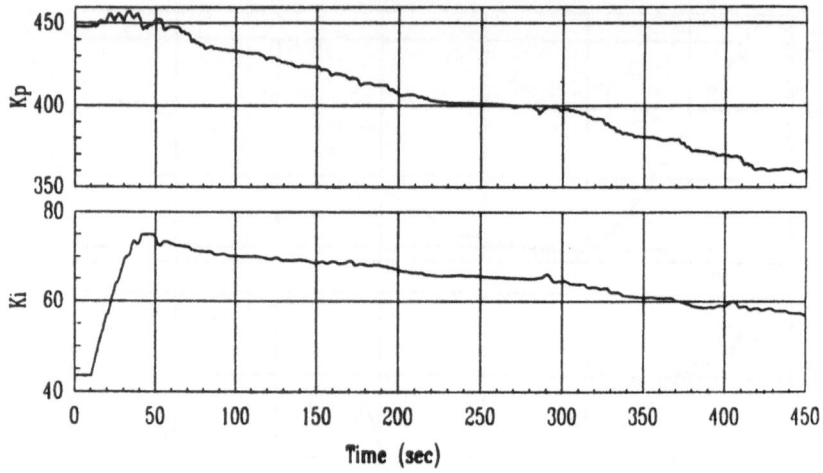

Fig. 12. Test Car A, 40 MPH

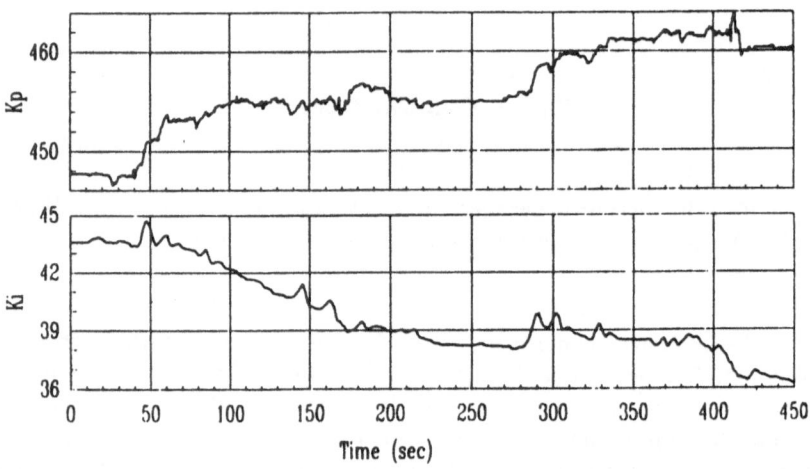

Fig. 13. Test Car B, 40 MPH

adaptive algorithm adjusts to weight changes to yield very similar performace in both cases.

In these tests, the adaptive controller demonstrated consistently good performance across vehicle lines and under all operating condition changes. The adaptive control has worked well in every vehicle tested and makes significant performance improvements in vehicles which have poor performance with a fixed gain PI control. For vehicles where the standard speed control already performs well, improvements from the adaptive algorithm are still significant at low speeds or at small throttle angles. Finally, the adaptive controller adjusts for load variations and produces consistent performance regardless of payload.

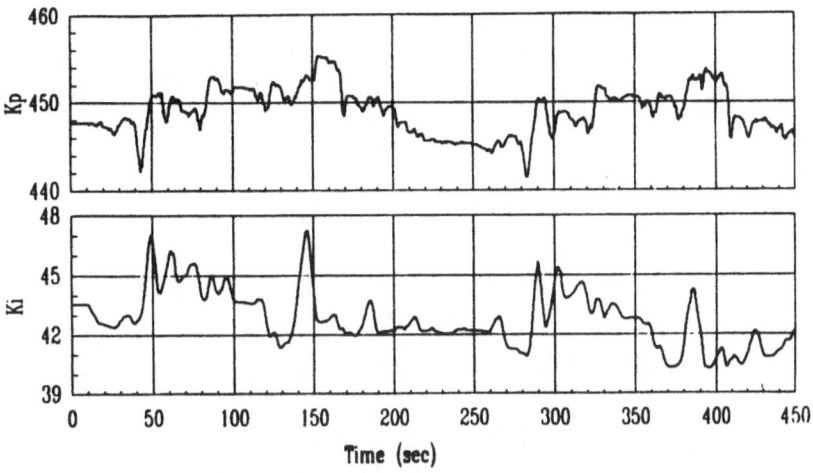

Fig. 14. Test Car C, 40 MPH

6   Conclusions

This paper has presented an adaptive algorithm to adjust the gains of a vehicle speed control system. By continuously adjusting the PI control gains, speed control performance can be optimized for each vehicle and operating condition. This helps to design a single speed control module without additional calibration or sacrifices in performance for certain car lines. It also allows improved performance for changing road conditions not possible with a fixed gain control or other types of adaptive control.

The results of vehicle testing confirm the performance improvements and robustness of the adaptive controller. Moreover, the adaptive control technique presented in this paper has been applied to solve other problems of electronically controlled automotive systems.

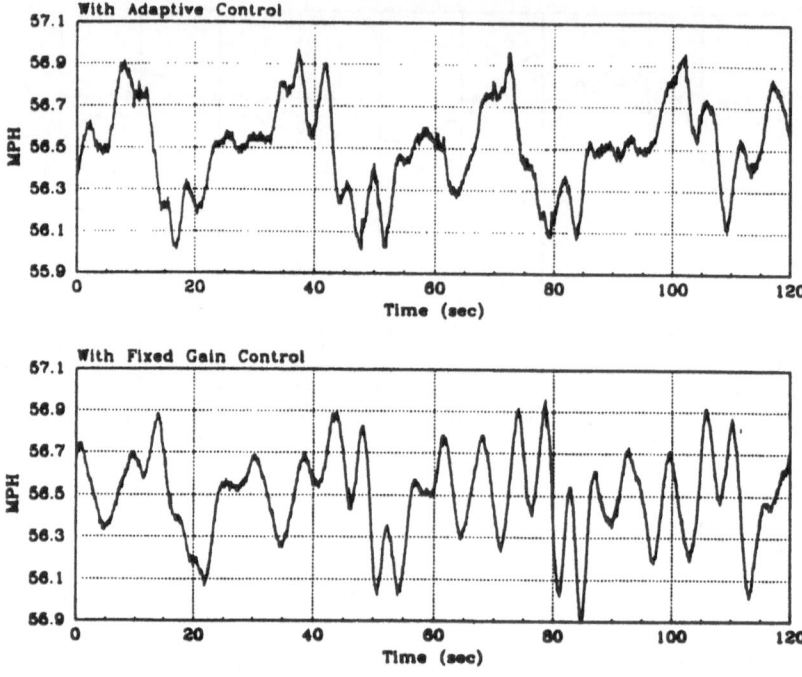

Fig. 15. Vehicle With High Frequency Road Disturbance

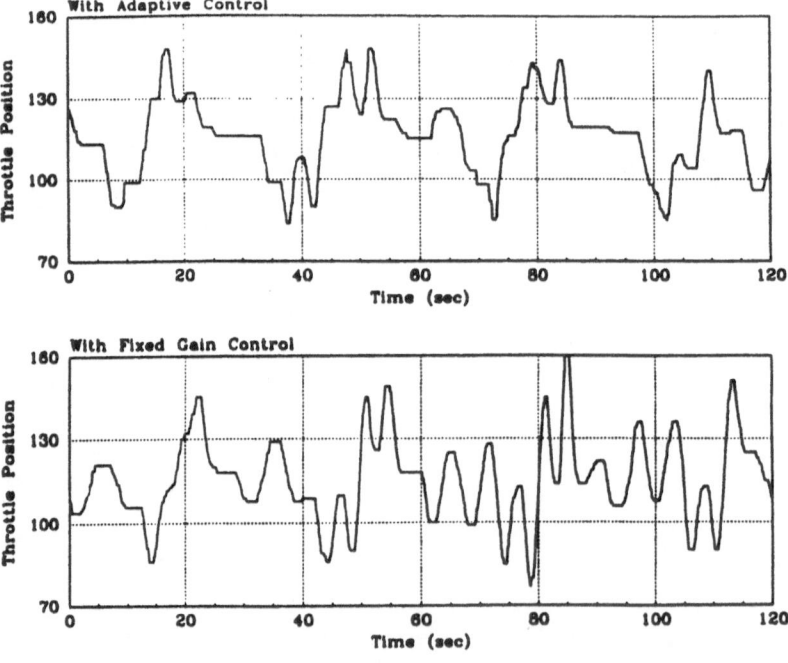

Fig. 16. Vehicle With High Frequency Road Disturbance

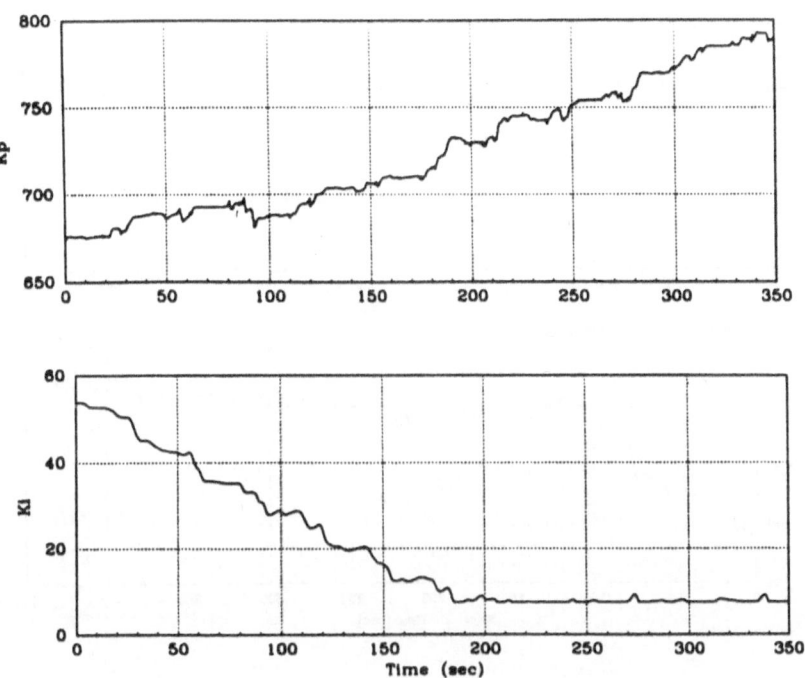

Fig. 17. Gains With High Frequency Disturbances

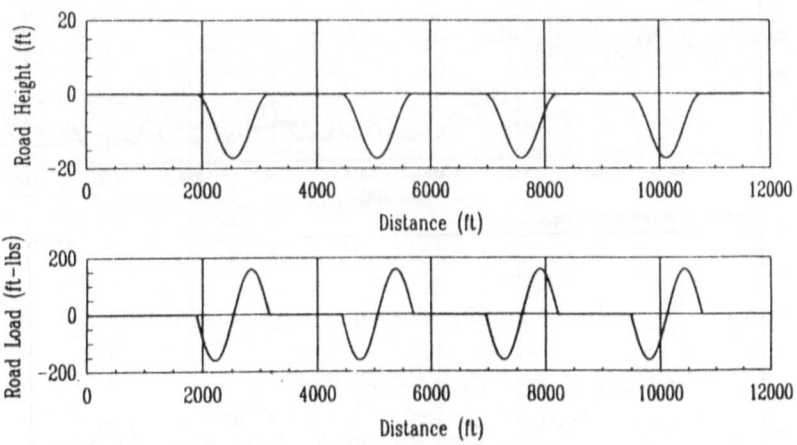

Fig. 18. Approximate Freeway Profile

20

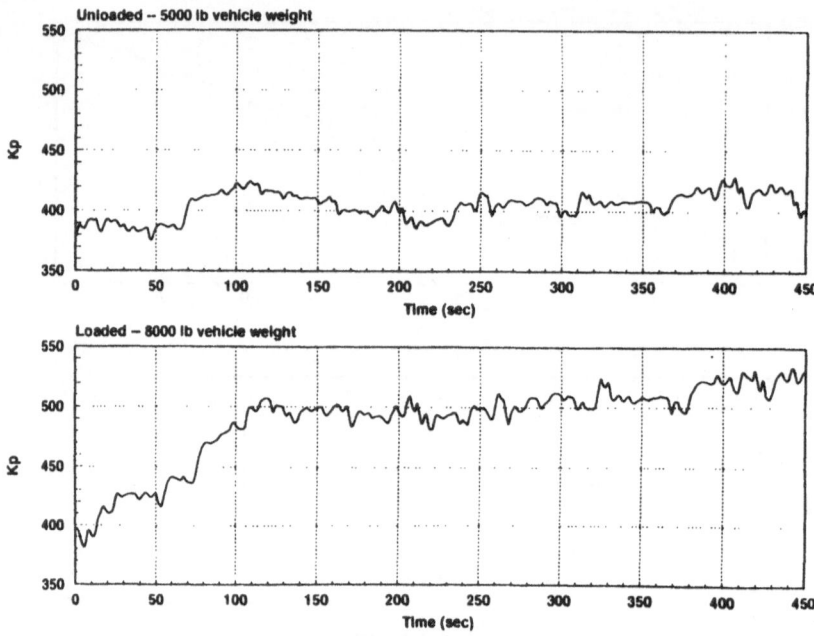

Fig. 19. Loaded and Unloaded Vehicle, Proportional Gains

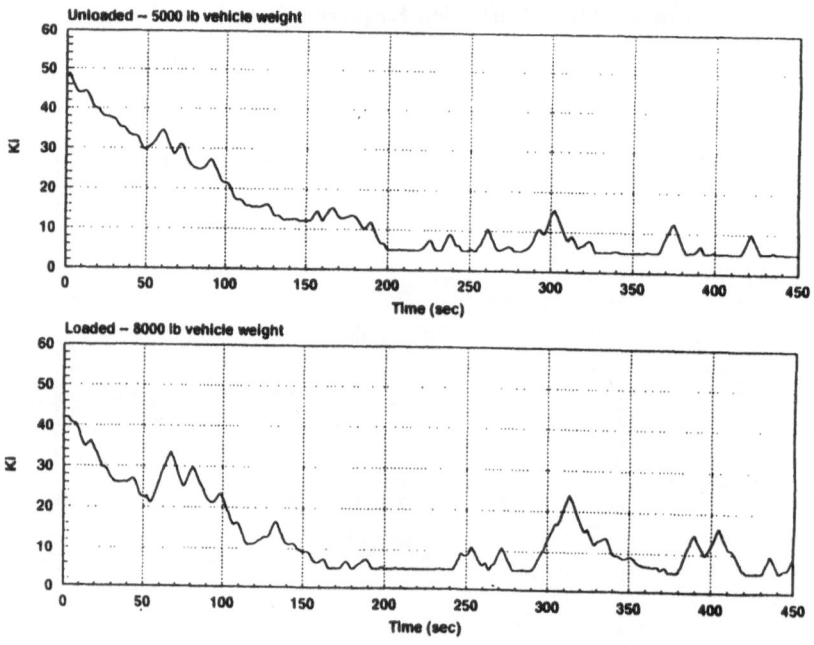

Fig. 20. Loaded and Unloaded Vehicle, Integral Gains

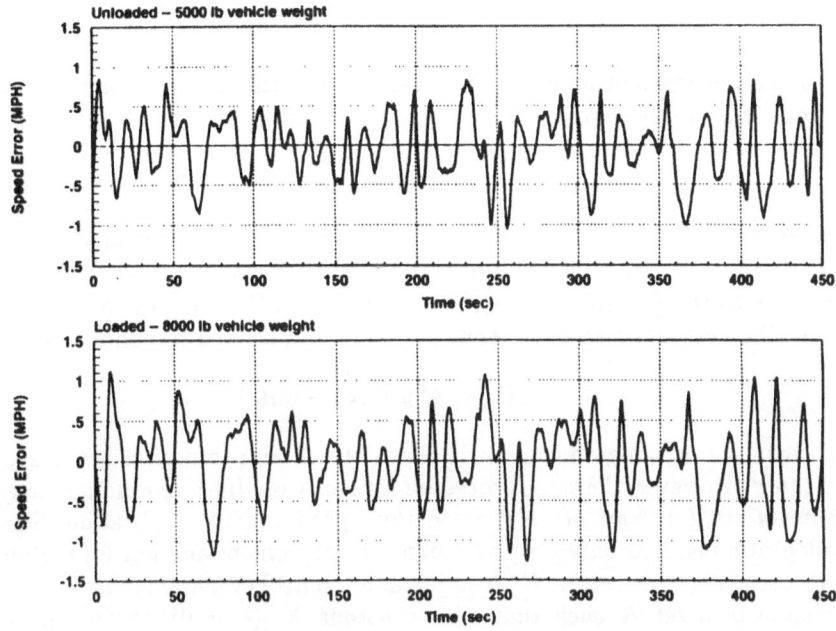

Fig. 21. Loaded and Unloaded Vehicle, Speed Errors

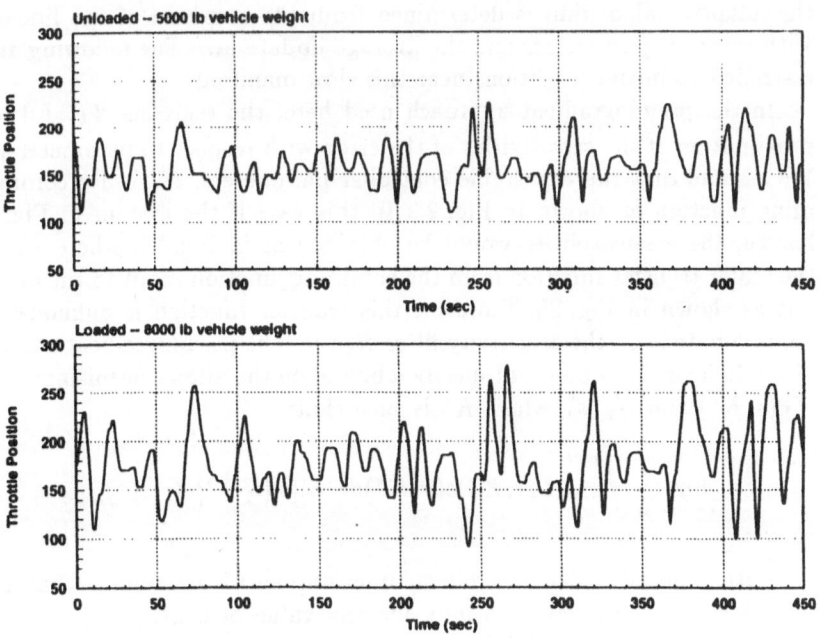

Fig. 22. Loaded and Unloaded Vehicle, Throttle Commands

APPENDIX

The pseudogradient adaptive approach used in this application relies upon the properties of slow adaptation. Such gradient tuning algorithms can be described by:

$$\dot{K}(t) = -\epsilon \Psi(t, K) e(t, K), \quad K \in \mathbb{R}^m \tag{7}$$

where K is a vector of tuneable parameters, $e(t, K)$ is an error signal and $\Psi(t, K)$ is the regressor which is an approximation of the sensitivity of the output with respect to the parameters, $\frac{\partial y(t,K)}{\partial K}$. In this derivation $e(t, K)$ will be defined as the difference between the output $y(t, K)$ and a desired response $y_m(t)$

$$e(t, K) \equiv y(t, K) - y_m(t), \tag{8}$$

so that the pseudogradient update law (7) will aim to minimize the average of $e(t, K)^2$. However, the same procedure is readily modified to minimize a weighted sum of $e(t, K)^2$ and $u(t, K)^2$ as in this speed control application. Since slow adaptation is used only for performance improvement and not for stabilization, we constrain the vector K of adjustable controller parameters, $k_1, ..., k_m$, to remain in a set $\bar{K}$ such that with constant K, ($\epsilon = 0$), the resulting linear system is stable. Our main tool in this analysis is an integral manifold, the so called *slow manifold* [14], that separates the fast linear plant and controller states from the slow parameter dynamics. To assure the existence of this manifold, we assume that $\Psi(t, K)$ and $e(t, K)$ are differentiable with respect to K and the input to the system, $r(t)$, is a uniformly bounded almost periodic function of time. Then applying the averaging theorem of Bogolubov, the stability of the adaptive algorithm is determined from the response of the linear system with constant parameters and the average update law. The following analysis is restricted to initial conditions near this slow manifold.

In the pseudogradient approach used here, the regressor $\Psi(t, K)$ is an approximation of the sensitivities of the error with respect to parameters, $\frac{\partial e(t,K)}{\partial K}$. We assume that the adjustable controller parameters, K, feed a common summing junction as shown in Fig. 23. In this case if the system in Fig. 23 were known, these sensitivities would be obtained as in Fig. 24, where $H_\Sigma(s, K)$ is the scalar transfer function from the summing junction input to the system output as shown in Fig. 25. However, this transfer function is unknown, and we approximate it by the sensitivity filter $H_\Psi(s)$. The pseudogradient stability condition to be derived here will specify a bound on the allowable mismatch between $H_\Sigma(s, K^*)$ and $H_\Psi(s)$, where K^* is such that:

$$\lim_{T \to \infty} \frac{1}{T} \int_t^{t+T} \Psi(\tau, K^*) e(\tau, K^*) d\tau = [\Psi(t, K^*) e(t, K^*)]_{\text{ave}} = 0. \tag{9}$$

With reduced order controllers, it is unrealistic to assume that the error $e(t, K)$ can be eliminated entirely for any value of controller parameters and plant variations. The remaining non-zero error is called the *tuned error*, $e^*(t) =$

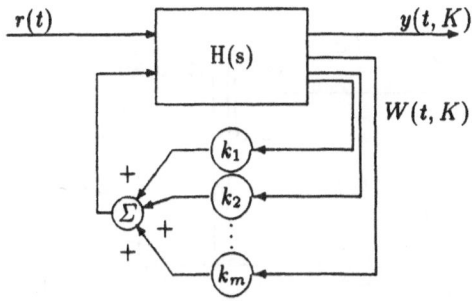

Fig. 23. Adjustable Linear System

$e(t, K^*)$. The error $e(t, K)$ is comprised of $e^*(t)$ and a term caused by the parameter error $\tilde{K} = K - K^*$. In the mixed (t,s) notation, the error $e(t, K)$ and the regressor $\Psi(t, K)$ are both obtained from the measured signals $W(t, K)$ as follows:

$$e(t, K) = H_\Sigma(s, K^*)[\tilde{K}^T(t)W(t, K)] + e^*(t) \tag{10}$$

$$\Psi(t, K) = H_\Psi(s)[W(t, K)]. \tag{11}$$

The error expression illustrated in Fig. 26 contains the exact sensitivity filter transfer function $H_\Sigma(s, K^*)$. The stability properties of this slowly adapting system are determined by examination of the average parameter update equation:

$$\dot{\tilde{K}}(t) = -\epsilon\{\Psi(t, K)V(t, K)^T\}_{ave}\tilde{K}(t) - \epsilon\Delta - \epsilon\{\Psi(t, K)e^*(t)\}_{ave}. \tag{12}$$

where we have again used $H_\Sigma(s, K^*)$ in the definition of the signal

$$V(t, K) \equiv H_\Sigma(s, K^*)[W(t, K)]. \tag{13}$$

The average swapping term Δ in (12) is

$$\Delta = \{H_\Sigma(s, K^*)[W(t, K)^T\tilde{K}(t)] - H_\Sigma(s, K^*)[W(t, K)^T]\tilde{K}(t)\}_{ave} \tag{14}$$

As shown in [14] and [15], for slow adaptation this term is small and does not affect the stability of (12) in a neighborhood of the equilibrium K^*. The stability of this equilibrium in the case when $e^*(t) = 0$ is established as follows:

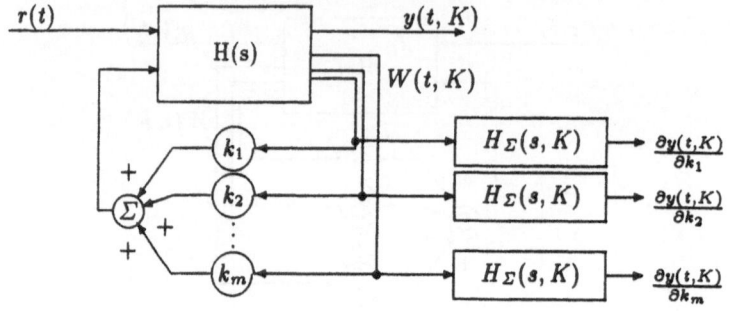

Fig. 24. Sensitivity System

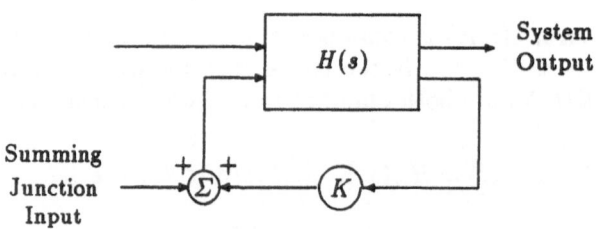

Fig. 25. $H_\Sigma(s, K)$

Lemma. *[15] Consider the average update system (12) with $\Delta(t) = 0$, $e(t, K^*) = 0$ and $K = K^*$. If*

$$\operatorname{Re}\lambda\{\Psi(t, K)V^T(t, K)\}_{ave} > 0 \tag{15}$$

then there exists an $\epsilon^* > 0$ such that $\forall \epsilon \in (0, \epsilon^*]$ the equilibrium $\tilde{K} = 0$ is uniformly asymptotically stable. □

To interpret the condition (15), we represent the regressor as

$$\Psi(t, K) = H_\Psi(s)H_\Sigma^{-1}(s, K^*)[V(t, K)] \tag{16}$$

so that the matrix in (15) can be visualized as in Fig. 27. Representing the almost periodic signal $V(t, K)$ as

$$V(t, K) = \sum_{i=-\infty}^{+\infty} v_i e^{\jmath \omega_i t} \tag{17}$$

a sufficient condition for (15) to hold is

$$\sum_{i=-\infty}^{+\infty} \text{Re}[H_\Psi(j\omega_i)H_\Sigma^{-1}(j\omega_i, K^*)]\text{Re}[v_i\bar{v}_i^T] > 0. \tag{18}$$

If the signal, $V(t)$, possesses an autocovariance, $R_V(z)$, it is persistently exciting (PE) *iff* $R_V(0) > 0$ [17]. It follows that $V(t,K)$ is PE *iff* $\sum_{i=-\infty}^{\infty} R_e[v_i\bar{v}_i^T] > 0$. Clearly if the sensitivity filter is exact, $H_\Psi(j\omega) = H_\Sigma(j\omega, K^*)$ and $V(t,K)$ is PE, then the sufficient stability condition in (18) holds When $H_\Psi(s)$ cannot be made exact, this stability condition is still satisfied when the pseudosensitivity filter is chosen such that $\text{Re}[H_\Psi(j\omega)H_\Sigma^{-1}(j\omega, K^*)] > 0$ for dominant frequencies, that is, the frequencies where $v_i\bar{v}_i^T$ is large. This condition serves as a guide for designing the pseudosensitivity filter $H_\Psi(s)$. As $H_\Sigma(s, K^*)$ is unknown beforehand, we use the a priori information about the closed loop system to design a filter $H_\Psi(s)$ such that in the dominant frequency range the following *pseudogradient condition* is satisfied for all plant and controller parameters of interest:

$$-90^\circ < \angle\{H_\Psi(j\omega)H_\Sigma^{-1}(j\omega, K^*)\} < 90^\circ \tag{19}$$

When (19) is satisfied, then the equilibrium K^* of the average update law (12) with $\Delta \equiv 0$, $e^*(t) \equiv 0$ is uniformly asymptotically stable. By Bogolubov's theorem and slow manifold analysis [14], this implies the local stability property of the actual adaptive system provided $e^*(t) \neq 0$ is sufficiently small. Since in this approach, both the controller structure and number of adjustable parameters are free to be chosen independently of plant order, there is no guarantee that $e^*(t)$ will be suffciently small. Although conservative bounds for $e^*(t)$ may be calculated [12], in practice since the design objective of the controller and Pseudogradient adaptive law is to minimize the average of $e(t,K)^2$, $e^*(t)$ is typically small.

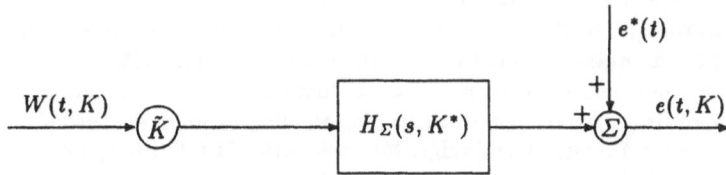

Fig. 26. Error Model

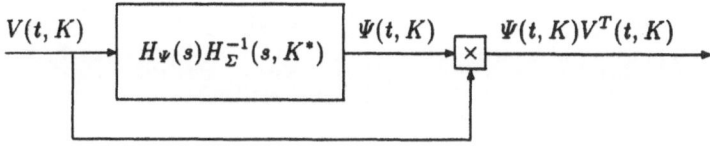

Fig. 27. Feedback Matrix

References

1. J. T. Ball, "Approaches and Trends in Automatic Speed Controls," SAE Tech. Paper #670195, 1967.
2. W. C. Follmer, "Electronic Speed Control," SAE Tech. Paper #740022, 1974.
3. S. J. Sobolak, "Simulation of the Ford Vehicle Speed Control System," SAE Tech. Paper #820777, 1982.
4. K. Nakamura, T. Ochiai, and K. Tanigawa, "Application of Microprocessor to Cruise Control System," *Proceedings of the IEEE Workshop on Automotive Application of Microprocessors*, pp. 37-44, 1982.
5. B. Chaudhure, R. J. Schwabel, and L. H. Voelkle, "Speed Control Integrated into the Powertrain Computer," SAE Tech. Paper #860480, 1986.
6. T. Tabe, H,. Takeuchi, M. Tsujii, and M. Ohba, "Vehicle Speed Control System Using Modern Control Theory," *Proceedings of the 1986 International Conference on Industrial Electronics, Control and Intrumentation*, (IECON'86), vol 1, pp. 365-370, 1986.
7. M. Uriuhara, T. Hattori, and S. Morida, "Development of Automatic Cruising Using Fuzzy Control System," *Journal of SAE of Japan*, vol. 42, no. 2, pp. 224-229, 1988.
8. M. Abate and N. Dosio, "Use of Fuzzy Logic for Engine Idle Speed Control," SAE Tech. Paper #900594, 1990.
9. T. Tsujii, H. Takeuchi, K. Oda, and M. Ohba, "Application of Self-Tuning to Automotive Cruise Control," *Proc. Amer. Contr. Conf.*, pp. 1843-1848, 1990.
10. G. Hong, and N. Collings, "Application of Self-Tuning Control," SAE Tech. Paper #900593, 1990.
11. P. V. Kokotovic, "Method of Sensitivity Points in the Investigation and Optimization of Linear Control Systems," *Automation and Remote Control*, vol 25, pp. 1670-1676, 1964.
12. D. S. Rhode, "Sensitivity Methods and Slow Adaptation," Ph.D. Thesis, University of Illinois at Urbana-Champaign, 1990.
13. D. S. Rhode and P. V. Kokotovic, "Parameter Convergence Conditions Independent of Plant Order," in *Proc. Amer. Contr. Conf.*, pp. 981-986, 1990.
14. B. D. Riedle, and P. V. Kokotovic, "Integral Manifolds of Slow Adaptation," *IEEE Trans. Automat. Contr.*, vol. 31, pp. 316-323, 1986.
15. P. V. Kokotovic, B. D. Riedle, and L. Praly, "On A Stability Criterion for Continuous Slow Adaptation," *Sys. Contr. Lett.*, vol. 6, pp. 7-14, 1985.
16. B. D. O. Anderson, R. R. Bitmead, C. R. Johnson, Jr., P. V. Kokotovic, R. L. Kosut, I. Mareels, L. Praly, and B. D. Riedle, *Stability of Adaptive Systems: Passivity and Averaging Analysis*, Cambridge, Massachusetts: M.I.T. Press, 1986.
17. S. Boyd and S. S. Sastry, "Necessary and Sufficient Conditions for Parameter Convergence in Adaptive Control," *Automatica*, vol. 22, no. 6, pp. 629-639, 1986.

This article was processed using the LaTeX macro package with LLNCS style

Modified Model Reference Adaptive Control: Design, Analysis and Performance Bounds*

Aniruddha Datta
Dept. of Electrical Engineering
Texas A&M University
College Station, TX 77843-3128
U.S.A.

Petros A. Ioannou
Dept. of Electrical Engineering-Systems
University of Southern California
Los Angeles, CA 90089-2563
U.S.A.

Abstract. Boundedness of all closed loop signals and asymptotic convergence of the tracking error to zero or to a residual set are the criteria that have been most commonly used to characterize the performance of model reference adaptive control schemes in the absence of persistently exciting signals. These criteria do not reveal the large transient oscillations often observed in simulations and do not exclude the possibility of bursting at steady state in the presence of small modelling errors. The purpose of this paper is to address the issue of performance by using two additional criteria to assess performance in the ideal and non-ideal situations. They are the mean square tracking error criterion and the L_∞ tracking error bound criterion. We use these criteria to examine the performance of a standard model reference adaptive controller and motivate the design of a modified scheme that can have an arbitrarily improved nominal performance in the ideal case and in the presence of bounded input disturbances. It is shown that for these cases the modified scheme can provide an arbitrarily improved zero-state transient performance and an arbitrary reduction in the size of possible bursts that may occur at steady state. As in every robust control design the nominal performance has to be traded off with robust stability and therefore the improvement in performance achieved by the proposed scheme is limited by the size of the unmodelled dynamics, as established in the paper. Another expected limitation to nominal performance is sensor noise which due to space limitation is not examined here.

* This work was supported by the National Science Foundation under Grants ECS-9119722 and ECS-9210726.

1 Introduction

The initial motivation behind the introduction of adaptive control was the need to design controllers for dynamical systems with large parametric uncertainty. A major breakthrough in adaptive control occurred in 1980 when several research groups [1]-[3] published results which showed that, under certain ideal conditions, it was possible to design globally stable adaptive systems.

Subsequently, it was shown [4, 5] that in the presence of modelling errors, such as bounded disturbances and/or unmodelled dynamics, an adaptive control scheme designed for stability in an ideal situation, could go unstable. The main cause of instability was the adaptive law for estimating the unknown parameters that made the closed loop system nonlinear and more susceptible to the effects of modelling errors. Consequently the main effort of the adaptive control research of the 80's [6]-[17] was to modify the adaptive laws for robustness. This led to the emergence of robust adaptive control that dominated the adaptive control research during the 80's. Several intelligent modifications were proposed which led to adaptive control schemes that counteract instabilities and guarantee signal boundedness and a tracking or regulation error that is of the order of the modelling error in the mean square sense for any given finite initial conditions [7, 8, 11]. These results however provide little or no information about convergence rates and transient behaviour. Indeed in [18] an example was presented to show that an asymptotically perfect tracking performance could go hand in hand with an arbitrarily poor transient behaviour. Furthermore there is no guarantee that the tracking will be of the order of the modelling error at steady state. Indeed it was shown using simple examples [19, 20, 21, 22] that phenomena such as chaos, large transient oscillations and bursting at steady state can occur, when modelling errors such as small bounded disturbances are present, without violating the boundedness properties of the modified schemes.

In an effort to eliminate some of these undesirable phenomena several researchers suggested the use of reference input signals that are dominantly rich [8, 23]. These signals are shown to lead to a high level of persistence of excitation relative to the level of the modelling error, that in turn guarantees exponential convergence for the tracking and parameter error to residual sets whose size is of the order of the modelling error. The use of dominantly rich signals however may destroy the tracking properties of the adaptive scheme especially in the case where the desired reference input signal is not a rich one. In another effort using averaging techniques and persistence of excitation [13] the performance of MRAC was analyzed for sufficiently small parameter errors. While this analysis improved the understanding of the behaviour of adaptive schemes locally, it did not provide much information about the global behaviour of adaptive schemes in the presence of large parametric uncertainties and in the absence of persistently exciting signals.

Recently Sun et. al. [24] proposed a novel approach to improve the performance of a non-adaptive model following scheme in the presence of parametric uncertainty by modifying the usual model reference control law with the addition of an extra term that compensates for the unknown parameters. This result was later extended to the adaptive case [25] and led to a modified control law referred to as a "modified certainty equivalence" design. It was intuitively argued that such an adaptive design would provide an improved transient performance, provided that the adaptation was slow. Simulation results showed that this was indeed the case.

In this paper, we first introduce two additional criteria to study the performance of standard by now robust MRAC schemes. They are the mean square tracking error criterion and the L_∞ tracking error bound. Our results using the first criterion reveal that the standard MRAC schemes may indeed have "bad" transient behaviour especially when the initial parameter error is large. The second criterion, on the other hand, provides us with a measure to quantify transient behaviour. Motivated from the results of Sun [25] and the swapping lemma we then propose a modified MRAC scheme which we show to have arbitrary zero-state performance improvement in terms of both these criteria in the ideal case and in the presence of bounded input disturbances. When unmodelled dynamics are present, the performance of the modified scheme has to be traded off with robust stability in a similar manner as in the non-adaptive case. In addition, the presence of high frequency sensor noise may impose limitations on the effectiveness of the modified scheme.

Related work can be found in [26] where a high gain switching MRAC scheme was proposed for obtaining arbitrarily good transient and steady state performance. The improvement in performance was obtained by modifying the standard model reference control objective to one of "approximate tracking." As a result, nonzero tracking errors remained at steady state. Our modified MRAC scheme, on the other hand, improves the transient performance while retaining the ideal steady state behaviour. Other contemporaneous work related to transient performance in adaptive control can be found in [27, 28]. A fundamental difference between these results and the one in this paper is that the former are aimed at *only estimating* transient signal bounds in model reference adaptive control, while the result here is concerned with *improvement* in the *zero-state* performance. In this context, it should be pointed out that after a preliminary version of this paper was presented [29], other approaches for improving the zero-state transient response of adaptive systems, for the ideal case, have also been reported in the literature [30].

The paper is organized as follows. Section 2 contains some mathematical preliminaries which form the backbone of our design and analysis in subsequent sections. In Section 3, we consider a standard MRAC scheme and analyze its performance using the mean square tracking error criterion and the L_∞ tracking error bound. In Section 4, we motivate, propose and analyze a modified MRAC scheme and demonstrate that it can be designed to arbitrarily improve the zero-state performance of the closed loop adaptive system in terms of both criteria considered. In Section 5, we show that the modified MRAC scheme can be designed to arbitrarily attenuate the effect of bounded input disturbances on the tracking error and therefore improve transients and reduce the size of possible bursts at steady state. In Section 6, we analyze the performance of the modified MRAC scheme in the presence of unmodelled dynamics. The tradeoff between performance improvement and robust stability is discussed. For clarity of presentation, the technical details are given in separate appendices. In Section 7, we carry out simulations on simple examples to demonstrate the efficacy of the modified MRAC scheme. Section 8 concludes the paper by summarizing the main results and outlining the directions for future research.

2 Mathematical Preliminaries

In this section, we give some definitions and lemmas which are useful in the stability and robustness analysis of adaptive control schemes.

Definition 2.1 *For any signal* $x : [0, \infty) \to R^n$, x_t *denotes the truncation of* x *to the interval* $[0, t]$ *and is defined as*

$$x_t(\tau) = \begin{cases} x(\tau) & \text{if } \tau \leq t \\ 0 & \text{otherwise} \end{cases} \tag{2.1}$$

Definition 2.2 *For any signal* $x : [0, \infty) \to R^n$, *and for any* $\delta \geq 0$, $t \geq 0$, $\|x_t\|_2^\delta$ *is defined as*

$$\|x_t\|_2^\delta \triangleq \left(\int_0^t e^{-\delta(t-\tau)} [x^T(\tau) x(\tau)] d\tau \right)^{\frac{1}{2}} \tag{2.2}$$

Remark 2.1 *The* $\|(.)_t\|_2^\delta$ *represents the exponentially weighted* L_2 *norm of the signal truncated to* $[0, t]$. *When* $\delta = 0$ *and* $t = \infty$, $\|(.)_t\|_2^\delta$ *becomes the usual* L_2 *norm and will be denoted by* $\|.\|_2$. *It can be shown that* $\|.\|_2^\delta$ *satisfies the usual properties of the vector norm.*

Definition 2.3 *Let* $H(s)$ *be a transfer function matrix whose entries are stable and proper. Then*

$$\|H(s)\|_\infty \triangleq \sup_\omega \left\{ \lambda_{max} \left[H^*(j\omega) H(j\omega) \right] \right\}^{\frac{1}{2}} \tag{2.3}$$

where $\lambda_{max}[.]$ *denotes the largest (necessarily real here) eigenvalue and* H^* *denotes the conjugate transpose of* H. *The* $\|.(s)\|_\infty$ *is the so-called* H_∞-*norm widely used in robust control [31].*

Definition 2.4 *Let* $H(s)$ *be a transfer function matrix whose entries are proper and analytic in* $Re[s] \geq -\delta/2$. *Then*

$$\|H(s)\|_\infty^\delta \triangleq \|H(s - \delta/2)\|_\infty \tag{2.4}$$

If in addition the entries of $H(s)$ *are strictly proper then*

$$\|H(s)\|_2^\delta \triangleq \left(\frac{1}{2\pi} \int_{-\infty}^\infty trace \left[H^*(j\omega - \delta/2) H(j\omega - \delta/2) \right] d\omega \right)^{\frac{1}{2}} \tag{2.5}$$

Definition 2.5 *Consider the signals* $x : [0, \infty) \to R^n$, $y : [0, \infty) \to R^+$ *and the set*

$$S(y) = \left\{ x : [0, \infty) \to R^n | \int_t^{t+T} x^T(\tau) x(\tau) d\tau \le \int_t^{t+T} y(\tau) d\tau + c \right\}$$

for some $c \ge 0$ *and* $\forall \, t, \, T \ge 0$. *We say that* x *is* y-small *in the mean if* $x \in S(y)$.

The following Lemmas play a key role in the robustness analysis of adaptive control schemes.

Lemma 2.1 *Let*

$$z = H(s)[u] = H * u$$

where $H(s)$ *is a transfer function matrix with causal and proper entries. If the entries of* $H(s)$ *are analytic in* $Re[s] \ge -\delta/2$ *for some* $\delta \ge 0$ *and* $u \in L_{2e}$ *then*

$$\|z_t\|_2^{\delta} \le \|H(s)\|_{\infty}^{\delta} \|u_t\|_2^{\delta} \tag{2.6}$$

If in addition the entries of $H(s)$ *are strictly proper then*

$$|z(t)| \le \|H(s)\|_2^{\delta} \|u_t\|_2^{\delta} \tag{2.7}$$

The proof of Lemma 2.1 can be found in [16, 32].

Lemma 2.2 *(Swapping Lemma) Let* $\tilde{\theta}, w : R^+ \to R^n$ *and let* $\tilde{\theta}$ *be differentiable. Let* $W(s)$ *be a proper stable rational transfer function with a minimal realization* $(A, \, b, \, c, \, d)$ *i.e.*

$$W(s) = c^T(sI - A)^{-1} b + d$$

Then

$$W(s)[\tilde{\theta}^T w] = \tilde{\theta}^T W(s)[w] + W_c(s)[(W_b(s)[w^T]) \dot{\tilde{\theta}}] \tag{2.8}$$

where

$$W_c(s) = -c^T(sI - A)^{-1}, \quad W_b(s) = (sI - A)^{-1} b$$

The proof of Lemma 2.2 can be found in [14, 15, 32].

Lemma 2.3 *[16] Let* $\tilde{\theta}, \, w : R^+ \to R^n$ *and* $\tilde{\theta}, \, w$ *be differentiable. Let* $W(s)$ *be a rational transfer function with relative degree* n^* *and with stable poles and zeros. Then*

$$\tilde{\theta}^T w = \Lambda_1(s, \alpha)[\dot{\tilde{\theta}}^T w + \tilde{\theta}^T \dot{w}]$$

$$+ \Lambda_0(s, \alpha) W^{-1}(s) \left[\tilde{\theta}^T W(s)[w] + W_c(s)[(W_b(s)[w^T]) \dot{\tilde{\theta}}] \right] \tag{2.9}$$

where $W_c(s)$, $W_b(s)$ *are as defined in Lemma 2.2,* $s\Lambda_1(s, \alpha) = 1 - \Lambda_0(s, \alpha)$, $\Lambda_0(s, \alpha) = \frac{\alpha^{n^*}}{(s+\alpha)^{n^*}}$ *for any arbitrary constant* $\alpha > 0$. *Furthermore for large* α

$$\|\Lambda_1(s, \alpha)\|_{\infty}^{\delta} \le \frac{c}{\alpha}, \quad \|\Lambda_0(s, \alpha) W^{-1}(s)\|_{\infty}^{\delta} \le c\alpha^{n^*} \tag{2.10}$$

for $\delta \ll 2\alpha$ *and some* $c \in R^+$.
The proof of Lemma 2.3 can be found in [16, 32].

Lemma 2.4 *If $u \in S(\mu)$ for some $\mu \geq 0$ then $\forall \delta > 0$, $\|u_t\|_2^\delta \in L_\infty$.*
The proof of Lemma 2.4 can be obtained by adapting the proof of Lemma 3.6 in [32].

3 Performance of a Standard MRAC Scheme

In this section, we consider a standard MRAC scheme and analyze its performance. The scheme consists of a model reference control algorithm and a parameter estimator using normalization and projection.

The plant to be controlled is described by the input-output pair u,y which are related by

$$y = G_0(s)[u] = k_p \frac{Z_0(s)}{R_0(s)}[u] \tag{3.1}$$

The control objective is to choose the control input u so that the output y of the plant tracks as closely as possible the output y_m of a stable reference model

$$y_m = W_m(s)[r] = k_m \frac{Z_m(s)}{R_m(s)}[r] \tag{3.2}$$

for any piecewise continuous, uniformly bounded reference input signal $r(t)$.

In order to meet the control objective, we make the following standard assumptions [7, 16] concerning the plant $G_0(s)$ and the reference model $W_m(s)$:

- (A1) $R_0(s)$ is a monic polynomial of degree n.
- (A2) $Z_0(s)$ is a monic Hurwitz polynomial of degree $m < n$.
- (A3) The constant k_p is known.[2]
- (A4) The relative degree $n^* = n - m$ is known.
- (A5) $Z_m(s)$, $R_m(s)$ are monic Hurwitz polynomials of degree q_m, p_m respectively with $p_m \leq n$, $n^* = p_m - q_m$ and $k_m > 0$.

We consider the following MRAC scheme that has been widely analyzed in the literature of adaptive control.
Control law:

$$u = \theta^T w + c_0 r, \ c_0 = \frac{k_m}{k_p} \tag{3.3}$$

where

$$w = [w_1^T, \ w_2^T, \ y]^T, \ \theta = [\theta_1^T, \ \theta_2^T, \ \theta_3]^T$$

$$w_1 = \frac{a(s)}{\Lambda(s)}[u], \ w_2 = \frac{a(s)}{\Lambda(s)}[y]$$

$$a(s) = [s^{n-2}, \ s^{n-3}, \ \cdots s, \ 1]^T$$

$$\Lambda(s) = Z_m(s)\lambda(s)$$

---

[2] This assumption is only for the sake of simplicity, and can be relaxed as in [7, 11, 16].

where $\lambda(s)$ is a monic Hurwitz polynomial of degree $n - 1 - q_m$.

Adaptive law:

$$\dot{\theta} = Pr[-\gamma\epsilon\phi], \ \theta(0) \in C_\theta \tag{3.4}$$

$$\epsilon = \frac{[\theta^T \phi + c_0 y - W_m(s)[u]]}{m^2} \tag{3.5}$$

$$\phi = W_m(s)[w] \tag{3.6}$$

$$\frac{d}{dt}m^2 = -\delta_0(m^2 - 1) + u^2 + y^2, \ m^2(0) = 1 \tag{3.7}$$

where $\gamma > 0$; $C_\theta = \{\theta \in R^{2n-1} | |\theta| \leq M_\theta\}$, $M_\theta > |\theta^*|$ is a design constant; θ^* is the desired controller parameter vector often referred to as the tuned parameter vector. The projection $Pr[.]$ operator is as defined in [33].

The design parameter $\delta_0 > 0$ is chosen so that $\lambda(s)$, $Z_m(s)$, $R_m(s)$ have all their roots in $Re[s] < -\frac{\delta_0}{2}$. The adaptive law (3.4) includes two rather standard by now modifications: a parameter projection and a special normalizing signal generated on line. Other modifications such as switching-σ, dead-zone etc. can also be used without changing the qualitative nature of the results of the paper. Similarly the projection sets can be changed especially when some apriori knowledge about θ^* is available such as lower bounds on each element etc.

Remark 3.1 *The signal m in (3.7) is a normalizing signal [9] and the quantity ϵ in (3.5) is the normalized estimation error [32]. For a detailed treatment of the theory underlying the design of the adaptive law (3.4)-(3.7), the reader is referred to [32].*

The following theorem describes the properties of (3.3), (3.4) when applied to the plant (3.1).

Theorem 3.1 *All the signals in the closed loop plant (3.1), (3.3), (3.4) are uniformly bounded and the tracking error $e_1(t) \triangleq y - y_m$ converges to zero as $t \to \infty$.*

Proof. The proof is given in [16] and is included in Appendix A for the sake of completeness.

Theorem 3.1 establishes boundedness $\forall t \geq 0$ and zero tracking error asymptotically with time. These two properties have been the main measures of performance in most MRAC schemes that are globally stable and do not employ any persistently exciting signals.

Two additional performance measures which as we will show are appropriate for MRAC schemes are the mean square (MS) tracking error defined as

$$E(t) \triangleq \frac{1}{t} \int_0^t e_1^2(z)dz, \ t > 0 \tag{3.8}$$

and the L_∞ tracking error bound defined as $\sup_{t\geq0} |e_1(t)|$. The MS criterion gives a measure of the average "energy" of the tracking error signal whereas the

L_∞ bound characterizes the pointwise in time behavior of the tracking error. We use these two measures to examine the performance of the standard MRAC scheme.

Theorem 3.2 *(Performance): Consider the closed loop plant (3.1), (3.3), (3.4). Then* $\forall\, t > 0$, *the tracking error* e_1 *satisfies*

$$\frac{1}{t}\int_0^t e_1^2(z)dz \le c\left[\frac{1}{\alpha^2}(\bar{m}^2+1) + \alpha^{2n^*}\bar{m}^2\frac{|\tilde{\theta}(0)|^2}{t}\right] \qquad (3.9)$$

where $\bar{m}$ *is an upper bound for the normalizing signal* $m(t)$, *i.e.* $m(t) \le \bar{m}$ $\forall\, t \ge 0$; $\tilde{\theta}(t) \triangleq \theta(t) - \theta^*$ *is the parameter error at time* t; $\alpha > 0$ *is an arbitrary large constant; and* c *is the generic symbol for a positive constant. Furthermore, we have*

$$\sup_{t \ge 0} |e_1(t)| \le c\bar{m} \qquad (3.10)$$

where $c > 0$ *depends on the choice of the reference model, the filters and* M_θ.

Proof. The proof is given in Appendix A.

Remark 3.2 *Throughout this paper* c *will be the generic symbol for positive constants. These constants can be explicitly calculated as in [16], [32] and such explicit calculations are necessary when analyzing quantitative robustness in adaptive control e.g. [16]. However, for the purpose of this paper, the exact values are unimportant and so we do not make any attempt to calculate them, especially since these calculations being very tedious would only unnecessarily contribute to the length and technicality of the paper.*

Remark 3.3 *The inequalities (3.9), (3.10) are developed under the assumption of zero initial conditions for the plant, reference model and filters. Non-zero initial conditions will introduce exponentially decaying to zero terms, which can be accounted for in the analysis, at the expense of additional tedious, yet simple algebraic manipulations. In this case, the use of a uniform[3] bound* $\bar{m}$ *on the normalizing signal* $m(t)$ *will lead to conservative estimates. Instead, the steps in the proof of Theorem 3.1 can be repeated, carrying the exponentially decaying to zero terms all throughout, leading to a bound on* $m(t)$ *which itself contains exponentially decaying to zero terms. Using such a bound, and repeating the steps in the proof of Theorem 3.2, with the exponentially decaying to zero terms carried all along, equation (3.9) becomes*

$$\frac{1}{t}\int_0^t e_1^2(z)dz \le (c+c_\epsilon)\left[\frac{1}{\alpha^2}(1+\frac{1}{t}c_\epsilon) + \alpha^{2n^*}\frac{1}{t}\right] + \frac{1}{t}c_\epsilon \qquad (3.11)$$

and equation (3.10) becomes

---

[3] Throughout this paper, the qualifier uniform will denote uniformity with respect to time, unless explicitly stated otherwise.

$$|e_1(t)| \le c + c_\epsilon + \varepsilon_t \tag{3.12}$$

where ε_t is a generic exponentially decaying to zero term that accounts for the decaying effect of the non-zero initial conditions on the various bounds while $c_\epsilon > 0$ is a generic constant which accounts for the non-decaying effect of the initial conditions (e.g. the non-decaying effect of the initial conditions on the L_2 bounds for $\frac{\tilde{\theta}^T \phi}{m}$, $\dot{\tilde{\theta}}$, etc.).

Equation (3.9) reveals that a large $|\tilde{\theta}(0)|$ may lead to a large mean square tracking error especially during the initial stages of adaptation. This may manifest itself as bad transient response, which is a characteristic of adaptive controllers observed quite often in simulations when the initial parameter error $\tilde{\theta}(0)$ happens to be large. Equation (3.10) on the other hand gives us a uniform bound for the tracking error e_1. Since $e_1(t) \to 0$ as $t \to \infty$, this bound may be conservative especially for large values of t. Nevertheless it does set a limit on the magnitude of possible oscillations that may occur during the transient phase.

Neither the mean square tracking error given by (3.9), nor the L_∞ tracking error bound given in (3.10) provide much insight as to how to improve performance. If a good apriori information is available about θ^*, then $\tilde{\theta}(0)$ can be made small resulting in a smaller mean-square tracking error. A smaller $\tilde{\theta}(0)$ will also result in a smaller $\bar{m}$ (see Appendix A, (A.14) onwards), possibly leading to an improved transient behaviour. However, the interest in adaptive control usually arises when the parametric uncertainty happens to be large.

In the next section we modify the control law (3.3) so that the performance, measured as either the mean square tracking error or the L_∞ tracking error bound, can be arbitrarily improved.

4 Performance of a Modified MRAC Scheme

In this section we propose a modified MRAC scheme that has the capability of improving performance by properly choosing a certain design parameter. The modification introduced is motivated from the swapping lemma and the results of Sun [25] and is explained as follows. Instead of the control law (3.3), consider the control law

$$u = \theta^T w + c_0 r - Q(s) \left[\epsilon m^2 + W_c(s) \left[(W_b(s)[w^T]) \dot{\tilde{\theta}} \right] \right] \tag{4.1}$$

where $Q(s)$ is a stable filter to be specified later, and $W_c(s)$, $W_b(s)$ are as obtained from Lemma 2.2 with $W(s) = W_m(s)$. Since the signals ϵ, m, w and $\tilde{\theta}$ are all available for measurement, and the transfer functions $W_c(s)$, $W_b(s)$ are known, the above control law is implementable. Now using (A.19) (Appendix A) and the Swapping Lemma, we obtain

$$\epsilon m^2 + W_c(s) \left[(W_b(s)[w^T]) \dot{\tilde{\theta}} \right] = \tilde{\theta}^T \phi + W_c(s) \left[(W_b(s)[w^T]) \dot{\tilde{\theta}} \right] = W_m(s) \left[\tilde{\theta}^T w \right]$$

This when substituted into (4.1) yields

$$u = \theta^{*T} w + c_0 r + [1 - Q(s)W_m(s)] [\tilde{\theta}^T w] \tag{4.2}$$

Combining (3.1) and (4.2), we obtain the expression

$$e_1 = \frac{1}{c_0} W_m(s) [1 - Q(s)W_m(s)] [\tilde{\theta}^T w] \tag{4.3}$$

for the tracking error.

Our objective here is to choose the filter $Q(s)$ so as to attenuate the effect of the parametric uncertainty on the tracking error e_1. To do so, we make use of the following lemma which shows that $Q(s)$ can be chosen to make $\|\frac{1}{c_0} W_m(s) [1 - Q(s)W_m(s)]\|_{\hat{A}}$ arbitrarily small.[4]

Lemma 4.1 Let $W_m(s)$ be strictly proper and minimum phase. Then

$$\inf_{Q(s)\in \hat{A}} \left\| \frac{1}{c_0} W_m(s) [1 - Q(s)W_m(s)] \right\|_{\hat{A}} = 0.$$

Furthermore, $Q(s) = \frac{W_m^{-1}(s)}{(\tau s+1)^{n^*}}$, $\tau > 0$ is an infimizing sequence in the sense that $\|\frac{1}{c_0} W_m(s) [1 - Q(s)W_m(s)]\|_{\hat{A}} \to 0$ as $\tau \to 0$.

Proof. Now substituting $Q(s) = \frac{W_m^{-1}(s)}{(\tau s+1)^{n^*}}$ into $\|\frac{1}{c_0} W_m(s) [1 - Q(s)W_m(s)]\|_{\hat{A}}$, and after some straight forward calculations, we can show that

$$\|\frac{1}{c_0} W_m(s) [1 - Q(s)W_m(s)]\|_{\hat{A}} \le \tau c$$

where $c > 0$ is a constant independent of τ, thereby proving the lemma.

In view of Lemma 4.1, we choose $Q(s) = \frac{W_m^{-1}(s)}{(\tau s+1)^{n^*}}$, $\tau > 0$. This leads to the control law

$$u = \theta^T w + c_0 r - \frac{W_m^{-1}(s)}{(\tau s + 1)^{n^*}} [\epsilon m^2 + W_c(s)[(W_b(s)[w^T])\dot{\tilde{\theta}}]] \tag{4.4}$$

where $\tau > 0$ is a design constant. The control law (4.4) will be used together with the adaptive law (3.4) to form a modified MRAC scheme.

A similar modified MRAC scheme was proposed in [25] where the modification to the control input was motivated from non-adaptive design considerations [24] and consequently did not include the $\dot{\tilde{\theta}}$ term in (4.4). The above derivation of the control law (4.4) was carried out assuming that the high frequency gain k_p is known. For the case of unknown k_p, a similar modification to the control law can be derived, where the estimation error and "swap terms" get replaced by a feedback term involving the tracking error. Since the steps involved are quite

---

[4] Here $\hat{A}$ is the set of all bounded-input bounded-output (BIBO) stable transfer functions and $\|.\|_{\hat{A}}$ is the corresponding L_∞ gain [36].

different from the approach taken here, the design and analysis of such a scheme is addressed in another paper [34].

It is clear from (3.1), (4.4) that

$$u = \theta^{*T} w + c_0 r + W_\tau(s)[\tilde{\theta}^T w] \tag{4.5}$$

and

$$e_1 = \frac{1}{c_0} W_m(s) W_\tau(s)[\tilde{\theta}^T w] \tag{4.6}$$

$$\text{where } W_\tau(s) \triangleq 1 - \frac{1}{(\tau s + 1)^{n^*}} \tag{4.7}$$

The following Theorem establishes the stability properties of the modified MRAC scheme when applied to the plant (3.1).

Theorem 4.1 *All the signals in the closed-loop plant (3.1), (3.4), (4.4) are uniformly bounded and the tracking error $e_1(t)$ converges to zero as $t \to \infty$. Furthermore the bound for the normalizing signal m is independent of $\tau \in (0, \tau_{max}]$ where $\tau_{max} > 0$ is any finite constant i.e. $m(t) \leq \bar{m} \ \forall \, t$ and $\forall \, \tau \in (0, \tau_{max}]$.*

Proof. The proof is given in Appendix B.

The performance of the closed loop system (3.1), (3.4), (4.4) is given by the following Theorem.

Theorem 4.2 *(Performance): Consider the closed loop plant (3.1), (3.4), (4.4) with $\tau \in (0, \tau_{max}]$ where $\tau_{max} > 0$ is any finite number. Then $\forall \, t > 0$, the tracking error e_1 satisfies*

$$\frac{1}{t} \int_0^t e_1^2(z) dz \leq \tau^2 c \left[\frac{1}{\alpha^2} (\bar{m}^2 + 1) + \alpha^{2n^*} \bar{m}^2 \frac{|\tilde{\theta}(0)|^2}{t} \right] \tag{4.8}$$

where $\bar{m}$ is as defined in Theorem 4.1 and $\alpha > 0$ is an arbitrary large constant. Furthermore, we have

$$\sup_{t \geq 0} |e_1(t)| \leq \tau c(\bar{m} + 1) \tag{4.9}$$

where $c > 0$ is independent of $\tau \in (0, \tau_{max}]$.

Proof. The proof is given in Appendix B.

Since $\bar{m}$ is *independent* of $\tau \in (0, \tau_{max}]$, it is clear from (4.8) that using the MRAC scheme (3.4), (4.4) we can make the zero-state mean square tracking error *arbitrarily* small by choosing τ to be small enough. This by itself is not sufficient to guarantee an improvement in the zero-state transient performance. However, from (4.9) it is clear that the uniform bound on the zero-state tracking error can also be made as small as we wish by reducing τ. Therefore we can arbitrarily improve the transient performance of the zero-state tracking error.

Remark 4.1 *Non-zero initial conditions on the states of the plant and filters introduce exponentially decaying to zero terms which do affect performance. However, the performance improvement results of this paper are stated for zero initial conditions. For non-zero initial conditions the bounds (4.8), (4.9) have to be modified as in Remark 3.3 by including exponentially decaying to zero terms. Employing such an analysis (4.8) becomes*

$$\frac{1}{t}\int_0^t e_1^2(z)dz \leq \tau^2(c+c_\epsilon)\left[\frac{1}{\alpha^2}\left\{1+\frac{1}{t}c_\epsilon\left(1+\frac{1}{\tau^{n^*}}\right)^2\right\}\right.$$

$$\left.+\alpha^{2n^*}\frac{1}{t}\left\{1+\left(1+\frac{1}{\tau^{n^*}}\right)^2 c_\epsilon\right\}\right]+\frac{1}{t}c_\epsilon \qquad (4.10)$$

and (4.9) becomes

$$|e_1(t)| \leq \tau(c+c_\epsilon)+c\left(1+\frac{1}{\tau^{n^*}}\right)\varepsilon_t \qquad (4.11)$$

From equation (4.11), it is clear that for non-zero initial conditions, the modified MRAC scheme results in an instantaneous bound on the tracking error whose value at $t = 0$ may actually be higher than that of the standard unmodified scheme. However, the terms that contribute to the possibly higher instantaneous performance bounds in (4.11) and (4.10) are themselves decaying, either exponentially fast or as $\frac{1}{t}$, so that the modified scheme will still lead to improved performance, once the transients due to non-zero initial conditions have become small. In this context, it should be pointed out that the modified MRAC scheme of this paper seeks to improve performance only by attenuating the effect of large parameter mismatch on the tracking error. No attempt, whatsoever, was made here to improve the response due to non-zero initial conditions. Indeed, the problem of improving the response due to non-zero initial conditions, which remain even when the parametric uncertainty disappears, i.e. even when we have the perfectly matched case, is not a problem specific to adaptive control, which is aimed primarily towards dealing with parametric uncertainty.

5 Performance in the Presence of Bounded Input Disturbances

In this section, we study the effect of bounded input disturbances on the stability and performance of the modified MRAC scheme. Instead of (3.1), we now consider the plant

$$y = G_0(s)\,[u+d] = k_p\frac{Z_0(s)}{R_0(s)}\,[u+d] \qquad (5.1)$$

where $d(t)$ is a bounded disturbance. The control law (4.4) and the adaptive law (3.4) are kept the same as before with the only modification that now (3.7) is replaced by

$$\frac{d}{dt}m^2 = -\delta_0\left(m^2 - \frac{1}{\alpha_0^2}\right) + u^2 + y^2, \quad m^2(0) = \frac{1}{\alpha_0^2}, \quad \alpha_0 > 0 \qquad (5.2)$$

The only difference between (3.7) and (5.2) is that the lower bound of m^2 in (3.7) is unity while in the case of (5.2), it is $\frac{1}{\alpha_0^2}$. Thus by choosing α_0 to be small in (5.2), we are able to increase the lower bound of m^2 so that the "normalized contribution" of d entering the adaptive law (3.4) can be made as small as we wish. This is mainly done for the sake of simpler analysis by avoiding the use of contradiction arguments [11] in establishing the boundedness of m when $\alpha_0^2 = 1$.

The following Theorem establishes the stability properties of the MRAC scheme (3.4), (4.4) when applied to the plant (5.1).

Theorem 5.1 *Let $\tau \in (0, \tau_{max}]$ where $\tau_{max} > 0$ is any finite number. Then $\exists \alpha_0^* > 0$ such that if $\alpha_0 < \alpha_0^*$ in (5.2), all the signals in the closed loop plant (5.1), (3.4), (4.4) are uniformly bounded. Furthermore, the normalizing signal $m(t)$ satisfies*

$$m^2(t) \leq \bar{m}^2 \qquad (5.3)$$

where $\bar{m} > 0$ is independent of $\tau \in (0, \tau_{max}]$.

Proof. The proof is given in Appendix C.

The performance of the closed loop system (5.1), (3.4), (4.4) is given by the following Theorem.

Theorem 5.2 *Consider the closed loop plant (5.1), (3.4), (4.4) with $\alpha_0 < \alpha_0^*$ where α_0^* is as defined in Theorem 5.1. Then $\forall\, t > 0$ the tracking error e_1 satisfies*

$$\frac{1}{t}\int_0^t e_1^2(z)dz \leq \tau^2 c\left[\left\{\frac{1}{\alpha^2}(\bar{m}^2 + 1) + d_0^2\right\}\right.$$
$$\left. + \alpha^{2n^*}\bar{m}^2\left(d_0^2 + \frac{|\tilde{\theta}(0)|^2}{t}\right)\right] \qquad (5.4)$$

where $\bar{m}$ is as defined in Theorem 5.1; $\alpha > 0$ is an arbitrary large constant; and $d_0 > 0$ is an upper bound for $d(t)$. If in addition $\dot{d} \in L_\infty$, then we have

$$\sup_{t\geq 0}|e_1(t)| \leq \tau c\left[(\bar{m} + 1)(d_0 + 1) + \sup_{t\geq 0}|\dot{d}(t)|\right] \qquad (5.5)$$

where $c > 0$ is independent of $\tau \in (0, \tau_{max}]$.

Proof. The proof is given in Appendix C.

From (5.3), (5.4), we see that, using the modified MRAC scheme, we are able to attenuate the effect of bounded disturbances on the mean square tracking error by simply reducing the value of τ. Furthermore, from (5.3), (5.5) we see that by decreasing τ, the uniform bound on the zero-state tracking error can be made as small as we wish. This decrease in the uniform bound is expected to bring about an improvement in the zero-state transient behaviour as also a reduction in the size of the possible bursts that may occur at steady state. Thus disturbance rejection and nominal performance improvement go hand in hand. This is a well known observation in linear feedback theory and here we have just shown that a similar observation is true for the modified MRAC scheme also.

In our foregoing analysis, the disturbance in the adaptive system was caused by a bounded disturbance at the plant input. Another interesting way in which a disturbance may appear in an adaptive scheme is when a forgetting factor such as a σ-modification is introduced into the adaptive law, mainly for the purpose of enhancing robustness. We now consider one such scheme and show how the use of the modified control law (4.4) can improve the performance.

For MRAC schemes using a fixed σ-modification [4, 11], the adaptive law (3.4) is replaced by

$$\dot{\theta} = -\gamma\epsilon\phi - \gamma\sigma\theta \tag{5.6}$$

where $\sigma > 0$ is a design constant and ϵ, ϕ are as defined in (3.5), (3.6). For such schemes the tracking error cannot be shown to converge to zero (even in the ideal case) [11]. In fact, in [22] it has been shown that an adaptive scheme using the standard control law (3.3) and a fixed σ-modification in the adaptive law, can exhibit bursting behaviour at steady state solely due to the forgetting factor σ which introduces the disturbance term $\gamma\sigma\theta^*$ in the adaptive law. We now show that by replacing the control law (3.3) with (4.4) and keeping the adaptive law (5.6) the same as before, we are able to reduce the mean square tracking error as also the L_∞ tracking error bound to as small a level as we wish by simply decreasing τ.

The following theorem establishes the stability properties of the MRAC scheme (4.4), (5.6) when applied to the plant (3.1).

Theorem 5.3 *Let $\tau \in (0, \tau_{max}]$ where $\tau_{max} > 0$ is any finite number. Then $\exists \sigma_0 > 0$ such that $\forall \sigma \in [0, \sigma_0)$, all the signals in the closed loop plant (3.1), (4.4), (5.6) are uniformly bounded. Furthermore, the normalizing signal m satisfies*

$$m^2(t) \leq \bar{m}^2 \tag{5.7}$$

where $\bar{m} > 0$ is independent of $\tau \in (0, \tau_{max}]$.

Proof. The proof is very similar to that of Theorem 4.1 and is, therefore, omitted. The only difference is that, this time instead of being in L_2, the signals $\frac{\tilde{\theta}^T \phi}{m}$, $\dot{\tilde{\theta}} \in S(c\sigma)$. Hence the same proof goes through provided σ_0 is small enough.

The performance of the closed loop system (3.1), (4.4), (5.6) is given by the following theorem.

Theorem 5.4 *Consider the closed loop plant (3.1), (4.4), (5.6) with $\sigma \in [0, \sigma_0)$ where σ_0 is as defined in Theorem 5.3. Then $\forall\, t\; >\; 0$ the tracking error e_1 satisfies*

$$\frac{1}{t}\int_0^t e_1^2(z)dz \leq \tau^2 c\left[\frac{1}{\alpha^2}(\bar{m}^2 + 1)\right.$$

$$\left.+\alpha^{2n^*}\bar{m}^2\left(\sigma + \frac{|\tilde{\theta}(0)|^2}{t}\right)\right] \tag{5.8}$$

where $\bar{m}$ is as defined in Theorem 5.3; and $\alpha\; >\; 0$ is an arbitrary large constant. Furthermore, we have

$$\sup_{t\geq 0}|e_1(t)| \leq \tau c\left[\bar{m}(\sigma + 1) + 1\right] \tag{5.9}$$

where $c\; >\; 0$ is independent of $\tau \in (0,\ \tau_{max}]$.

Proof. The proof is very similar to that of Theorem 4.2 and is, therefore, omitted.

From (5.7) and (5.8), it is clear that, using the modified control law (4.4), the mean square tracking error can be made arbitrarily small by reducing τ. Also from (5.9), it follows that the uniform bound on the zero-state tracking error can be made as small as we wish. This decrease in the uniform bound is expected to not only improve the zero-state transient behaviour but also bring about a reduction in the size of the bursts that may possibly occur at steady state.

The performance improvement achieved by the modified MRAC scheme seems to be too good to be true. One may wonder what we lose for such an improvement in performance. The answer is a well known fact in feedback control; performance improvement has to be traded off with robust stability. In addition, the presence of high frequency sensor noise may impose limitations on the bandwidth of the high pass compensator in (4.4) so that, in practice, τ cannot be chosen to be arbitrarily small. The trade-off between performance improvement and robust stability is established in the following section where the modified MRAC scheme is analyzed in the presence of unmodelled dynamics. Due to space limitations, the corresponding analysis in the presence of sensor noise is omitted.

6 Performance in the Presence of Unmodelled Dynamics

Instead of (3.1), let us now consider the plant

$$y = G_0(s)\left[1 + \mu\Delta_m(s)\right][u] \tag{6.1}$$

where $\mu\; >\; 0$ is a small constant; $G_0(s)$ is as defined in (3.1) and represents the modelled part of the plant; and $\Delta_m(s)$ is a multiplicative uncertainty[5] that satisfies

---

[5] Here the unmodelled dynamics are assumed to be of the multiplicative type. This is without any loss of generality, and other kinds of uncertainty characterizations such as additive or stable factor perturbations [35] can be used just as well.

- (A6) There exists a known $\delta^* > 0$ and a stable rational transfer function $W(s)$ of relative degree $< n^*$ having all its poles and zeros in $Re[s] < -\frac{\delta^*}{2}$ such that $\|W(s)\Delta_m(s)\|_\infty^{\delta^*} < \infty$.

The adaptive law (3.4) and the control law (4.4) are kept the same as before with the only modification that now the design parameter δ_0 in (3.7) is chosen to satisfy $\delta_0 < \delta^*$, in addition to the earlier requirements placed on its selection for the ideal case.

The following Theorem establishes the stability properties of the MRAC scheme (3.4), (4.4) when applied to the plant (6.1).

Theorem 6.1 Let $\tau \in (\tau_{min}, \tau_{max}]$ where $\tau_{max} > \tau_{min} > 0$ are any finite numbers. Then $\exists\ \mu^* > 0$ such that $\forall\ \mu \in [0,\ \mu^*)$, all the signals in the closed loop plant (6.1), (3.4), (4.4) are uniformly bounded. Furthermore, the normalizing signal $m(t)$ satisfies

$$m^2(t) \leq \bar{m}^2 \triangleq \frac{c}{A^2}\left[1 + \frac{\frac{cB^2}{A^2}(c_\eta + 1)}{\left(1 - ce^{\frac{c}{A^2}B^2 c_\eta}\right)}e^{\frac{cB^2}{A^2}(1 - c_\eta)}\right] \tag{6.2}$$

$$A \triangleq 1 - \mu^* c\left(1 + \frac{1}{(\tau_{min})^{n^*}}\right) - c_\alpha$$

$$B \triangleq 1 + \mu^* c\left(1 + \frac{1}{(\tau_{min})^{n^*}}\right)$$

where $c_\alpha \in (0,\ 1)$ is some constant and $c_\eta = c\mu$ for some $c > 0$.

Proof. The proof is given in Appendix D.

From the proof of Theorem 6.1 (Appendix D, (D.18)), we see that as τ_{min} is decreased, μ^* becomes smaller and smaller. Thus the performance improvement that we obtained in Section 4 was achieved at the expense of robust stability. In other words, the modified MRAC scheme demonstrates, in an adaptive control context, the tradeoff between improving performance and guaranteeing robust stability. Also from (D.18), it is clear that as τ_{min} approaches zero, then so does μ^*. Thus, in a practical situation, τ_{min} cannot be made arbitrarily small because otherwise the resulting MRAC scheme may be unstable. Hence arbitrary performance improvement is no longer possible. Instead, as we shall see, the amount of possible performance improvement depends on the size of the unmodelled dynamics.

Till now, we have been using the mean square tracking error and the L_∞ tracking error bound to characterize performance. However, in the presence of unmodelled dynamics, calculation of the L_∞ tracking error bound involves the use of induced L_∞ norms which are not easy to calculate [36]. To facilitate the calculation of the L_∞ tracking error bound, we make the following additional assumption concerning $\Delta_m(s)$:

- (A7) There exists a stable minimum phase rational transfer function $W_1(s)$ of relative degree $< n^*$ such that $\|W_1(s)\Delta_m(s)\|_A < \infty$.

The performance of the closed loop system (6.1), (3.4), (4.4) is given by the following Theorem.

Theorem 6.2 *Consider the closed loop plant (6.1), (3.4), (4.4) with $\tau \in (\tau_{min}, \tau_{max}]$ and $\mu \in [0, \mu^*)$ where μ^* is as defined in Theorem 6.1. Then $\forall\, t > 0$ the tracking error e_1 satisfies*

$$
\frac{1}{t}\int_0^t e_1^2(z)dz \leq \frac{\tau^2 c}{\left[1 - \mu^* c\left(1 + \frac{1}{(\tau_{min})^{n^*}}\right)\right]^2}
$$

$$
\left[\frac{1}{\alpha^2}\frac{\left\{1 + \mu^{*2}\left(1 + \frac{1}{(\tau_{min})^{2n^*}}\right)\right\}\bar{m}^2 + 1}{\left\{1 - \mu^* c\left(1 + \frac{1}{(\tau_{min})^{n^*}}\right)\right\}^2}\right.
$$

$$
\left. + \alpha^{2n^*}\bar{m}^2\left(c_\eta + \frac{|\tilde{\theta}(0)|^2}{t}\right)\right] + \frac{\mu^{*2} c}{\left[1 - \mu^* c\left(1 + \frac{1}{(\tau_{min})^{n^*}}\right)\right]} \tag{6.3}
$$

where $\bar{m}$, c_η are as defined in Theorem 6.1; and $\alpha > 0$ is an arbitrary large constant. Furthermore, we have

$$
\sup_{t\geq 0}|e_1(t)| \leq \frac{\tau c}{\left[1 - \mu^* c\left(1 + \frac{1}{(\tau_{min})^{n^*}}\right)\right]}\left[\left\{1 + \mu^*\left(1 + \frac{1}{(\tau_{min})^{n^*}}\right)\right\}\bar{m} + 1\right]
$$

$$
+ \frac{\mu^* c}{\left[1 - \mu^* c\left(1 + \frac{1}{(\tau_{min})^{n^*}}\right)\right]} \tag{6.4}
$$

where the allowable μ^* for (6.4) to hold may be smaller than the one for (6.3) (see Appendix D, (D.30)).

Proof. The proof is given in Appendix D.

From (6.3),(6.4), we see that if $\tau \in (\tau_{min}, \tau_{max}]$, then a reduction in τ within this interval will lead to an improved tracking performance. However, τ must be chosen to be greater than τ_{min}. Since robust stability considerations prevent us from choosing $\tau < \tau_{min}$, arbitrary zero-state performance improvement is no longer possible. Nevertheless if, as τ_{min} is decreased, the ratio $\frac{\mu^*}{(\tau_{min})^{n^*}}$ can be maintained less than some pre-specified constant, then the right hand side of (6.3), (6.4) as also the expression for $\bar{m}$ in (6.2) no longer depends on τ_{min}. As a result, we can decrease τ_{min} and thereby permit a smaller choice of τ leading to further improvement in the zero-state tracking performance. The extent to which τ_{min} can be decreased while keeping the ratio $\frac{\mu^*}{(\tau_{min})^{n^*}}$ less than some pre-specified constant, depends on how small a μ^* is acceptable to us. This, in turn, depends on the amount of unmodelled dynamics that our design is required to tolerate. If, for instance, there are no unmodelled dynamics, then μ^* equals zero is acceptable and, in this case, we can make τ_{min} as small as we wish, leading to arbitrary zero-state performance improvement. Thus the amount of zero-state performance improvement possible depends on the size of the unmodelled dynamics. The "smaller" the size of the unmodelled dynamics, the larger the *possible* zero-state performance improvement. In the limit, as the unmodelled dynamics disappear, arbitrary zero-state performance improvement becomes, once again, possible. The above is not a peculiarity of our proposed modified MRAC scheme. Indeed, it is a fundamental characteristic of any feed-back control system.

In the foregoing discussion, we have considered only unmodelled dynamics. In a realistic situation, however, high frequency sensor noise will also be present. This may impose limitations on the bandwidth of the high pass compensator in (4.4) which, in turn, would prevent us from choosing an arbitrarily small value for τ. This is a well known tradeoff in every robust control design.

7   Examples and Simulations

In this section, we first demonstrate the efficacy of the modified MRAC scheme in improving performance by using a simple first order example. The plant to be controlled is described by

$$y = \frac{1}{s-1}[1 + \Delta_m(s)][u + d] \tag{7.1}$$

where $\Delta_m(s) = -\frac{2\mu s}{\mu s+1}$, $\mu > 0$ is a multiplicative uncertainty and $d(t)$ is a bounded input disturbance. The reference model is

$$y_m = \frac{1}{s+1}[r] \tag{7.2}$$

Thus $\theta^* = -2.0$. In all the simulations for this example, $C_\theta = [-4, 4]$, $\delta_0 = 0.1$, $\gamma = 1$, $\theta(0) = -3.5$, $c_0 = 1$. The initial conditions on the states of the plant and reference model were all set equal to zero. In Figs. 1 through 4, u and e represent the plant input and tracking error respectively while 'tau' represents the parameter τ. Also in Fig. 5, yp represents the plant output.

We first simulated the ideal case i.e. with $d = \mu = 0$. We chose $r(t) = 1.0$ and obtained the plots shown in Fig. 1 with the standard and modified ($\tau = 0.5$ and $\tau = 0.01$) schemes. From this figure, it is clear that the modified scheme gives a much better transient tracking performance than the standard one. Furthermore, as predicted by the theory, a decrease in τ brings about an improvement in the transient behaviour. The control inputs in Fig. 1 are certainly not impulsive. So, although the modified scheme provides excellent transient behaviour, there are no impulsive controls. The situation, however, is likely to be different when the initial conditions are non-zero.

Next, we simulated the plant in the presence of the bounded disturbance $d(t) = 0.1\mathrm{Sin}0.5t$. The parameter α_0 in (5.2) was chosen to be 1.0. With $r(t) = 1.0$, we obtained the plots shown in Fig. 2. For the modified scheme, τ was chosen to be 0.5. From Fig. 2, it is clear that the modified scheme not only gives a better transient tracking performance as compared to the standard one, but it

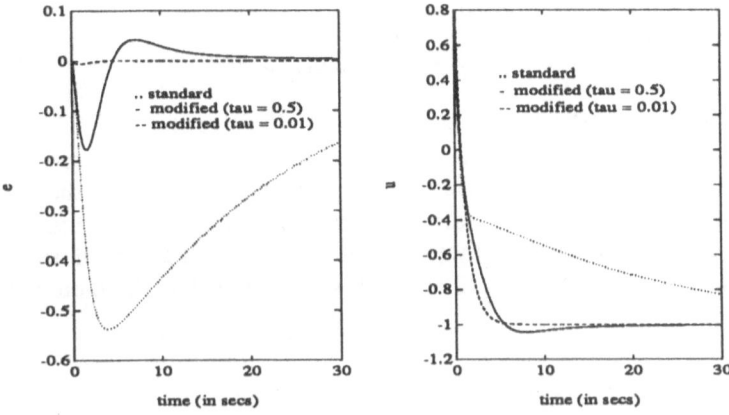

Fig. 1. Responses for the Ideal Case ($r(t) = 1.0$)

also attenuates the effect of the input disturbance on the output tracking error. This agrees with our analysis in Section 5.

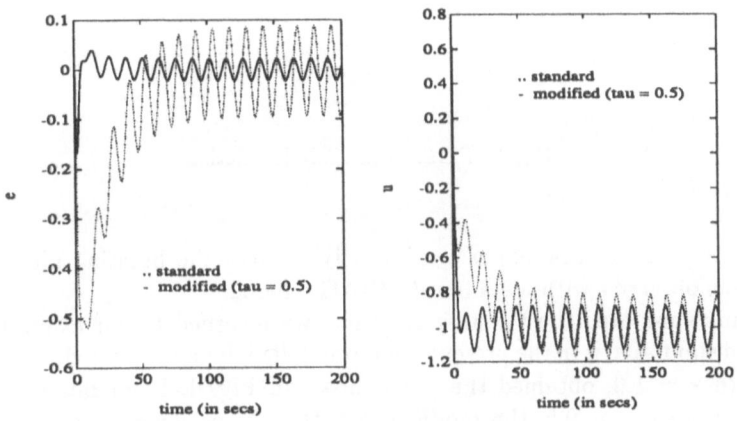

Fig. 2. Responses with an Input Disturbance ($d(t) = 0.1\mathrm{Sin}0.5t$, $r(t) = 1.0$)

Having demonstrated that the modified MRAC scheme does provide good transient behaviour and disturbance rejection, we simulated a bursting example. We considered the same plant as the one simulated in [22] i.e.

$$y = \frac{s+2}{s^2 - s - 2}[u] \tag{7.3}$$

The reference model was chosen as

$$y_m = \frac{1}{s + a_m} r, \ a_m > 0 \tag{7.4}$$

We first considered the MRAC scheme (3.3), (5.6) applied to (7.3). Choosing $\Lambda(s) = (s + 1)$, $r = 0$, $\gamma = 1$, $\sigma = 0.02$, $\delta_0 = 1.9$, $a_m = 1$, $y(0) = 1.0$ and $\theta(0) = [0, 0, 0]^T$, we obtained the bursting phenomena shown by the dotted line in Fig. 3. Keeping all design parameters and initial conditions fixed, we then replaced the control law (3.3) by (4.4) and chose $\tau = 0.1$ to obtain the response shown by the solid line in Fig. 3. From Fig. 3 we see that the modified MRAC

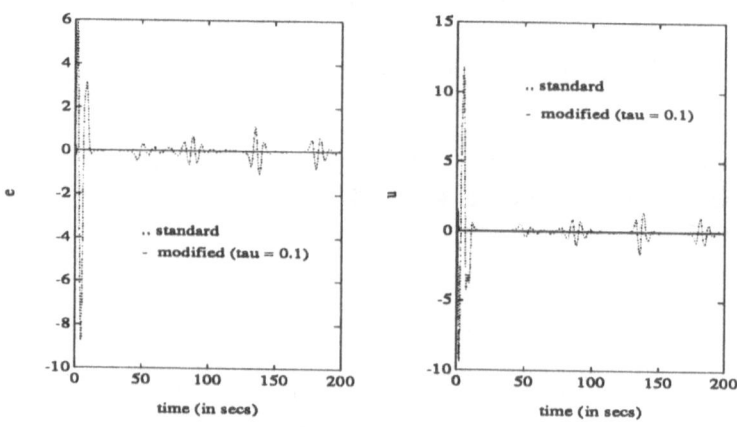

Fig. 3. Bursts and their Reduction

scheme proposed in this paper significantly reduces the bursting phenomenon which was observed with a standard MRAC scheme.

Having simulated the bursting example, we returned to our example (7.1), (7.2) and simulated it in the presence of unmodelled dynamics. We chose $\mu = 0.01$ and, with $r = 1.0$, obtained the plots shown in Fig. 4. From this figure, it is clear that, with $\tau = 0.5$, the modified MRAC scheme gives a better transient performance as compared to the standard one. Reducing the value of τ from 0.5 to 0.1 further improves the transient performance. The value of τ was then further reduced to 0.01 and led to the instability phenomena shown in Fig. 5. This confirms our observation that, in the presence of unmodelled dynamics, τ cannot be chosen to be too small or else instability will result.

8 Conclusion

In this paper we first introduced the mean square tracking error criterion and the L_∞ tracking error bound criterion to examine the performance of a standard

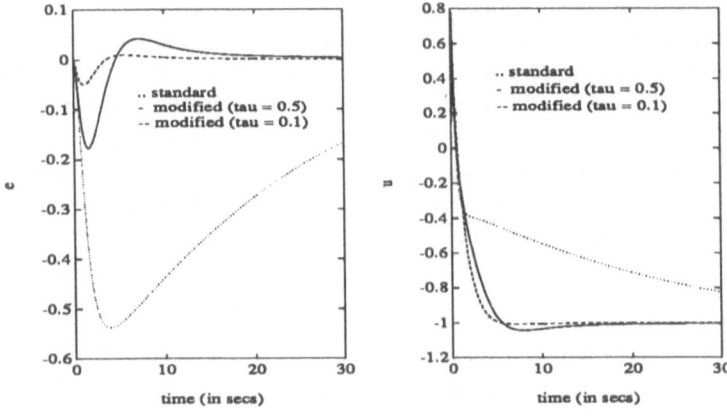

Fig. 4. Responses in the Presence of Unmodelled Dynamics ($\mu = 0.01$, $r(t) = 1.0$)

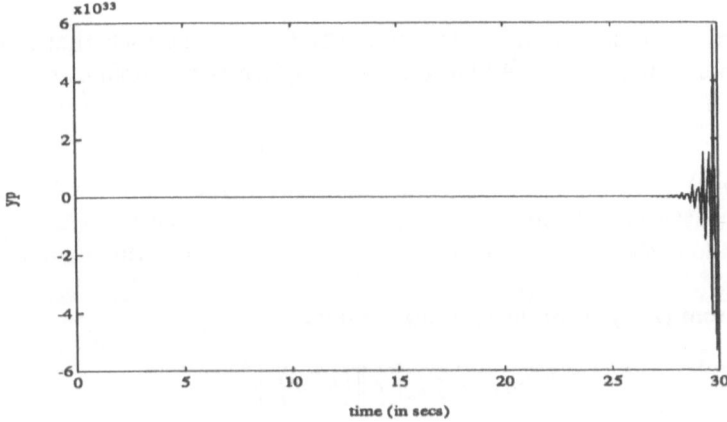

Fig. 5. Instability Using Modified MRAC ($\mu = 0.01$, $r(t) = 1.0$, $\tau = 0.01$)

robust MRAC scheme. We showed that the standard scheme may exhibit bad transient behaviour especially when the initial parameter error is large. We then proposed a modified scheme whose zero-state performance can be arbitrarily improved in the presence of large parametric uncertainties and bounded input disturbances. The performance improvement is in terms of both the mean square tracking error and the L_∞ tracking error bound and the latter is shown to imply an improved zero-state transient behaviour and smaller possible bursts at steady state. In the presence of unmodelled dynamics arbitrary zero-state performance improvement can no longer be achieved without affecting robust

48

stability. Moreover, the presence of high frequency sensor noise also prevents us from obtaining arbitrarily improved zero-state performance. These tradeoffs, between performance on one hand, and robustness and sensor noise on the other, are well known in feedback control systems and are therefore not surprising.

The MRAC schemes considered in this paper are based on the assumption that the plant high frequency gain k_p is known apriori. The extension of these results to allow uncertainty in k_p requires a somewhat different approach to the problem and is presented in [34].

APPENDIX A

Proof of Theorem 3.1: From (3.1), (3.3), we can show that

$$u = \frac{R_0(s)}{k_p Z_0(s)} \frac{W_m(s)}{c_0} \left[\tilde{\theta}^T w + c_0 r \right] \tag{A.1}$$

$$y = \frac{W_m(s)}{c_0} \left[\tilde{\theta}^T w + c_0 r \right] \tag{A.2}$$

where $\tilde{\theta} = \theta - \theta^*$ is the parameter error. Let $\delta_1 > 0$ be such that $Z_0(s)$ has all its roots in $\text{Re}[s] < -\frac{\delta_1}{2}$. Let $\delta \in (0, \ min[\delta_0, \delta_1])$ and define the fictitious signal $m_f(t)$ by

$$m_f(t) = 1 + \|u_t\|_2^\delta + \|y_t\|_2^\delta \tag{A.3}$$

Clearly $m(t) \leq m_f(t) \ \forall \ t \geq 0$. From (A.2), (A.1) and Lemma 2.1, it can be shown that the boundedness of $m_f(t)$ would guarantee the boundedness of all the closed loop signals. Accordingly, we proceed to establish the boundedness of $m_f(t)$.

Now from (A.1)-(A.3), using (2.6), we obtain

$$m_f(t) \leq c + c \left\| \left(\tilde{\theta}^T w \right)_t \right\|_2^\delta \tag{A.4}$$

Now using (2.9) with $W(s) = W_m(s)$ and $\Lambda_0(s, \ \alpha) = \frac{\alpha^{n^*}}{(s+\alpha)^{n^*}}$, $\alpha > \frac{\delta}{2}$, we obtain

$$\tilde{\theta}^T w = \Lambda_1(s, \ \alpha) \left[\dot{\tilde{\theta}}^T w + \tilde{\theta}^T \dot{w} \right]$$

$$+ \Lambda_0(s, \ \alpha) W_m^{-1}(s) \left[\tilde{\theta}^T \phi + W_c(s)[(W_b(s)[w^T])\dot{\tilde{\theta}}] \right] \tag{A.5}$$

$$\text{Now } w = H(s)[u \ y]^T + [0 \ 0 \ y]^T \tag{A.6}$$

$$\text{where } H(s) = \begin{bmatrix} \frac{a(s)}{\Lambda(s)} & 0 \\ 0 & \frac{a(s)}{\Lambda(s)} \\ 0 & 0 \end{bmatrix}$$

$$\text{Thus } |w(t)| \leq c m_f(t) + |y(t)| \tag{A.7}$$

From (A.2), it follows that

$$|y(t)| \le c \left[\left\| \left(\tilde{\theta}^T w \right)_t \right\|_2^\delta + c \right]$$
$$\le c \left[\|w_t\|_2^\delta + c \right] \text{ (since } \tilde{\theta} \text{ is bounded)} \tag{A.8}$$

Also from (A.6), we obtain

$$\|w_t\|_2^\delta \le c m_f \tag{A.9}$$

Combining (A.7), (A.8), (A.9), we obtain

$$|w(t)| \le c m_f(t) + c \tag{A.10}$$

Also from (A.6) we obtain

$$\|\dot{w}_t\|_2^\delta \le c m_f(t) + \|\dot{y}_t\|_2^\delta$$
$$\le c m_f(t) + c \left[\left\| \left(\tilde{\theta}^T w \right)_t \right\|_2^\delta + c \right] \text{ (using (A.2))}$$
$$\le c m_f(t) + c \left[m_f + c \right] \text{ (using (A.9))}$$
$$\le c m_f + c \tag{A.11}$$

Now $\left| W_b(s)[w^T] \right| \le c \|w_t\|_2^\delta$ (using (2.7))
$$\le c m_f \text{ (using (A.9))} \tag{A.12}$$

Combining (A.5), (A.10), (A.11), (A.12), we obtain

$$\left\| \left(\tilde{\theta}^T w \right)_t \right\|_2^\delta \le c + \frac{c}{\alpha} m_f(t) + c \alpha^{n^*} \left\| \left(\tilde{\theta}^T \phi \right)_t \right\|_2^\delta$$
$$+ \left(\frac{c}{\alpha} + c \alpha^{n^*} \right) \left\| \left(|\dot{\tilde{\theta}}| m_f \right)_t \right\|_2^\delta \tag{A.13}$$

Combining (A.4) and (A.13), we obtain

$$m_f(t) \le c + \frac{c}{\alpha} m_f(t) + c \alpha^{n^*} \left\| \left(\tilde{\theta}^T \phi \right)_t \right\|_2^\delta + \left(\frac{c}{\alpha} + c \alpha^{n^*} \right) \left\| \left(|\dot{\tilde{\theta}}| m_f \right)_t \right\|_2^\delta$$

Choose α large enough so that $\frac{c}{\alpha} < 1$. Then

$$m_f(t) \le c + c \alpha^{n^*} \left\| \left(\frac{\tilde{\theta}^T \phi}{m} m_f \right)_t \right\|_2^\delta + \left(\frac{c}{\alpha} + c \alpha^{n^*} \right) \left\| \left(|\dot{\tilde{\theta}}| m_f \right)_t \right\|_2^\delta \tag{A.14}$$

The adaptive law (3.4) guarantees that $\frac{\tilde{\theta}^T \phi}{m}$, $\dot{\tilde{\theta}} \in L_2$ [32] (also see (A.23), (A.27)). This fact together with the Gronwall Lemma [36] implies that $m_f \in L_\infty$ so that $m \in L_\infty$ and all the closed loop signals are bounded.

Now from (A.2)

$$e_1 = \frac{W_m(s)}{c_0} [\tilde{\theta}^T w] \tag{A.15}$$

From the Swapping Lemma, and the boundedness of m we can show that $e_1 \in L_2$. Furthermore since $\tilde{\theta}^T w \in L_\infty$ and $W_m(s)$ is strictly proper, it follows that $\dot{e}_1 \in L_\infty$ so that $e_1(t)$ is uniformly continuous. Hence, by Barbalat's Lemma [37], it follows that $e_1(t) \to 0$ as $t \to \infty$ and the proof is complete.

Proof of Theorem 3.2: From (A.2), the tracking error $e_1 = y - y_m$ satisfies

$$e_1 = \frac{1}{c_0} W_m(s)[\tilde{\theta}^T w] \tag{A.16}$$

so that

$$\int_0^t e_1^2(z)dz \le c \int_0^t \left(\tilde{\theta}^T w\right)^2 dz \tag{A.17}$$

We now proceed to obtain an expression for $\int_0^t \left(\tilde{\theta}^T w\right)^2 dz$.

It can be shown [11] that θ^* satisfies

$$W_m(s)[u] - \theta^{*T}\phi = c_0 y \tag{A.18}$$

and

$$\epsilon = \frac{\tilde{\theta}^T \phi}{m^2}, \quad \tilde{\theta} \triangleq \theta - \theta^* \tag{A.19}$$

Let us now consider the positive definite function

$$V = \frac{\tilde{\theta}^T \tilde{\theta}}{2\gamma} = \frac{|\tilde{\theta}|^2}{2\gamma} \tag{A.20}$$

Then it can be shown [33, 32] that along the solution of (3.4) we have

$$\dot{V} = \frac{\tilde{\theta}^T}{\gamma} Pr[-\gamma \epsilon \phi]$$

$$\le -\epsilon^2 m^2 = -\frac{(\tilde{\theta}^T \phi)^2}{m^2} \tag{A.21}$$

i.e. $V, \tilde{\theta} \in L_\infty$ and $\epsilon m, \frac{\tilde{\theta}^T \phi}{m} \in L_2$. Furthermore

$$\int_0^t \epsilon^2 m^2 dz = \int_0^t \frac{(\tilde{\theta}^T \phi)^2}{m^2} dz \le V_0 - V_t \tag{A.22}$$

where $V_0 \triangleq V\left(\tilde{\theta}(0)\right)$ and $V_t = V\left(\tilde{\theta}(t)\right)$. Since $V_t \ge 0$, it follows that

$$\int_0^t \epsilon^2 m^2 dz = \int_0^t \frac{(\tilde{\theta}^T \phi)^2}{m^2} dz \le V_0 = \frac{|\tilde{\theta}(0)|^2}{2\gamma} \tag{A.23}$$

It can be shown [32] that

$$|\dot{\tilde{\theta}}|^2 \le \gamma^2 \frac{|\phi|^2}{m^2} \epsilon^2 m^2 \tag{A.24}$$

Now $\phi = W_m(s) H_1(s) \begin{bmatrix} u \\ y \end{bmatrix}$

where $H_1(s) \triangleq \begin{bmatrix} \frac{a(s)}{A(s)} & 0 \\ 0 & \frac{a(s)}{A(s)} \\ 0 & 1 \end{bmatrix}$

Thus, using (2.7)

$$\frac{|\phi|^2}{m^2} \leq c \qquad (A.25)$$

Substituting (A.25) in (A.24) we obtain

$$|\dot{\tilde{\theta}}|^2 \leq c\epsilon^2 m^2 \qquad (A.26)$$

so that $\displaystyle\int_0^t |\dot{\tilde{\theta}}|^2 dz \leq c|\tilde{\theta}(0)|^2 \qquad (A.27)$

Now using (2.9) with $W(s) = W_m(s)$ and $\Lambda_0(s, \alpha) = \frac{\alpha^{n^*}}{(s+\alpha)^{n^*}}$, $\alpha > 0$, we obtain

$$\tilde{\theta}^T w = \Lambda_1(s, \alpha)\left[\dot{\tilde{\theta}}^T w + \tilde{\theta}^T \dot{w}\right]$$
$$+\Lambda_0(s, \alpha)W_m^{-1}(s)\left[\tilde{\theta}^T \phi + W_c(s)\left(\{W_b(s)[\omega^T]\}\dot{\tilde{\theta}}\right)\right]$$

so that

$$\left\|\left(\tilde{\theta}^T w\right)_t\right\|_2 \leq \frac{c}{\alpha}\left[\left\|\left(\dot{\tilde{\theta}}^T w\right)_t\right\|_2 + \left\|\left(\tilde{\theta}^T \dot{w}\right)_t\right\|_2\right]$$
$$+c\alpha^{n^*}\left[\left\|\left(\frac{\tilde{\theta}^T \phi}{m}m\right)_t\right\|_2 + \left\|\left(|\dot{\tilde{\theta}}|m\right)_t\right\|_2\right]$$
$$\leq \frac{c}{\alpha}\left[\left\|\left(\dot{\tilde{\theta}}^T w\right)_t\right\|_2 + \left\|\left(\tilde{\theta}^T \dot{w}\right)_t\right\|_2\right]$$
$$+c\alpha^{n^*}\bar{m}\left[\left\|\left(\frac{\tilde{\theta}^T \phi}{m}\right)_t\right\|_2 + \left\|\left(|\dot{\tilde{\theta}}|\right)_t\right\|_2\right] \qquad (A.28)$$

Now $|\dot{\tilde{\theta}}| \leq \gamma|\epsilon m|\frac{|\phi|}{m}$

$= \gamma\frac{|\tilde{\theta}^T \phi|}{m}\frac{|\phi|}{m}$ (since $\epsilon = \frac{\tilde{\theta}^T \phi}{m^2}$)

$\leq c$ (since $\frac{\phi}{m} \in L_\infty$) $\qquad (A.29)$

From (A.10) we have $|w| \leq cm_f + c \leq c(\bar{m} + 1)$.

Thus $(\|w_t\|_2)^2 \leq c(\bar{m}^2 + 1)t \qquad (A.30)$

Again from (A.1), (A.2), we obtain

$$\|u_t\|_2 \le c\|w_t\|_2 + c\|r_t\|_2$$
$$\|y_t\|_2 \le c\|w_t\|_2 + c\|r_t\|_2$$
$$\|\dot{y}_t\|_2 \le c\|w_t\|_2 + c\|r_t\|_2$$

Since $\dot{w} = sH(s)[u\ y]^T + [0\ 0\ \dot{y}]^T$, using (A.30), it follows that

$$(\|\dot{w}_t\|_2)^2 \le c(\bar{m}^2 + 1)t \tag{A.31}$$

Combining (A.28), (A.29), (A.30), (A.31) we obtain

$$\int_0^t (\tilde{\theta}^T w)^2 dz \le \frac{c}{\alpha^2}(\bar{m}^2 + 1)t$$
$$+c\alpha^{2n^*}\bar{m}^2 \left[\int_0^t \frac{(\tilde{\theta}^T \phi)^2}{m^2} dz + \int_0^t |\dot{\tilde{\theta}}|^2 dz\right] \tag{A.32}$$

Combining (A.17), (A.23), (A.27), (A.32), we obtain

$$\int_0^t e_1^2(z)dz \le c\left[\frac{1}{\alpha^2}(\bar{m}^2 + 1)t + \alpha^{2n^*}\bar{m}^2|\tilde{\theta}(0)|^2\right] \tag{A.33}$$

from which (3.9) follows.

To obtain (3.10), we once again start from (A.16). Using (2.7) we have

$$|e_1(t)| \le \left\|\frac{1}{c_0}W_m(s)\right\|_2^\delta \left\|(\tilde{\theta}^T w)_t\right\|_2^\delta$$
$$\le \left\|\frac{1}{c_0}W_m(s)\right\|_2^\delta c\|w_t\|_2^\delta \ (\text{ since } \tilde{\theta} \in L_\infty)$$
$$\le c\bar{m} \ (\text{using (A.9)})$$

and the proof is complete.

APPENDIX B

Proof of Theorem 4.1: From (3.1), (4.4), we can show that

$$u = \frac{R_0(s)}{k_p Z_0(s)} \frac{W_m(s)}{c_0} \left[W_\tau(s)[\tilde{\theta}^T w] + c_0 r\right] \tag{B.1}$$

$$y = \frac{W_m(s)}{c_0} \left[W_\tau(s)[\tilde{\theta}^T w] + c_0 r\right] \tag{B.2}$$

where $W_\tau(s)$ is given by (4.7). Let $\tau \in (0, \tau_{max}]$ and let $\delta_1 > 0$ be such that $Z_0(s)$ has all its roots in $\text{Re}[s] < -\frac{\delta_1}{2}$. Let $\delta \in (0, min[\delta_0, \delta_1, \frac{1}{\tau_{max}}])$ and define the fictitious signal $m_f(t)$ by

$$m_f(t) = 1 + \|u_t\|_2^\delta + \|y_t\|_2^\delta \tag{B.3}$$

Now for $\delta < \dfrac{1}{\tau}$, $\left\|\dfrac{\tau s}{\tau s+1}\right\|_\infty^\delta = 1$ and $\left\|\dfrac{1}{\tau s+1}\right\|_\infty^\delta < 2$

Thus $\|W_\tau(s)\|_\infty^\delta = \left\|\dfrac{(\tau s+1)^{n^*}-1}{(\tau s+1)^{n^*}}\right\|_\infty^\delta$

$$= \left\|\dfrac{\tau s\left[(\tau s+1)^{n^*-1}+(\tau s+1)^{n^*-2}+\cdots+1\right]}{(\tau s+1)^{n^*}}\right\|_\infty^\delta$$

$< 2^{n^*} - 1$, a constant independent of τ

Hence starting with (B.1), (B.2) instead of (A.1), (A.2) we can repeat the steps in the proof of Theorem 3.1 to conclude that all the closed loop signals are uniformly bounded and $e_1(t) \to 0$ as $t \to \infty$.

The explicit bound for the signal m_f can be obtained from (A.14), which arises in this case also, by using the Gronwall Lemma and the fact that $\dfrac{\tilde{\theta}^T\phi}{m}$, $\dot{\tilde{\theta}} \in L_2$. As in the proof of Theorem 3.1, the L_2-bounds on $\dfrac{\tilde{\theta}^T\phi}{m}$ and $\dot{\tilde{\theta}}$ are obtained by exploiting only the properties of the adaptive law (3.4) and are, therefore, independent of the control law used, let alone be dependent on τ. Since none of the generic constants in (A.14) are dependent on $\tau \in (0, \tau_{max}]$, it can be shown that the bound on m_f, and hence that on m, is independent of $\tau \in (0, \tau_{max}]$.

Proof of Theorem 4.2: From (B.2) the tracking error e_1 satisfies

$$e_1 = \frac{1}{c_0}W_\tau(s)W_m(s)[\tilde{\theta}^T w] \tag{B.4}$$

But $W_\tau(s) = \dfrac{(\tau s+1)^{n^*}-1}{(\tau s+1)^{n^*}}$

$$= \dfrac{\tau s\left[(\tau s+1)^{n^*-1}+(\tau s+1)^{n^*-2}+\cdots+1\right]}{(\tau s+1)^{n^*}}$$

$$= \tau s W_{1\tau}(s)$$

where $W_{1\tau}(s) \triangleq \left[\dfrac{1}{\tau s+1}+\dfrac{1}{(\tau s+1)^2}+\cdots+\dfrac{1}{(\tau s+1)^{n^*}}\right]$ (B.5)

Thus from (B.4)

$$e_1 = \frac{1}{c_0}sW_m(s)\tau W_{1\tau}(s)[\tilde{\theta}^T w] \tag{B.6}$$

Since $\|W_{1\tau}(s)\|_\infty \le n^*$, and $W_m(s)$ is strictly proper, we have

$$\int_0^t e_1^2(z)dz \le \tau^2 c \int_0^t (\tilde{\theta}^T w)^2 dz \tag{B.7}$$

Equation (4.8) can now be obtained by repeating the steps ((A.18) through (A.33)) in the proof of Theorem 3.2.

To obtain (4.9), we once again start from (B.6). Using (2.7) we have

$$|e_1(t)| \leq \tau \left\| \frac{1}{c_0} W_m(s) W_{1\tau}(s) \right\|_2^\delta \left[\left\| \left(\dot{\tilde{\theta}}^T w \right)_t \right\|_2^\delta + \left\| \left(\tilde{\theta}^T \dot{w} \right)_t \right\|_2^\delta \right]$$

$$\leq \tau \| W_{1\tau}(s) \|_\infty^\delta \left\| \frac{1}{c_0} W_m(s) \right\|_2^\delta [c\bar{m} + c]$$

(since $\tilde{\theta}$, $\dot{\tilde{\theta}} \in L_\infty$, $\|w_t\|_2^\delta \leq cm_f$, $\|\dot{w}_t\|_2^\delta \leq cm_f + c$)

so that the proof is complete.

APPENDIX C

In order to prove Theorems 5.1 and 5.2, we need the following preliminary Lemma which essentially establishes the properties of (3.4) *in the presence of the disturbance d.*

Lemma C.1 *The adaptive law (3.4) guarantees that*

$$\int_t^{t+T} \frac{(\tilde{\theta}^T \phi)^2}{m^2} dz \leq c_\eta T + \frac{|\tilde{\theta}(t)|^2}{\gamma} \quad \forall\, t,\, T \geq 0 \tag{C.1}$$

$$\int_t^{t+T} |\dot{\tilde{\theta}}|^2 dz \leq c \left[c_\eta T + \frac{|\tilde{\theta}(t)|^2}{\gamma} \right] \quad \forall\, t,\, T \geq 0 \tag{C.2}$$

where $c_\eta \triangleq \frac{\eta^2}{m^2}$ and $\eta = \frac{[\Lambda(s) - \theta_1^{*T} a(s)]}{\Lambda(s)} W_m(s)[d]$.

Proof. It can be shown [11] that θ^* satisfies

$$W_m(s)[u] - \theta^{*T} \phi + \eta = c_0 y \tag{C.3}$$

$$\text{where } \eta \triangleq \frac{[\Lambda(s) - \theta_1^{*T} a(s)]}{\Lambda(s)} W_m(s)[d] \tag{C.4}$$

Using (C.3) in (3.5), we obtain

$$\epsilon = \frac{\tilde{\theta}^T \phi + \eta}{m^2} \tag{C.5}$$

Consider the positive definite function

$$V = \frac{\tilde{\theta}^T \tilde{\theta}}{2\gamma} = \frac{|\tilde{\theta}|^2}{2\gamma}$$

Then, it can be shown that along the solution of (3.4), we have

$$\dot{V} \le -\epsilon \tilde{\theta}^T \phi$$

$$= -\tilde{\theta}^T \phi \left(\frac{\tilde{\theta}^T \phi + \eta}{m^2} \right) \quad (\text{ using (C.5)})$$

$$\le -\frac{(\tilde{\theta}^T \phi)^2}{2m^2} + \frac{\eta^2}{2m^2} \quad (\text{ completing squares })$$

so that

$$\int_t^{t+T} \frac{(\tilde{\theta}^T \phi)^2}{m^2} dz \le c_\eta T + \frac{|\tilde{\theta}(t)|^2}{\gamma} \quad \forall\, t,\, T \ge 0 \tag{C.6}$$

Similarly,

$$\int_t^{t+T} \epsilon^2 m^2 dz \le c_\eta T + \frac{|\tilde{\theta}(t)|^2}{\gamma} \quad \forall\, t,\, T \ge 0 \tag{C.7}$$

so that $\frac{\tilde{\theta}^T \phi}{m}$, $\epsilon m \in \mathcal{S}(c_\eta)$.

It can be shown that $|\dot{\tilde{\theta}}|^2 \le c\epsilon^2 m^2 \frac{|\phi|^2}{m^2}$

$$\text{Since } \phi = W_m(s)H_1(s) \begin{bmatrix} u \\ y \end{bmatrix}$$

from (2.7) it follows that $|\phi|^2 \le cm^2$.

$$\text{Thus } |\dot{\tilde{\theta}}|^2 \le c\epsilon^2 m^2 \tag{C.8}$$

Combining (C.8) with (C.7), the proof is complete.

Proof of Theorem 5.1: From (5.1), (4.4), we can show that

$$u = \frac{R_0(s)}{c_0 k_p Z_0(s)} W_m(s)W_\tau(s)[\tilde{\theta}^T w] + \frac{\left[\theta_2^{*T} a(s) + \theta_3^* \Lambda(s)\right]}{c_0 \Lambda(s)} W_m(s)[d]$$

$$- \frac{R_0(s)}{c_0 k_p Z_0(s)(\tau s + 1)^{n^*}} \frac{\Lambda(s) - \theta_1^{*T} a(s)}{\Lambda(s)} W_m(s)[d]$$

$$+ \frac{R_0(s)}{k_p Z_0(s)} W_m(s)[r] \tag{C.9}$$

$$y = \frac{1}{c_0} W_m(s)W_\tau(s)[\tilde{\theta}^T w] + \frac{1}{c_0} W_m(s)W_\tau(s)\frac{\left[\Lambda(s) - \theta_1^{*T} a(s)\right]}{\Lambda(s)}[d]$$

$$+ W_m(s)[r] \tag{C.10}$$

Let $\tau \in (0,\, \tau_{max}]$ and let $\delta_1 > 0$ be such that $Z_0(s)$ has all its roots in $\text{Re}[s] < -\frac{\delta_1}{2}$. Let $\delta \in (0,\, min[\delta_0,\, \delta_1,\, \frac{1}{\tau_{max}}])$ and define the fictitious signal $m_f(t)$ by

$$m_f(t) = \frac{1}{\alpha_0} + \|u_t\|_2^\delta + \|y_t\|_2^\delta \tag{C.11}$$

Then, from (C.9)-(C.11), using (2.6), we obtain

$$m_f(t) \le \frac{1}{\alpha_0} + c + c \left\| \left(\tilde{\theta}^T w \right)_t \right\|_2^\delta \tag{C.12}$$

Now using (2.9) with $W(s) = W_m(s)$ and $\Lambda_0(s, \alpha) = \frac{\alpha^{n^*}}{(s+\alpha)^{n^*}}$, $\alpha > \frac{\delta}{2}$, we obtain

$$\tilde{\theta}^T w = \Lambda_1(s, \alpha) \left[\dot{\tilde{\theta}}^T w + \tilde{\theta}^T \dot{w} \right]$$
$$+ \Lambda_0(s, \alpha) W_m^{-1}(s) \left[\tilde{\theta}^T \phi + W_c(s) \left[(W_b(s)[w^T]) \dot{\tilde{\theta}} \right] \right] \tag{C.13}$$

$$\text{Now } w = H(s)[u \ y]^T + [0 \ 0 \ y]^T \tag{C.14}$$
$$\Rightarrow |w(t)| \le c m_f + |y(t)| \tag{C.15}$$

From (C.10), it follows that

$$|y(t)| \le c m_f + c \tag{C.16}$$

Combining (C.15), (C.16) we obtain

$$|w(t)| \le c m_f + c \tag{C.17}$$

Also from (C.14), we obtain

$$\|\dot{w}_t\|_2^\delta \le c m_f + \|\dot{y}_t\|_2^\delta$$
$$\le c m_f + (c m_f + c) \ \text{(using (C.10))}$$
$$\le c m_f + c \tag{C.18}$$

$$\text{Now } \left| W_b(s)[w^T] \right| \le c \|w_t\|_2^\delta \ \text{(using (2.7))}$$
$$\le c m_f \ \text{(using (C.14))} \tag{C.19}$$

Combining (C.13), (C.17), (C.18), (C.19), we obtain

$$\left\| \left(\tilde{\theta}^T w \right)_t \right\|_2^\delta \le c + \frac{c}{\alpha} m_f(t) + c \alpha^{n^*} \left\| \left(\tilde{\theta}^T \phi \right)_t \right\|_2^\delta$$
$$+ \left(\frac{c}{\alpha} + c \alpha^{n^*} \right) \left\| \left(|\dot{\tilde{\theta}}| m_f \right)_t \right\|_2^\delta \tag{C.20}$$

Combining (C.12) and (C.20), we obtain

$$m_f(t) \le \frac{1}{\alpha_0} + c + \frac{c}{\alpha} m_f$$
$$+ c \alpha^{n^*} \left\| \left(\tilde{\theta}^T \phi \right)_t \right\|_2^\delta + \left(\frac{c}{\alpha} + c \alpha^{n^*} \right) \left\| \left(|\dot{\tilde{\theta}}| m_f \right)_t \right\|_2^\delta$$

Choose α large enough so that $\frac{c}{\alpha} < 1$. Then

$$m_f(t) \le \frac{c}{\alpha_0} + c + c\alpha^{n^*} \left\| \left(\frac{\tilde{\theta}^T \phi}{m} m_f \right)_t \right\|_2^\delta$$
$$+ \left(\frac{c}{\alpha} + c\alpha^{n^*} \right) \left\| \left(|\dot{\tilde{\theta}}| m_f \right)_t \right\|_2^\delta$$

Squaring both sides

$$m_f^2(t) \le \frac{c}{\alpha_0^2} + c + c\alpha^{2n^*} \int_0^t \left(\frac{\tilde{\theta}^T \phi}{m} \right)^2 m_f^2(z) e^{-\delta(t-z)} dz$$
$$+ \left(\frac{c}{\alpha^2} + c\alpha^{2n^*} \right) \int_0^t |\dot{\tilde{\theta}}|^2 m_f^2(z) e^{-\delta(t-z)} dz$$

or $m_f^2(t) \le \frac{c}{\alpha_0^2} + c + \int_0^t \left[c\frac{(\tilde{\theta}^T \phi)^2}{m^2} + c|\dot{\tilde{\theta}}|^2 \right] e^{-\delta(t-z)} m_f^2(z) dz$

Define $\gamma_0^2 = c\frac{(\tilde{\theta}^T \phi)^2}{m^2} + c|\dot{\tilde{\theta}}|^2$ (C.21)

Then using the Bellman Gronwall Lemma [36], we have

$$m_f^2(t) \le \left[\frac{c}{\alpha_0^2} + c \right] \left[1 + \int_0^t e^{-\delta(t-z)} \gamma_0^2(z) e^{\int_z^t \gamma_0^2(y) dy} dz \right]$$ (C.22)

From (C.1), (C.2) and (C.21), it follows that $\gamma_0^2 \in \mathcal{S}(cc_\eta)$ where $c_\eta = \frac{\eta^2}{m^2}$. Since $m^2 \ge \frac{1}{\alpha_0^2}$ and $\eta \in L_\infty$, it follows that by choosing α_0 in (5.2) to be small enough we can make c_η as small as we wish. Thus, using Lemma 2.4, it follows that $m_f \in L_\infty$ provided α_0 in (5.2) is chosen to be small enough, i.e. less than some constant α_0^*. Furthermore, from (C.22), we obtain $m_f^2 \le \bar{m}^2 = c$. This completes the proof.

Proof of Theorem 5.2: From (C.10), the tracking error e_1 satisfies

$$e_1 = \frac{1}{c_0} W_\tau(s) W_m(s) [\tilde{\theta}^T w] + \frac{1}{c_0} W_m(s) W_\tau(s) \frac{\left[\Lambda(s) - \theta_1^{*T} a(s) \right]}{\Lambda(s)} [d]$$ (C.23)

from which we obtain

$$\int_0^t e_1^2(z) dz \le \tau^2 c \left[\int_0^t \left(\tilde{\theta}^T w \right)^2 dz + d_0^2 t \right]$$ (C.24)

Now using (2.9) with $W(s) = W_m(s)$ and $\Lambda_0(s, \alpha) = \frac{\alpha^{n^*}}{(s+\alpha)^{n^*}}$, $\alpha > 0$, we obtain

$$\tilde{\theta}^T w = \Lambda_1(s, \alpha) \left[\dot{\tilde{\theta}}^T w + \tilde{\theta}^T \dot{w} \right] + \Lambda_0(s, \alpha) W_m^{-1}(s) \left[\tilde{\theta}^T \phi + W_c(s) \left[(W_b(s)[w^T]) \dot{\tilde{\theta}} \right] \right]$$

so that

$$\left\|\left(\tilde{\theta}^T w\right)_t\right\|_2 \le \frac{c}{\alpha}\left[\left\|\left(\dot{\tilde{\theta}}^T w\right)_t\right\|_2 + \left\|\left(\tilde{\theta}^T \dot{w}\right)_t\right\|_2\right]$$

$$+ca^{n^*}\left[\left\|\left(\frac{\tilde{\theta}^T \phi}{m}m\right)_t\right\|_2 + \left\|\left(|\dot{\tilde{\theta}}|m\right)_t\right\|_2\right]$$

$$\le \frac{c}{\alpha}\left[\left\|\left(\dot{\tilde{\theta}}^T w\right)_t\right\|_2 + \left\|\left(\tilde{\theta}^T \dot{w}\right)_t\right\|_2\right]$$

$$+ca^{n^*}\bar{m}\left[\left\|\left(\frac{\tilde{\theta}^T \phi}{m}\right)_t\right\|_2 + \left\|\left(|\dot{\tilde{\theta}}|\right)_t\right\|_2\right] \qquad (C.25)$$

Now $|\dot{\tilde{\theta}}| \le \gamma|\epsilon m|\dfrac{|\phi|}{m}$

$$= \gamma\frac{|\tilde{\theta}^T \phi + \eta|}{m}\frac{|\phi|}{m} \text{ (since } \epsilon = \frac{\tilde{\theta}^T \phi + \eta}{m^2}\text{)}$$

$$\le c \text{ (since } \tfrac{\phi}{m}, \tfrac{\eta}{m} \in L_\infty) \qquad (C.26)$$

From (C.17) we have $|w| \le cm_f + c \le c(\bar{m} + 1)$.

$$\text{Thus } (\|w_t\|_2)^2 \le c(\bar{m}^2 + 1)t \qquad (C.27)$$

Again from (C.9), (C.10), we obtain

$$\|u_t\|_2 \le c\|w_t\|_2 + c\|r_t\|_2 + c\|d_t\|_2$$
$$\|y_t\|_2 \le c\|w_t\|_2 + c\|r_t\|_2 + c\|d_t\|_2$$
$$\|\dot{y}_t\|_2 \le c\|w_t\|_2 + c\|r_t\|_2 + c\|d_t\|_2$$

Since $\dot{w} = sH(s)[u\ y]^T + [0\ 0\ \ddot{y}]^T$, using (C.27), it follows that

$$(\|\dot{w}_t\|_2)^2 \le c\left[\bar{m}^2 + 1\right]t \qquad (C.28)$$

Combining (C.25), (C.26), (C.27), (C.28) we obtain

$$\int_0^t \left(\tilde{\theta}^T w\right)^2 dz \le \frac{c}{\alpha^2}\left[\bar{m}^2 + 1\right]t$$

$$+ca^{2n^*}\bar{m}^2\left[\int_0^t \frac{(\tilde{\theta}^T \phi)^2}{m^2}dz + \int_0^t |\dot{\tilde{\theta}}|^2 dz\right] \qquad (C.29)$$

Combining (C.1), (C.2), (C.24), (C.29) we obtain

$$\int_0^t e_1^2(z)dz \le \tau^2 c\left[\left\{\frac{1}{\alpha^2}(\bar{m}^2 + 1) + d_0^2\right\}t\right.$$

$$\left. +\alpha^{2n^*}\bar{m}^2\left(c_\eta t + |\tilde{\theta}(0)|^2\right)\right]$$

Since $c_\eta = cd_0^2$ for some $c > 0$, we obtain (5.4) by dividing both sides of the above inequality by t.

Equation (5.5) can be obtained by essentially repeating the steps used in arriving at (4.9) (Appendix B, Proof of Theorem 4.2) and using the fact that $\dot{d} \in L_\infty$.

APPENDIX D

In order to prove Theorems 6.1 and 6.2, we need the following preliminary Lemma which essentially establishes the properties of (3.4) *in the presence of the unmodelled dynamics.*

Lemma D.1 *The adaptive law (3.4) guarantees that*

$$\int_t^{t+T} \frac{(\tilde{\theta}^T \phi)^2}{m^2} dz \leq c_\eta T + \frac{|\tilde{\theta}(t)|^2}{\gamma} \quad \forall\, t,\, T \geq 0 \tag{D.1}$$

$$\int_t^{t+T} |\dot{\tilde{\theta}}|^2 dz \leq c \left[c_\eta T + \frac{|\tilde{\theta}(t)|^2}{\gamma} \right] \quad \forall\, t,\, T \geq 0 \tag{D.2}$$

where $c_\eta \triangleq \frac{\eta^2}{m^2}$ *and* $\eta = \mu \frac{[\Lambda(s) - \theta_1^{*T} a(s)]}{\Lambda(s)} W_m(s) \Delta_m(s)[u]$.

Proof. It can be shown [11] that θ^* satisfies

$$W_m(s)[u] - \theta^{*T}\phi + \eta = c_0 y \tag{D.3}$$

$$\text{where } \eta \triangleq \mu \frac{[\Lambda(s) - \theta_1^{*T} a(s)]}{\Lambda(s)} W_m(s) \Delta_m(s)[u] \tag{D.4}$$

Using (D.3) in (3.5), we obtain

$$\epsilon = \frac{\tilde{\theta}^T \phi + \eta}{m^2} \tag{D.5}$$

The rest of the proof is an exact repitition of the steps in the proof of Lemma C.1 (Appendix C).

Proof of Theorem 6.1: From (6.1), (4.4) we can show that

$$u = \frac{R_0(s)}{c_0 k_p Z_0(s)} W_m(s) W_r(s) \left[\tilde{\theta}^T w \right] + \mu \frac{\left[\theta_2^{*T} a(s) + \theta_3^* \Lambda(s) \right]}{c_0 \Lambda(s)} W_m(s) \Delta_m(s)[u]$$

$$- \mu \frac{R_0(s)}{c_0 k_p Z_0(s)} \frac{\Lambda(s) - \theta_1^{*T} a(s)}{\Lambda(s)} \frac{W_m(s) \Delta_m(s)}{(\tau s + 1)^{n^*}} [u] + \frac{R_0(s)}{k_p Z_0(s)} W_m(s)[r] \tag{D.6}$$

$$y = \frac{1}{c_0} W_m(s) W_r(s) \left[\tilde{\theta}^T w \right] + \mu \frac{\left[\theta_2^{*T} a(s) + \theta_3^* \Lambda(s) \right]}{c_0 \Lambda(s)} W_m(s) \Delta_m(s)[y]$$

$$- \mu \frac{R_0(s)}{c_0 k_p Z_0(s)} \frac{\left[\Lambda(s) - \theta_1^{*T} a(s) \right]}{\Lambda(s)} W_m(s) \frac{\Delta_m(s)}{(\tau s + 1)^{n^*}} [y] + W_m(s)[r] \tag{D.7}$$

Let $\tau \in (\tau_{min}, \tau_{max}]$ and let $\delta_1 > 0$ be such that $Z_0(s)$ has all its roots in $Re[s] < -\frac{\delta_1}{2}$. Let $\delta \in (0, \; min[\delta_0, \; \delta_1, \; \frac{1}{\tau_{max}}])$ and define the fictitious signal $m_f(t)$ by

$$m_f(t) = 1 + \|u_t\|_2^\delta + \|y_t\|_2^\delta \tag{D.8}$$

Then from (D.6)-(D.8), using (2.6), we obtain

$$m_f(t) \leq c + c\mu \left(1 + \frac{1}{(\tau)^{n^*}}\right) m_f + c \left\|\left(\tilde{\theta}^T w\right)_t\right\|_2^\delta$$

Let $\mu \in [0, \; \mu_1^*)$ where μ_1^* will be chosen later in the proof. Then

$$m_f(t) \leq c + c\mu_1^* \left(1 + \frac{1}{(\tau_{min})^{n^*}}\right) m_f + c \left\|\left(\tilde{\theta}^T w\right)_t\right\|_2^\delta \tag{D.9}$$

Now, using (2.9) with $W(s) = W_m(s)$ and $\Lambda_0(s, \; \alpha) = \frac{\alpha^{n^*}}{(s+\alpha)^{n^*}}$, $\alpha > \frac{\delta}{2}$, we obtain

$$\tilde{\theta}^T w = \Lambda_1(s, \; \alpha)\left[\dot{\tilde{\theta}}^T w + \tilde{\theta}^T \dot{w}\right]$$

$$+\Lambda_0(s, \; \alpha)W_m^{-1}(s)\left[\tilde{\theta}^T \phi + W_c(s)\left[(W_b(s)[w^T])\dot{\tilde{\theta}}\right]\right] \tag{D.10}$$

$$\text{Now } w = H(s)[u \; y]^T + [0 \; 0 \; y]^T \tag{D.11}$$

$$\Rightarrow |w(t)| \leq cm_f + |y(t)| \tag{D.12}$$

From (D.7), it follows that

$$|y(t)| \leq c\left[1 + \mu_1^*\left(1 + \frac{1}{(\tau_{min})^{n^*}}\right)\right] m_f + c \tag{D.13}$$

Combining (D.12), (D.13) we obtain

$$|w(t)| \leq c\left[1 + \mu_1^*\left(1 + \frac{1}{(\tau_{min})^{n^*}}\right)\right] m_f + c \tag{D.14}$$

Also from (D.11), we obtain

$$\|\dot{w}_t\|_2^\delta \leq cm_f + \|\dot{y}_t\|_2^\delta$$

$$\leq cm_f + c\left[1 + \mu_1^*\left(1 + \frac{1}{(\tau_{min})^{n^*}}\right)\right] m_f + c \text{ (using (D.7))}$$

$$\leq c\left[1 + \mu_1^*\left(1 + \frac{1}{(\tau_{min})^{n^*}}\right)\right] m_f + c \tag{D.15}$$

$$\text{Now } |W_b(s)[w^T]| \leq c\|w_t\|_2^\delta \text{ (using (2.7))}$$

$$\leq cm_f \text{ (using (D.11))} \tag{D.16}$$

Combining (D.10), (D.14), (D.15), (D.16), we obtain

$$\left\|\left(\tilde{\theta}^T w\right)_t\right\|_2^\delta \le c + \frac{c}{\alpha}\left[1 + \mu_1^*\left(1 + \frac{1}{(\tau_{min})^{n^*}}\right)\right] m_f + c\alpha^{n^*}\left\|\left(\tilde{\theta}^T \phi\right)_t\right\|_2^\delta$$
$$+ \left\{\frac{c}{\alpha}\left[1 + \mu_1^*\left(1 + \frac{1}{(\tau_{min})^{n^*}}\right)\right] + c\alpha^{n^*}\right\}\left\|\left(|\dot{\tilde{\theta}}|m_f\right)_t\right\|_2^\delta \tag{D.17}$$

Combining (D.9) and (D.17), we obtain

$$m_f(t) \le c + c\mu_1^*\left(1 + \frac{1}{(\tau_{min})^{n^*}}\right) m_f$$
$$+ \frac{c}{\alpha}\left[1 + \mu_1^*\left(1 + \frac{1}{(\tau_{min})^{n^*}}\right)\right] m_f + c\alpha^{n^*}\left\|\left(\tilde{\theta}^T \phi\right)_t\right\|_2^\delta$$
$$+ \left\{\frac{c}{\alpha}\left[1 + \mu_1^*\left(1 + \frac{1}{(\tau_{min})^{n^*}}\right)\right] + c\alpha^{n^*}\right\}\left\|\left(|\dot{\tilde{\theta}}|m_f\right)_t\right\|_2^\delta$$

Now choose μ_1^* small enough and α large enough so that

$$A \triangleq 1 - c\mu_1^*\left(1 + \frac{1}{(\tau_{min})^{n^*}}\right) - \frac{c}{\alpha}\left[1 + \mu_1^*\left(1 + \frac{1}{(\tau_{min})^{n^*}}\right)\right] > 0 \tag{D.18}$$

Then $m_f(t) \le \dfrac{1}{A}\left[c + c\alpha^{n^*}\left\|\left(\tilde{\theta}^T \phi\right)_t\right\|_2^\delta\right.$
$$\left. + \left\{\frac{c}{\alpha}\left[1 + \mu_1^*\left(1 + \frac{1}{(\tau_{min})^{n^*}}\right)\right] + c\alpha^{n^*}\right\}\left\|\left(|\dot{\tilde{\theta}}|m_f\right)_t\right\|_2^\delta\right]$$

Squaring both sides

$$m_f^2(t) \le \frac{c}{A^2} + \frac{c}{A^2}\alpha^{2n^*}\int_0^t \left(\frac{\tilde{\theta}^T \phi}{m}\right)^2 m_f^2(z)e^{-\delta(t-z)}dz$$
$$+ \frac{1}{A^2}\left[\frac{c}{\alpha}\left\{1 + \mu_1^*\left(1 + \frac{1}{(\tau_{min})^{n^*}}\right)\right\} + c\alpha^{n^*}\right]^2\int_0^t |\dot{\tilde{\theta}}|^2 m_f^2(z)e^{-\delta(t-z)}dz$$

or $m_f^2(t) \le \dfrac{c}{A^2} + \displaystyle\int_0^t \gamma_0^2(z)m_f^2(z)e^{-\delta(t-z)}dz$

where

$$\gamma_0^2 = \frac{1}{A^2}\left[c\alpha^{2n^*}\frac{\left(\tilde{\theta}^T \phi\right)^2}{m^2}\right.$$
$$\left. + \left(\frac{c}{\alpha}\left\{1 + \mu_1^*\left(1 + \frac{1}{(\tau_{min})^{n^*}}\right)\right\} + c\alpha^{n^*}\right)^2|\dot{\tilde{\theta}}|^2\right] \tag{D.19}$$

Then using the Bellman Gronwall Lemma [36], we have

$$m_f^2(t) \leq \frac{c}{A^2} \left[1 + \int_0^t e^{-\delta(t-z)} \gamma_0^2(z) e^{\int_z^t \gamma_0^2(y)dy} dz \right] \tag{D.20}$$

From (D.1), (D.2) and (D.19), it follows that $\gamma_0^2 \in \mathcal{S}(cc_\eta)$. From (D.4) and (2.7), it is also clear that $c_\eta = c\mu$ for some $c > 0$ so that c_η can be made as small as we wish by restricting μ to be small. Thus, using Lemma 2.4, it follows that $\exists\, \mu^* > 0, \mu^* \leq \mu_1^*$ such that $\forall\, \mu \in [0, \mu^*)$, $m_f \in L_\infty$. Furthermore, from (D.20) and Lemma 2.4, we can obtain the bound for m^2 given in (6.2) and, therefore, the proof is complete.

Proof of Theorem 6.2: From (D.7)

$$e_1 = \mu \frac{\left[\theta_2^{*T} a(s) + \theta_3^* \Lambda(s) \right]}{c_0 \Lambda(s)} W_m(s) \Delta_m(s)[e_1]$$

$$- \mu \frac{R_0(s)}{c_0 k_p Z_0(s)} \frac{\left[\Lambda(s) - \theta_1^{*T} a(s) \right]}{\Lambda(s)} \frac{W_m(s) \Delta_m(s)}{(\tau s + 1)^{n^*}} [e_1]$$

$$+ \frac{1}{c_0} W_m(s) W_r(s) \left[\tilde{\theta}^T w \right] + \mu \frac{\left[\theta_2^{*T} a(s) + \theta_3^* \right]}{c_0 \Lambda(s)} W_m(s) \Delta_m(s) W_m(s)[r]$$

$$- \mu \frac{R_0(s)}{c_0 k_p Z_0(s)} \frac{\left[\Lambda(s) - \theta_1^{*T} a(s) \right]}{\Lambda(s)} \frac{W_m(s) \Delta_m(s)}{(\tau s + 1)^{n^*}} W_m(s)[r] \tag{D.21}$$

from which we obtain

$$\|e_{1t}\|_2 \leq \mu^* c \|e_{1t}\|_2 + c \frac{\mu^*}{(\tau_{min})^{n^*}} \|e_{1t}\|_2 + c\tau \left\| \left(\tilde{\theta}^T w \right)_t \right\|_2 + \mu^* c \|r_t\|_2$$

From (D.18)

$$1 - \mu^* c \left(1 + \frac{1}{(\tau_{min})^{n^*}} \right) > 0 \tag{D.22}$$

so that

$$\int_0^t e_1^2(z)dz \leq \frac{\tau^2 c}{\left[1 - \mu^* c \left(1 + \frac{1}{(\tau_{min})^{n^*}} \right) \right]^2} \int_0^t \left(\tilde{\theta}^T w \right)^2 dz$$

$$+ \frac{\mu^{*2} c}{\left[1 - \mu^* c \left(1 + \frac{1}{(\tau_{min})^{n^*}} \right) \right]^2} t \tag{D.23}$$

Now using (2.9) with $W(s) = W_m(s)$ and $\Lambda_0(s, \alpha) = \frac{\alpha^{n^*}}{(s+\alpha)^{n^*}}$, $\alpha > 0$, we obtain

$$\tilde{\theta}^T w = \Lambda_1(s, \alpha) \left[\dot{\tilde{\theta}}^T w + \tilde{\theta}^T \dot{w} \right] + \Lambda_0(s, \alpha) W_m^{-1}(s) \left[\tilde{\theta}^T \phi + W_c(s) \left[(W_b(s)[w^T]) \dot{\tilde{\theta}} \right] \right]$$

so that $\left\|\left(\tilde{\theta}^T w\right)_t\right\|_2 \leq \dfrac{c}{\alpha}\left[\left\|\left(\dot{\tilde{\theta}}^T w\right)_t\right\|_2 + \left\|\left(\tilde{\theta}^T \dot{w}\right)_t\right\|_2\right]$

$$+ c\alpha^{n^*}\left[\left\|\left(\frac{\tilde{\theta}^T \phi}{m} m\right)_t\right\|_2 + \left\|\left(|\dot{\tilde{\theta}}|m\right)_t\right\|_2\right]$$

$$\leq \frac{c}{\alpha}\left[\left\|\left(\dot{\tilde{\theta}}^T w\right)_t\right\|_2 + \left\|\left(\tilde{\theta}^T \dot{w}\right)_t\right\|_2\right]$$

$$+ c\alpha^{n^*}\bar{m}\left[\left\|\left(\frac{\tilde{\theta}^T \phi}{m}\right)_t\right\|_2 + \left\|\left(|\dot{\tilde{\theta}}|\right)_t\right\|_2\right] \qquad (D.24)$$

Now $|\dot{\tilde{\theta}}| \leq \gamma|\epsilon m|\dfrac{|\phi|}{m}$

$$= \gamma\frac{|\tilde{\theta}^T \phi + \eta|}{m}\frac{|\phi|}{m} \quad \text{(since } \epsilon = \frac{\tilde{\theta}^T \phi + \eta}{m^2}\text{)}$$

$$\leq c \quad \text{(since } \frac{\phi}{m}, \frac{\eta}{m} \in L_\infty\text{)} \qquad (D.25)$$

From (D.7), it follows that

$$|y| \leq c\left[1 + \mu^*\left(1 + \frac{1}{(\tau_{min})^{n^*}}\right)\right]m_f + c$$

Since $w = H(s)[u\ y]^T + [0\ 0\ y]^T$ we have

$$|w| \leq c\left[1 + \mu^*\left(1 + \frac{1}{(\tau_{min})^{n^*}}\right)\right]m_f + c$$

Thus

$$(\|w_t\|_2)^2 \leq c\left[\left\{1 + \mu^{*2}\left(1 + \frac{1}{(\tau_{min})^{2n^*}}\right)\right\}\bar{m}^2 + 1\right]t \qquad (D.26)$$

Again from (D.6), (D.7), we obtain

$$\|u_t\|_2 \leq c\|w_t\|_2 + \mu^*c\left(1 + \frac{1}{(\tau_{min})^{n^*}}\right)\|u_t\|_2 + c\|r_t\|_2$$

$$\|y_t\|_2 \leq c\|w_t\|_2 + \mu^*c\left(1 + \frac{1}{(\tau_{min})^{n^*}}\right)\|y_t\|_2 + c\|r_t\|_2$$

$$\|\dot{y}_t\|_2 \leq c\|w_t\|_2 + \mu^*c\left(1 + \frac{1}{(\tau_{min})^{n^*}}\right)\|\dot{y}_t\|_2 + c\|r_t\|_2$$

Since $\dot{w} = sH(s)[u\ y]^T + [0\ 0\ \dot{y}]^T$, using (D.26), it follows that

$$(\|\dot{w}_t\|_2)^2 \leq \frac{c\left[\left\{1 + \mu^{*2}\left(1 + \frac{1}{(\tau_{min})^{2n^*}}\right)\right\}\bar{m}^2 + 1\right]}{\left[1 - \mu^*c\left(1 + \frac{1}{(\tau_{min})^{n^*}}\right)\right]^2}t \qquad (D.27)$$

Combining (D.24), (D.25), (D.26), (D.27) we obtain

$$\int_0^t \left(\tilde{\theta}^T w\right)^2 dz \le \frac{c}{\alpha^2} \frac{1}{\left[1 - \mu^* c \left(1 + \frac{1}{(\tau_{min})^{n^*}}\right)\right]^2}$$

$$\left[\left\{1 + \mu^{*2}\left(1 + \frac{1}{(\tau_{min})^{2n^*}}\right)\right\}\bar{m}^2 + 1\right]t$$

$$+ c\alpha^{2n^*}\bar{m}^2\left[\int_0^t \frac{\left(\tilde{\theta}^T \phi\right)^2}{m^2} dz + \int_0^t |\dot{\tilde{\theta}}|^2 dz\right] \qquad (D.28)$$

Combining (D.1), (D.2), (D.23), (D.28) we obtain

$$\int_0^t e_1^2(z)dz \le \frac{\tau^2 c}{\left[1 - \mu^* c\left(1 + \frac{1}{(\tau_{min})^{n^*}}\right)\right]^2}$$

$$\left[\frac{1}{\alpha^2}\frac{\left\{1 + \mu^{*2}\left(1 + \frac{1}{(\tau_{min})^{2n^*}}\right)\right\}\bar{m}^2 + 1}{\left\{1 - \mu^* c\left(1 + \frac{1}{(\tau_{min})^{n^*}}\right)\right\}^2}t\right.$$

$$\left.+ \alpha^{2n^*}\bar{m}^2\left(c_\eta t + |\tilde{\theta}(0)|^2\right)\right] + \frac{\mu^{*2}c}{\left[1 - \mu^* c\left(1 + \frac{1}{(\tau_{min})^{n^*}}\right)\right]^2}t$$

from which (6.3) follows.

To obtain (6.4), we once again start from (D.21). Using (A7) we have

$$|e_1(t)| \le \mu^* c\left[1 + \frac{1}{(\tau_{min})^{n^*}}\right]\sup_{t \ge 0}|e_1(t)| + \left|\frac{\tau s}{c_0}W_{1\tau}(s)W_m(s)\left[\tilde{\theta}^T w\right]\right| + \mu^* c$$

$$(\text{since } W_\tau(s) = \tau s W_{1\tau}(s))$$

$$\le \mu^* c\left[1 + \frac{1}{(\tau_{min})^{n^*}}\right]\sup_{t \ge 0}|e_1(t)|$$

$$+ \tau\left\|W_{1\tau}(s)\frac{W_m(s)}{c_0}\right\|_2^\delta\left\|\left(\dot{\tilde{\theta}}^T w + \tilde{\theta}^T \dot{w}\right)_t\right\|_2^\delta + \mu^* c \text{ (using (2.7))}$$

$$\le \mu^* c\left[1 + \frac{1}{(\tau_{min})^{n^*}}\right]\sup_{t \ge 0}|e_1(t)|$$

$$+ \tau\|W_{1\tau}(s)\|_\infty^\delta\left\|\frac{W_m(s)}{c_0}\right\|_2^\delta\left[c\|w_t\|_2^\delta + c\|\dot{w}_t\|_2^\delta\right] + \mu^* c$$

$$(\text{since } \tilde{\theta}, \dot{\tilde{\theta}} \in L_\infty)$$

Using (D.11) and (D.15), we obtain

$$|e_1(t)| \le \mu^* c\left[1 + \frac{1}{(\tau_{min})^{n^*}}\right]\sup_{t \ge 0}|e_1(t)|$$

$$+\tau c \left[\left\{ 1 + \mu^* \left(1 + \frac{1}{(\tau_{min})^{n^*}} \right) \right\} \bar{m} + 1 \right] + \mu^* c \qquad \text{(D.29)}$$

Assume that μ^* is chosen small enough[6] so that

$$1 - \mu^* c \left(1 + \frac{1}{(\tau_{min})^{n^*}} \right) > 0 \qquad \text{(D.30)}$$

Then (6.4) follows from (D.29).

References

1. A. S. Morse, "Global Stability of Parameter Adaptive Control Systems," *IEEE Trans. on Automat. Contr.*, Vol. AC-25, 433-439, June 1980.
2. K. S. Narendra, Y. H. Lin, and L. S. Valavani, "Stable Adaptive Controller Design — Part II: Proof of Stability," *IEEE Trans. on Automat. Contr.*, Vol. AC-25, 440-448, June 1980.
3. G. C. Goodwin, P. J. Ramadge and P. E. Caines, "Discrete time multivariable adaptive control," *IEEE Trans. on Automat. Contr.*, Vol. AC-25, 449-456, June 1980.
4. P. A. Ioannou and P. V. Kokotovic, *Adaptive Systems with Reduced Models*, Berlin: Springer-Verlag, 1983.
5. C. E. Rohrs, L. Valavani, M. Athans and G. Stein, " Robustness of Continuous-time Adaptive Control Algorithms in the Presence of Unmodelled Dynamics," *IEEE Trans. on Automat. Contr.*, Vol. AC-30, No. 9, 881-889, Sept. 1985.
6. P. A. Ioannou and P. V. Kokotovic, "Instability Analysis and the Improvement of Robustness of Adaptive Control," *Automatica*, Vol. 20, No. 5, Sept. 1984.
7. P. A. Ioannou and K. S. Tsakalis, "A Robust Direct Adaptive Controller," *IEEE Trans. on Automat. Contr.*, Vol. AC-31, No. 11, 1033-1043, Nov. 1986.
8. K. S. Narendra and A. M. Annaswamy, *Stable Adaptive Systems*. Prentice Hall, 1989.
9. L. Praly, "Robust Model Reference Adaptive Controllers – Part 1: Stability Analysis," in *Proc. 23rd IEEE Conf. Decision Contr.*, 1984.
10. G. Kreisselmeier, "A Robust Indirect Adaptive Control Approach," *Int. Journal of Control*, Vol. 43, No. 1, 161-175, 1986.
11. P. Ioannou and J. Sun, "Theory and Design of Robust Direct and Indirect Adaptive Control Schemes," *Int. Journal of Control*, Vol. 47, No. 3, 775-813, 1988.
12. R. H. Middleton, G. C. Goodwin, D. J. Hill and D. Q. Mayne, "Design Issues in Adaptive Control," *IEEE Trans. on Automat. Contr.*, Vol. AC-33, No. 1, 50-58, Jan., 1988.
13. B. D. O. Anderson, R. R. Bitmead, C. R. Johnson, Jr., P. V. Kokotovic, R. L. Kosut, I. Mareels, L. Praly and B. D. Riedle, *Stability of Adaptive Systems: Passivity and Averaging Analysis*. Cambridge, MA: M. I. T. Press, 1986.
14. S. Sastry and M. Bodson, *Adaptive Control: Stability, Convergence and Robustness*. Prentice Hall, 1989.

---

[6] Equation (D.30) does not follow from (D.22). This is because though they qualitatively appear to be the same, the generic constants in (D.30) involve $\|.\|_A$ norms while those in (D.22) involve H_∞ norms and for a given stable transfer function, the former is an upper bound for the latter [36].

66

15. P. A. Ioannou and K. S. Tsakalis, "Time and Frequency Domain Bounds in Robust Adaptive Control," *Proceedings of the American Control Conference*, 1988.
16. K. S. Tsakalis, "Robustness of Model Reference Adaptive Controllers: An Input-Output Approach," *IEEE Trans. on Automat. Contr.*, Vol. AC-37, No. 5, 556-565, May 1992.
17. J. M. Krause, P. P. Khargonekar and G. Stein, "Robust Adaptive Control: Stability and Asymptotic Performance," *IEEE Trans. on Automat. Contr.*, Vol. AC-37, No. 3, 316-331, Mar., 1992.
18. Z. Zang and R. R. Bitmead, "Transient Bounds for Adaptive Control Systems," *Proc. 29th IEEE Conf. Decision Contr.*, 2724-2729, 1990.
19. I. M. Y. Mareels and R. R. Bitmead, "Nonlinear Dynamics in Adaptive Control: Chaotic and Periodic Stabilization," *Automatica*, Vol. 22, 641-655, 1986.
20. B. E. Ydstie, "Bifurcation and Complex Dynamics in Adaptive Control Systems," in *Proc. 25th IEEE Conf. Decision Contr.*, 1986.
21. M. P. Golden and B. E. Ydstie, "Chaos and Strange Attractors in Adaptive Control Systems," in *Proc. IFAC World Congress*, Munich, Vol. 10, 127-132, 1987.
22. L. Hsu and R. R. Costa, "Bursting Phenomena in Continuous-Time Adaptive Systems with a σ-Modification," *IEEE Trans. on Automat. Contr.*, Vol. AC-32, No. 1, 84-86, Jan., 1987.
23. P. A. Ioannou and G. Tao, "Dominant Richness and Improvement of Performance of Robust Adaptive Control," *Automatica*, Vol. 25, No. 2, 287-291, March 1989.
24. J. Sun, A. Olbrot and M. Polis, "Robust Stabilization and Robust Performance Using Model Reference Control and Modelling Error Compensation, " *IEEE Trans. on Automat. Contr.*, Vol. AC-39, No. 3, 630-635, 1994.
25. J. Sun, "A Modified Model Reference Adaptive Control Scheme for Improved Transient Performance," *IEEE Trans. on Automat. Contr.*, Vol. AC-38, No. 8, 1255-1259, Aug. 1993.
26. D. E. Miller and E. J. Davison, "An Adaptive Controller Which Provides an Arbitrarily Good Transient and Steady State Response," *IEEE Trans. on Automat. Contr.*, Vol. AC-36, No. 1, 68-81, Jan., 1991.
27. B. E. Ydstie, "Transient Performance and Robustness of Direct Adaptive Control," *IEEE Trans. on Automat. Contr.*, Vol.37, No. 8, 1091-1105, Aug. 1992.
28. R. Ortega, "On Morse's New Adaptive Controller: Parameter Convergence and Transient Performance," *IEEE Trans. on Automat. Contr.*, Vol. AC-38, No. 8, 1191-1202, Aug. 1993.
29. A. Datta and P. A. Ioannou, "Performance Improvement Versus Robust Stability in Model Reference Adaptive Control," *Proceedings of the 30th IEEE Conf. on Decision and Control*, 748-753, Dec. 1991.
30. M. Krstic, P. V. Kokotovic and I. Kanellakopoulos, "Transient Performance Improvement with a New Class of Adaptive Controllers," *Systems & Control Letters*, Vol. 21, 451-461, 1993.
31. B. A. Francis, *A Course in H_∞ Control Theory*, Berlin: Springer Verlag, 1987.
32. P. A. Ioannou and A. Datta, "Robust Adaptive Control: A Unified Approach," *Proceedings of the IEEE*, Vol. 79, No. 12, 1736-1768, December 1991.
33. G. C. Goodwin and D. Q. Mayne, "A Parameter Estimation Perspective of Continuous Time Model Reference Adaptive Control," *Automatica*, Vol. 23, No. 1, 57-70, 1987.
34. A. Datta and M. T. Ho, "On Modifying Model Reference Adaptive Control Schemes for Performance Improvement," *IEEE Trans. on Automat. Contr.*, Vol. 39, No. 9, Sept. 1994 (to appear).

35. M. Vidyasagar, *Control System Synthesis: A Factorization Approach.* MIT Press, Cambridge, 1985.
36. C. A. Desoer and M. Vidyasagar, *Feedback Systems: Input-Output Properties.* Academic Press, New York, 1975.
37. V. M. Popov, *Hyperstability of Control Systems.* Springer-Verlag, New York, 1973.

This article was processed using the LaTeX macro package with LLNCS style

Advances in Adaptive Nonlinear Control*

Ioannis Kanellakopoulos

Department of Electrical Engineering, UCLA, Los Angeles, CA 90024-1594.

Abstract. We present a brief account of recent advances in adaptive nonlinear control. Using simple examples, we illustrate design tools like adaptive backstepping, overparametrization, tuning functions, and nonlinear damping, and the new properties that they bestow upon the resulting adaptive systems, such as boundedness without adaptation and systematic transient performance improvement through trajectory initialization.

1 Introduction

Over the last few years, adaptive control of nonlinear systems has emerged as an exciting new research area, which has witnessed rapid and impressive developments. The first results [21, 27, 23, 1, 28, 6, 22], indicated that there were two factors contributing to the difficulty of the problem:

- the *uncertainty level*, defined as the number of integrators by which the uncertain parameters precede the control in the system equations, and
- the *growth rate* of the nonlinearities.

For some time, it appeared that global stability and tracking results could be obtained

- for arbitrary uncertainty level only if the growth rate of the nonlinearities was restricted to be essentially linear, and
- for arbitrary nonlinearities only if the uncertainty level was restricted to be zero or one.

During the 1990 Grainger lectures [12], however, a new design procedure [7] was presented, which guaranteed global stability and tracking for arbitrary uncertainty level and for arbitrary nonlinearity growth rate, provided that

- the *nonlinear complexity*, defined as the number of state variables that enter in the nonlinearities, is restricted to that of *strict-feedback systems*, i.e., systems whose nonlinearities depend only on state-variables that are "fed back."

---

* This work was supported in part by NSF through Grant RIA ECS-9309402 and in part by UCLA through the SEAS Dean's Fund.

nonlinear complexity ↑			
	$\dot{x}_1 = x_2$ $\dot{x}_2 = x_3$ $\dot{x}_3 = u + \theta\varphi(x_1, x_2, x_3)$	$\dot{x}_1 = x_2$ $\dot{x}_2 = x_3 + \theta\varphi(x_1, x_2, x_3)$ $\dot{x}_3 = u$	$\dot{x}_1 = x_2 + \theta\varphi(x_1, x_2, x_3)$ $\dot{x}_2 = x_3$ $\dot{x}_3 = u$
	$\dot{x}_1 = x_2$ $\dot{x}_2 = x_3$ $\dot{x}_3 = u + \theta\varphi(x_1, x_2)$	$\dot{x}_1 = x_2$ $\dot{x}_2 = x_3 + \theta\varphi(x_1, x_2)$ $\dot{x}_3 = u$	$\dot{x}_1 = x_2 + \theta\varphi(x_1, x_2)$ $\dot{x}_2 = x_3$ $\dot{x}_3 = u$
	$\dot{x}_1 = x_2$ $\dot{x}_2 = x_3$ $\dot{x}_3 = u + \theta\varphi(x_1)$	$\dot{x}_1 = x_2$ $\dot{x}_2 = x_3 + \theta\varphi(x_1)$ $\dot{x}_3 = u$	$\dot{x}_1 = x_2 + \theta\varphi(x_1)$ $\dot{x}_2 = x_3$ $\dot{x}_3 = u$
	0	1	2
	uncertainty level →		

Fig. 1. Examples of systems with different nonlinear complexities and uncertainty levels.

The chart presented in Figure 1 illustrates the nonlinear complexity vs. uncertainty level tradeoff in the absence of any restrictions on the growth rates of nonlinearities:

- For the systems in the first two columns as well as for the bottom system in the third column, global stability and tracking can be achieved. All these systems are strict-feedback systems.
- For the middle system in the third column, only regional stability and tracking are achievable. This is an example of a pure-feedback system.
- For the top system in the third column, only local results exist to date.

The reason for the nomenclature "strict-feedback systems" is illustrated in Figure 2, which contains the block-diagram representation of the system

$$\begin{aligned}\dot{x}_1 &= x_2 + \theta^{\mathrm{T}}\varphi_1(x_1)\\ \dot{x}_2 &= x_3 + \theta^{\mathrm{T}}\varphi_2(x_1, x_2)\\ \dot{x}_3 &= u + \theta^{\mathrm{T}}\varphi_3(x_1, x_2, x_3)\,,\end{aligned} \tag{1.1}$$

where θ is an unknown constant parameter vector.

Adaptive backstepping led to a multitude of global stability and tracking results for large classes of nonlinear systems [7, 8, 3, 9, 10, 11, 13, 17, 18, 19, 24, 29]. The latter results are all Lyapunov-based, i.e., the design procedure achieves the desired objectives by constructing a suitable Lyapunov function and rendering its derivative nonpositive. Estimation-based adaptive controllers for similar classes of nonlinear systems have also been recently developed [15, 16].

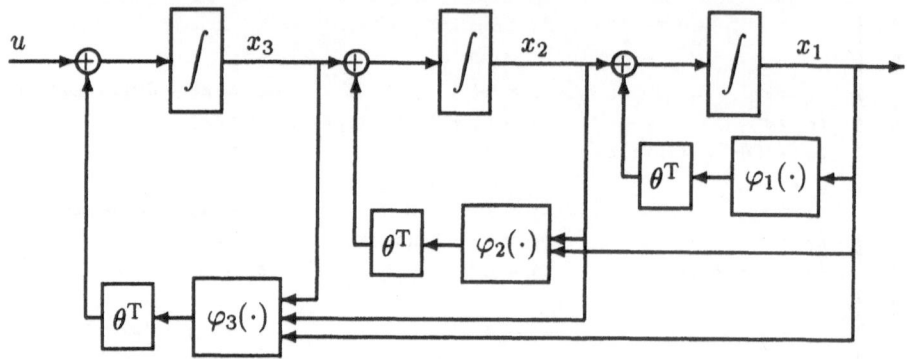

Fig. 2. Block diagram of the strict-feedback system (1.1). The nonlinearities depend only on variables that are "fed back".

In this paper we use some simple examples to illustrate several design tools such as adaptive backstepping, overparametrization, tuning functions, and κ-terms, which are used as building blocks for the construction of systematic design procedures. Section 2 deals with adaptive backstepping and discusses overparametrized schemes as well as schemes using tuning functions. In Section 3 we discuss the recently introduced κ-terms, and show how these terms and trajectory initialization can be combined with existing design procedures to yield adaptive controllers which guarantee boundedness without adaptation and without knowledge of bounds on the unknown parameters, and systematic improvement of transient tracking performance with the choice of design parameters. Section 4 deals deals with output-feedback schemes for systems whose nonlinearities depend only on the measured output. In Section 5, we state the general nonlinear results without proving them. Finally, in Section 6 we illustrate the application of these design procedures to adaptive control of linear systems.

2 Adaptive Backstepping

To illustrate adaptive backstepping, let us start with the simplest nonlinear system:

$$\dot{x} = u + \theta\varphi(x)\,, \tag{2.1}$$

whose block diagram is given in Figure 3. To achieve regulation of $x(t)$, we design an adaptive controller.

If θ were known, the control

$$u = -\theta\varphi(x) - c_1 x\,, \quad c_1 > 0 \tag{2.2}$$

would render the derivative of the Lyapunov function $V_0(x) = \frac{1}{2}x^2$ negative definite: $\dot{V} = -c_1 x^2$. Of course the control law (2.2) can not be implemented,

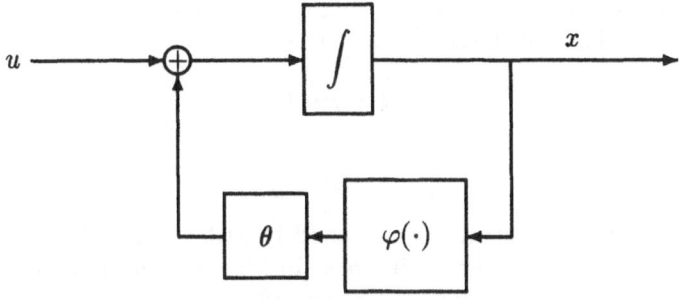

Fig. 3. The system (2.1).

since θ is unknown. Instead, one can employ its *certainty-equivalence* form in which θ is replaced by an estimate $\hat{\theta}$:

$$u = -\hat{\theta}\varphi(x) - c_1 x \,. \tag{2.3}$$

Substituting (2.3) into (2.2), we obtain

$$\dot{x} = -c_1 x + \tilde{\theta}\varphi(x) \,, \tag{2.4}$$

where $\tilde{\theta}$ is the *parameter error*:

$$\tilde{\theta} = \theta - \hat{\theta} \,. \tag{2.5}$$

The derivative of V_0 becomes

$$\dot{V}_0 = -c_1 x^2 + \tilde{\theta} x \varphi(x) \,. \tag{2.6}$$

Since the second term is indefinite and contains the unknown parameter error $\tilde{\theta}$, we can not conclude anything about the stability of (2.2). Therefore, we augment V_0 with a quadratic term in the parameter error $\tilde{\theta}$ to obtain the Lyapunov function:

$$V_1(x, \tilde{\theta}) = \frac{1}{2}x^2 + \frac{1}{2\gamma}\tilde{\theta}^2 \,, \tag{2.7}$$

where $\gamma > 0$ is the *adaptation gain*. The derivative of this function is

$$\begin{aligned}
\dot{V}_1 &= x\dot{x} + \frac{1}{\gamma}\tilde{\theta}\dot{\tilde{\theta}} \\
&= -c_1 x^2 + \tilde{\theta} x \varphi(x) + \frac{1}{\gamma}\tilde{\theta}\dot{\tilde{\theta}} \\
&= -c_1 x^2 + \tilde{\theta}\left[x\varphi(x) + \frac{1}{\gamma}\dot{\tilde{\theta}}\right] \,.
\end{aligned} \tag{2.8}$$

The second term is still indefinite and contains $\tilde{\theta}$ as a factor. However, the situation is much better than in (2.6), because we now have the dynamics of

$\dot{\tilde{\theta}} = -\dot{\hat{\theta}}$ at our disposal. With the appropriate choice of $\dot{\hat{\theta}}$ we can cancel the indefinite term. Thus, we choose the update law

$$\dot{\hat{\theta}} = -\dot{\tilde{\theta}} = \gamma\, x\varphi(x)\,, \tag{2.9}$$

which yields

$$\dot{V_1} = -c_1 x^2 \le 0\,. \tag{2.10}$$

The resulting adaptive system consists of (2.1) with the control (2.3) and the update law (2.9), and is shown in Figure 4. In Figure 5, this system is redrawn in its closed-loop form consisting of (2.4) and (2.9), namely

$$\dot{x} = -c_1 x + \tilde{\theta}\varphi(x)$$
$$\dot{\tilde{\theta}} = -\gamma\, x\varphi(x)\,. \tag{2.11}$$

Because $\dot{V_1} \le 0$, the equilibrium $x = 0$, $\tilde{\theta} = 0$ of (2.11) is globally stable. In addition, the desired regulation property $\lim_{t\to\infty} x(t) = 0$ follows from (2.10). The adaptive nonlinear controller which guarantees these properties is given by (2.4) and (2.9):

$$u = -c_1 x - \hat{\theta}\varphi(x)$$
$$\dot{\hat{\theta}} = \gamma\, x\varphi(x)\,. \tag{2.12}$$

One may think that the above adaptive design is so straightforward because (2.1) is a first-order system. In fact, this is due to the matching condition: the terms containing unknown parameters in (2.1) are in the span of the control, i.e., they can be directly cancelled by u when θ is known. To illustrate this point, let us consider the following second-order system, where again the uncertain term is "matched" by the control u:

$$\dot{x}_1 = x_2 + \varphi_1(x_1)$$
$$\dot{x}_2 = \theta\varphi_2(x) + u\,. \tag{2.13}$$

If θ were known, we would be able to use the change of coordinates

$$z_1 = x_1\,, \quad z_2 = x_2 + c_1 x_1 + \varphi_1(x_1)\,, \tag{2.14}$$

and then form the Lyapunov function

$$V_c(x) = \frac{1}{2}x_1^2 + \frac{1}{2}\left(x_2 + c_1 x_1 + \varphi_1(x_1)\right) = \frac{1}{2}z_1^2 + \frac{1}{2}z_2^2\,, \tag{2.15}$$

whose derivative is rendered negative definite

$$\dot{V}_c = -c_1 z_1^2 - c_2 z_2^2 \tag{2.16}$$

by the control

$$u = -c_2 z_2 - z_1 + \left(c_1 + \frac{\partial\varphi_1}{\partial x_1}\right)(x_2 + \varphi_1) - \theta\varphi_2\,. \tag{2.17}$$

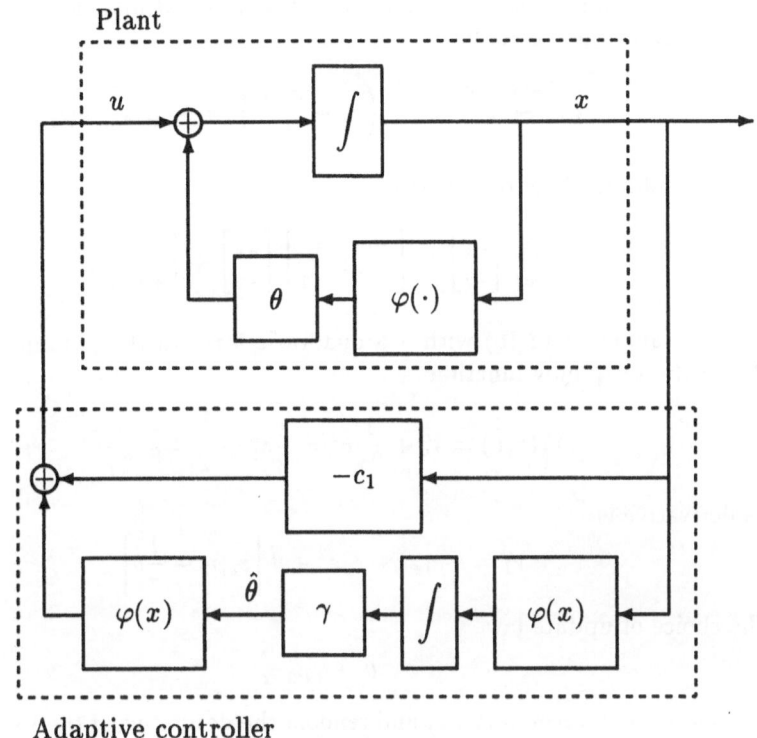

Plant

Adaptive controller

Fig. 4. The closed-loop adaptive system (2.11).

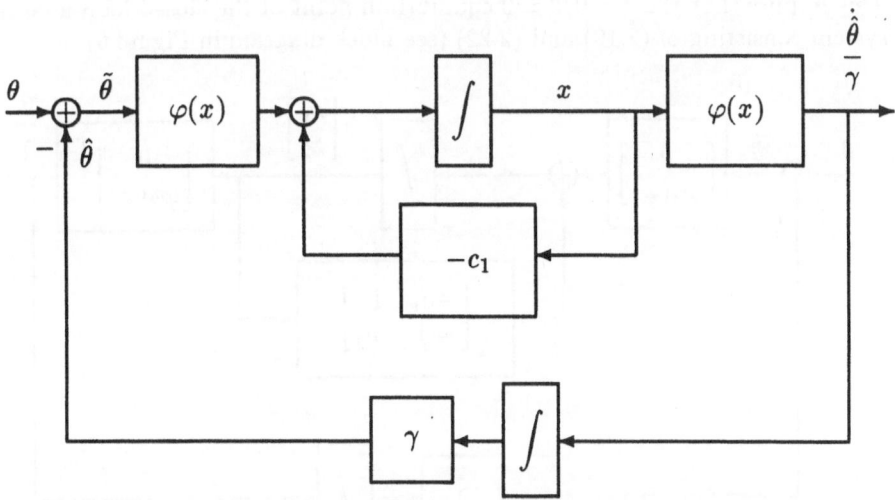

Fig. 5. An equivalent representation of (2.11).

Since θ is unknown, we again replace it with its estimate $\hat{\theta}$ in (2.17) to obtain the implemented control law:

$$u = -c_2 z_2 - x_1 + \left(c_1 + \frac{\partial \varphi_1}{\partial x_1}\right)(x_2 + \varphi_1) - \hat{\theta}\varphi_2 \,. \tag{2.18}$$

which results in the error system:

$$\frac{d}{dt}\begin{bmatrix} z_1 \\ z_2 \end{bmatrix} = \begin{bmatrix} -c_1 & 1 \\ -1 & -c_2 \end{bmatrix}\begin{bmatrix} z_1 \\ z_2 \end{bmatrix} + \begin{bmatrix} 0 \\ \varphi_2 \end{bmatrix}\tilde{\theta} \,. \tag{2.19}$$

Then we augment (2.16) with a a quadratic term in the parameter error $\tilde{\theta}$ to obtain the Lyapunov function:

$$V_1(z, \tilde{\theta}) = V_c + \frac{1}{2\gamma}\tilde{\theta}^2 = \frac{1}{2}z_1^2 + \frac{1}{2}z_2^2 + \frac{1}{2\gamma}\tilde{\theta}^2 \,. \tag{2.20}$$

Its derivative is

$$\dot{V}_1 = -c_1 z_1^2 - c_2 z_2^2 + \tilde{\theta}\left[z_2\varphi_2 - \frac{1}{\gamma}\dot{\hat{\theta}}\right]. \tag{2.21}$$

The choice of update law

$$\dot{\hat{\theta}} = \gamma\varphi_2 z_2 \tag{2.22}$$

eliminates the $\tilde{\theta}$-term in (2.21) and renders the derivative of the Lyapunov function (2.20) nonpositive:

$$\dot{V}_1 = -c_1 z_1^2 - c_2 z_2^2 \leq 0 \,. \tag{2.23}$$

This implies that the $z = 0, \tilde{\theta} = 0$ equilibrium point of the closed-loop adaptive system consisting of (2.19) and (2.22) (see block diagram in Figure 6)

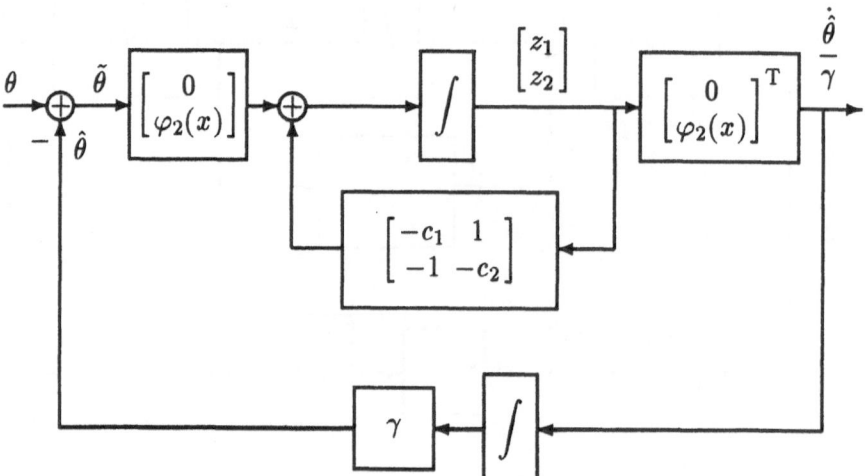

Fig. 6. The closed-loop adaptive system (2.24).

$$\frac{d}{dt} \begin{bmatrix} z_1 \\ z_2 \end{bmatrix} = \begin{bmatrix} -c_1 & 1 \\ -1 & -c_2 \end{bmatrix} \begin{bmatrix} z_1 \\ z_2 \end{bmatrix} + \begin{bmatrix} 0 \\ \varphi_2(x) \end{bmatrix} \tilde{\theta}$$

$$\dot{\tilde{\theta}} = -\gamma \begin{bmatrix} 0 & \varphi_2 \end{bmatrix} \begin{bmatrix} z_1 \\ z_2 \end{bmatrix} \tag{2.24}$$

is globally stable and, in addition, $x(t) \to 0$ as $t \to \infty$.

The adaptive design in the above examples was simple because of the matching: the parametric uncertainty was in the span of the control. We now move to the more general case of *extended matching*, where the parametric uncertainty enters the system one integrator before the control does:

$$\dot{x}_1 = x_2 + \theta\varphi(x_1) \tag{2.25a}$$

$$\dot{x}_2 = u. \tag{2.25b}$$

We use this example to introduce *adaptive backstepping*.

Let us first note that the system (2.25) is just the system (2.1) augmented by an integrator: in its block diagram, given in Figure 7, the scalar system (2.1) of Figure 3 appears in the dashed box.

Step 1. To design an adaptive controller for (2.25), we will exploit the fact that an adaptive controller is known for the subsystem in the dashed box. Indeed, if x_2 were the control input, then (2.25a) would be identical to (2.1), and the corresponding adaptive controller and Lyapunov function would be given by (2.12) and (2.7). Of course x_2 is just a state variable and not the control. Nevertheless, as its "desired value" we prescribe

$$x_{2 \text{ des}} = -c_1 x_1 - \hat{\theta}\varphi(x_1) \triangleq \alpha_1(x_1, \hat{\theta}). \tag{2.26}$$

Let $z_1 = x_1$ and let z_2 be the deviation of x_2 from its desired value:

$$z_2 = x_2 - x_{2 \text{ des}} = x_2 - \alpha_1(x_1, \hat{\theta}) = x_2 + c_1 x_1 + \hat{\theta}\varphi(x_1). \tag{2.27}$$

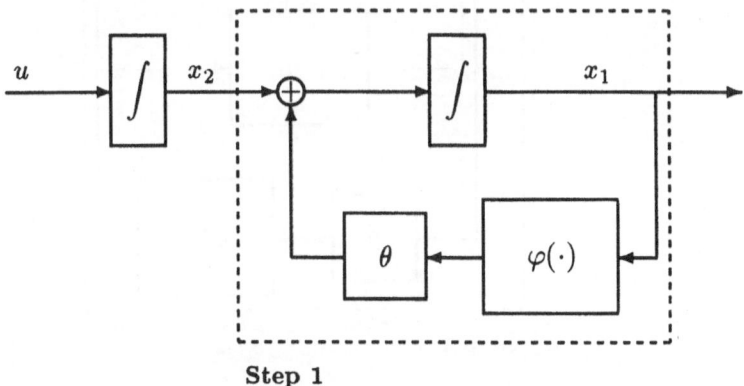

Step 1

Fig. 7. The block diagram of system (2.25).

We call x_2 a *virtual control*, and its desired value $\alpha_1(x_1, \hat{\theta})$ a *stabilizing function*. The variable z_2 is the corresponding *error variable*. Now we rewrite the system (2.25) in the (z_1, z_2)-coordinates in which it takes on a more convenient form, as illustrated in Figures 8 and 9. Starting from (2.25) and Figure 7, we add and subtract the stabilizing function $\alpha_1(x, \hat{\theta})$ to the $\dot{x}_1$-equation as shown in Figure 8. Then we use $\alpha_1(x)$ as the feedback control inside the dashed box and "backstep" $-\alpha_1(x)$ through the integrator, as in Figure 9.

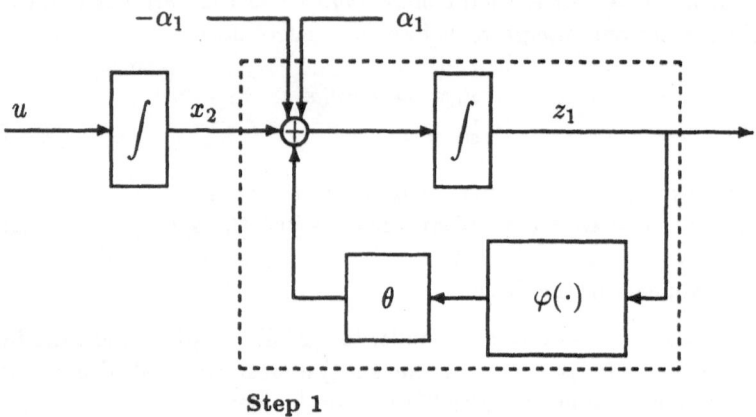

Fig. 8. Introducing $\alpha_1(x, \hat{\theta})$ as the desired value for x_2.

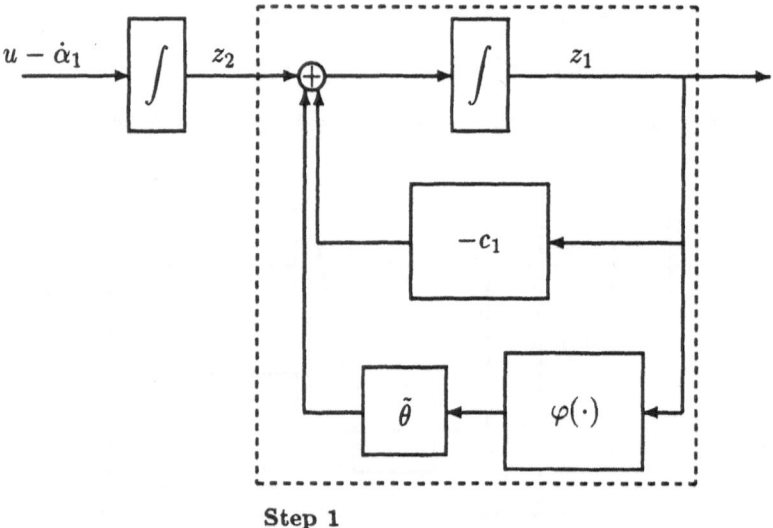

Fig. 9. Closing the feedback loop in the dashed box with $+\alpha_1$ and "backstepping" $-\alpha_1$ through the integrator.

In the new coordinates (z_1, z_2) the system is expressed as

$$\dot{z}_1 = \theta\varphi + \left[x_2 + c_1 x_1 + \hat{\theta}\varphi\right] - c_1 x_1 - \hat{\theta}\varphi = -c_1 z_1 + z_2 + \tilde{\theta}\varphi \quad (2.28a)$$

$$\dot{z}_2 = \dot{x}_2 - \dot{\alpha}_1 = u - \frac{\partial \alpha_1}{\partial x_1}(x_2 + \theta\varphi) - \frac{\partial \alpha_1}{\partial \hat{\theta}}\dot{\hat{\theta}}$$

$$= u - \frac{\partial \alpha_1}{\partial x_1}x_2 - \hat{\theta}\frac{\partial \alpha_1}{\partial x_1}\varphi - \tilde{\theta}\frac{\partial \alpha_1}{\partial x_1}\varphi - \frac{\partial \alpha_1}{\partial \hat{\theta}}\dot{\hat{\theta}}. \quad (2.28b)$$

A key feature of adaptive backstepping is that we don't use a differentiator to implement the time derivative $\dot{\alpha}_1$ in (2.28b); since $\alpha_1(x, \hat{\theta})$ is a known function, we compute its time derivative analytically as $\dot{\alpha}_1 = \frac{\partial \alpha_1}{\partial x}\dot{x} + \frac{\partial \alpha_1}{\partial \hat{\theta}}\dot{\hat{\theta}}$.

The first Lyapunov function is chosen from (2.7) as

$$V_1(z_1, \hat{\theta}) = \frac{1}{2}z_1^2 + \frac{1}{2\gamma}\tilde{\theta}^2, \quad (2.29)$$

Its derivative is

$$\dot{V}_1 = z_1 z_2 - c_1 z_1^2 + \tilde{\theta}\left(\varphi z_1 - \frac{1}{\gamma}\dot{\hat{\theta}}\right). \quad (2.30)$$

Another key feature of adaptive backstepping with tuning functions is that the choice of update law for $\hat{\theta}$ is *postponed* until the next step.

Step 2. We are now in the position to select a Lyapunov function for the system (2.25). Let us try to achieve this by augmenting the already existing Lyapunov function $V_1(z_1, \hat{\theta})$ with a quadratic term in the error variable z_2:

$$V_2 = V_1 + \frac{1}{2}z_2^2 = \frac{1}{2}z_1^2 + \frac{1}{2}z_2^2 + \frac{1}{2\gamma}\tilde{\theta}^2. \quad (2.31)$$

In view of (2.30) and (2.28b), the derivative of V_2 is

$$\dot{V}_2 = z_1 z_2 - c_1 z_1^2 + \tilde{\theta}\left(\varphi z_1 - \frac{1}{\gamma}\dot{\hat{\theta}}\right)$$

$$+ z_2\left[u - \frac{\partial \alpha_1}{\partial x_1}x_2 - \hat{\theta}\frac{\partial \alpha_1}{\partial x_1}\varphi - \tilde{\theta}\frac{\partial \alpha_1}{\partial x_1}\varphi - \frac{\partial \alpha_1}{\partial \hat{\theta}}\dot{\hat{\theta}}\right]$$

$$= -c_1 z_1^2 + \tilde{\theta}\left[\varphi z_1 - z_2\frac{\partial \alpha_1}{\partial x_1}\varphi - \frac{1}{\gamma}\dot{\hat{\theta}}\right]$$

$$+ z_2\left[z_1 + u - \frac{\partial \alpha_1}{\partial x_1}x_2 - \hat{\theta}\frac{\partial \alpha_1}{\partial x_1}\varphi - \frac{\partial \alpha_1}{\partial \hat{\theta}}\dot{\hat{\theta}}\right]. \quad (2.32)$$

In the last equation, all the terms containing $\tilde{\theta}$ have been grouped together. To eliminate them, the update law is chosen as

$$\dot{\hat{\theta}} = \gamma\left(\varphi z_1 - \frac{\partial \alpha_1}{\partial x_1}\varphi z_2\right). \quad (2.33)$$

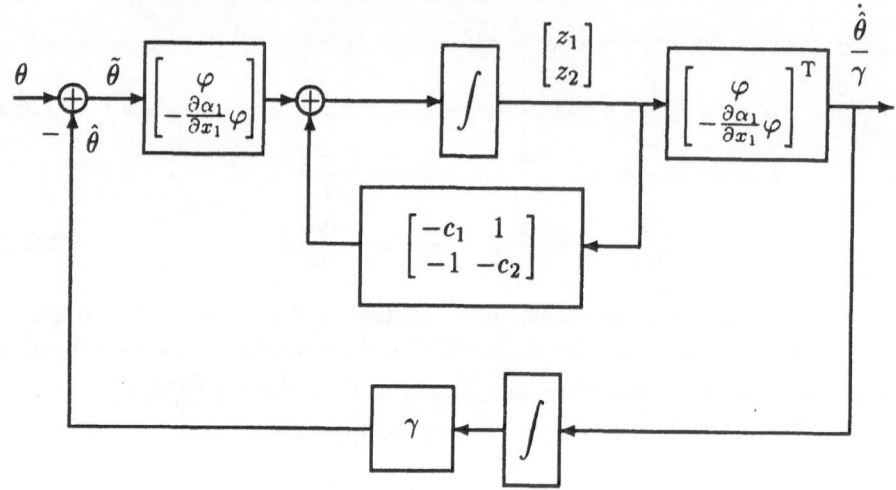

Fig. 10. The closed-loop adaptive system (2.37).

Then, the last bracketed term in (2.32) will be rendered equal to $-c_2 z_2^2$ with the control

$$u = -z_1 - c_2 z_2 + \frac{\partial \alpha_1}{\partial x_1} x_2 + \hat{\theta} \frac{\partial \alpha_1}{\partial x_1} \varphi + \frac{\partial \alpha_1}{\partial \hat{\theta}} \dot{\hat{\theta}}, \qquad (2.34)$$

where for $\dot{\hat{\theta}}$ we use the analytical expression of the update law (2.33). Substituting the expressions (2.33) and (2.34) into (2.32) we obtain:

$$\dot{V}_2 = -c_1 z_1^2 - c_2 z_2^2 \leq 0, \qquad (2.35)$$

and the error system becomes

$$\begin{aligned}
\dot{z}_1 &= -c_1 z_1 + z_2 + \tilde{\theta} \varphi \\
\dot{z}_2 &= -c_2 z_2 - z_1 - \tilde{\theta} \frac{\partial \alpha_1}{\partial x_1} \varphi \\
\dot{\hat{\theta}} &= \gamma \left(\varphi z_1 - \frac{\partial \alpha_1}{\partial x_1} \varphi z_2 \right).
\end{aligned} \qquad (2.36)$$

The matrix form of this system (see block diagram in Figure 10),

$$\begin{aligned}
\frac{d}{dt} \begin{bmatrix} z_1 \\ z_2 \end{bmatrix} &= \begin{bmatrix} -c_1 & 1 \\ -1 & -c_2 \end{bmatrix} \begin{bmatrix} z_1 \\ z_2 \end{bmatrix} + \begin{bmatrix} \varphi \\ -\frac{\partial \alpha_1}{\partial x_1} \varphi \end{bmatrix} \tilde{\theta} \\
\dot{\hat{\theta}} &= \gamma \begin{bmatrix} \varphi & -\frac{\partial \alpha_1}{\partial x_1} \varphi \end{bmatrix} \begin{bmatrix} z_1 \\ z_2 \end{bmatrix},
\end{aligned} \qquad (2.37)$$

makes its properties more visible:

- the constant system matrix has negative terms along its diagonal, while its off-diagonal terms are skew-symmetric, and
- the vector that multiplies the parameter error in the $\dot{z}$-equation is used in the update law for the parameter estimate.

The stability properties of (2.37) follow from (2.31) and (2.35), which establish that $z_1, z_2, \vartheta_1, \vartheta_2$ are bounded, and $z \to 0$ as $t \to \infty$. Since $z_1 = x_1$, x_1 is also bounded and converges to zero. The boundedness of x_2 then follows from the boundedness of α_1 (defined in (2.26)) and the fact that $x_2 = z_2 + \alpha_1$. Using (2.34) we conclude that the control u is also bounded. Finally, we note that the regulation of z and x_1 does *not* imply the regulation of x_2: from $z_2 = x_2 - \alpha_1$ and (2.26) we see that $x_2 + \hat{\theta}_1 \varphi(0)$ will converge to zero. Thus, x_2 is not guaranteed to converge to zero unless $\varphi(0) = 0$. However, x_2 will converge to a constant value:

$$\lim_{t \to \infty} = -\theta \varphi(0) \stackrel{\Delta}{=} x_2^e. \tag{2.38}$$

This can be seen from (2.25a): since x_1 and $\dot{x}_1$ converge to zero,[1] so does $x_2 + \theta \varphi(0)$.

The design procedure illustrated in the above example can be generalized to systems of arbitrary order which can be transformed into the strict-feedback (5.1) or the partially strict-feedback (5.2) canonical form. The number of design steps required is equal to the relative degree ρ of the system. In its original version [7] which used *overparametrization*, this procedure generated at each step a new stabilizing function α_i and a *new parameter estimate* ϑ_i is generated. As a result, if a system contains p unknown parameters, the overparametrized adaptive controller may employ as many as $p\,\rho$ parameter estimates. A schematic representation of this design procedure is given in Figure 11.

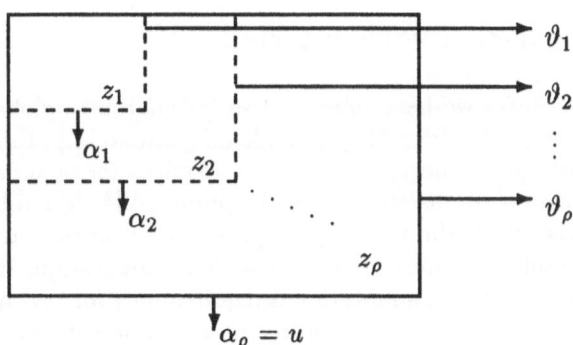

Fig. 11. The design procedure for overparametrized schemes.

The increase in the number of parameter estimates caused by overparametrization can be an undesirable feature, since it rapidly increases the dynamic order of the resulting adaptive controller. Soon after the design procedure of [7] was developed, the number of parameter estimates required was reduced in half [3]. Subsequently, the need for overparametrization was eliminated with the introduction of *tuning functions* [13], which require only as many estimates as there

---

[1] The fact that $\dot{x}_1 \to 0$ follows from the uniform continuity (since $\ddot{x}_1$ is bounded) and integrability (since $\int_0^t \dot{x}_1(s)ds = x_1(t) - x_1(0) \to -x_1(0)$ as $t \to \infty$) of $\dot{x}_1$.

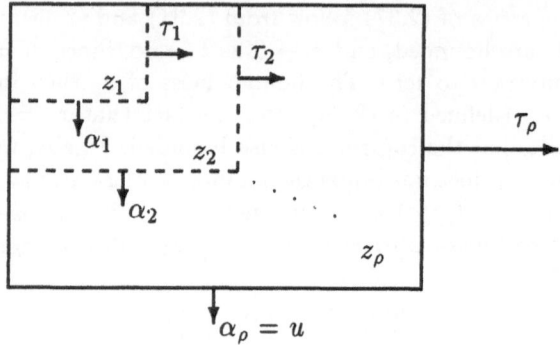

Fig. 12. The design procedure with tuning functions.

are unknown parameters. The tuning functions scheme can also be generalized to systems of arbitrary order which can be transformed into the strict-feedback (5.1) or the partially strict-feedback (5.2) canonical form. The number of design steps required is again equal to the relative degree ρ of the system. At each step, a new stabilizing function α_i and a new tuning function ϑ_i is generated. At the final step, the stabilizing function α_ρ becomes the control law u and the tuning function τ_ρ becomes the update law $\dot{\hat{\theta}} = \Gamma \tau_\rho$. A schematic representation of this design procedure is given in Figure 12.

3 Adaptive design with κ-terms

The design procedures we have presented so far guarantee global stability and asymptotic tracking. However, they provide no guarantees in terms of transient performance and, furthermore, adaptation is required for boundedness and stability. In this section we discuss a recently proposed [4, 5] modification of the design procedures which allows the systematic improvement of transient performance and guarantees boundedness even without adaptation. Hence, with the addition of these "κ-terms", adaptation is required only for asymptotic tracking. The κ-terms were instrumental in the recent development of estimation-based schemes for adaptive control of nonlinear systems [15, 16].

Disturbance-induced instability Consider the scalar nonlinear system depicted in Figure 13:

$$\dot{x} = u + \Delta(t)\varphi(x), \tag{3.39}$$

where $\varphi(x)$ is a known smooth nonlinearity, and $\Delta(t)$ is an exponentially decaying disturbance:

$$\Delta(t) = \Delta(0)e^{-kt}. \tag{3.40}$$

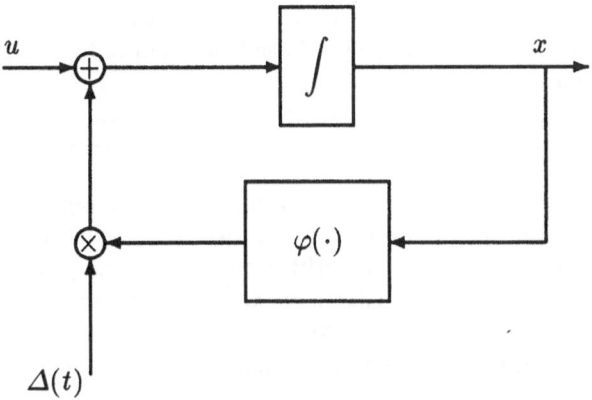

Fig. 13. The perturbed system (3.39).

Because of its exponentially decaying nature, one might be tempted to ignore the disturbance $\Delta(t)$ and use the simple control $u = -cx$, which results in the closed-loop system

$$\dot{x} = -cx + \Delta(0)e^{-kt}\varphi(x) . \tag{3.41}$$

While this design may be satisfactory in the case where $\varphi(x)$ is bounded by a constant or even a linear function of x, it is inadequate if $\varphi(x)$ is allowed to be any smooth nonlinear function. For example, consider the case $\varphi(x) = x^2$, which yields

$$\dot{x} = -cx + \Delta(0)e^{-kt}x^2 . \tag{3.42}$$

The solution $x(t)$ of this system can be calculated explicitly using the change of variable $w = 1/x$:

$$\dot{w} = -\frac{1}{x^2}\dot{x} = c\frac{1}{x} - \Delta(0)e^{-kt} = cw - \Delta(0)e^{-kt} , \tag{3.43}$$

which yields

$$w(t) = \left[w(0) - \frac{\Delta(0)}{c+k}\right]e^{ct} + \frac{\Delta(0)}{c+k}e^{-kt} . \tag{3.44}$$

The substitution $w = 1/x$ gives

$$x(t) = \frac{x(0)(c+k)}{[c+k - \Delta(0)x(0)]e^{ct} + \Delta(0)x(0)e^{-kt}} . \tag{3.45}$$

From (3.45) we see that the behavior of the closed-loop system (3.42) depends critically on the initial conditions $\Delta(0), x(0)$:

(i) If $\Delta(0)x(0) < c+k$, the solutions $x(t)$ are bounded and converge asymptotically to zero.

(ii) The situation changes dramatically when $\Delta(0)x(0) > c + k > 0$. The solutions $x(t)$ which start from such initial conditions not only diverge to infinity, but do so in *finite time:*

$$x(t) \to \infty \text{ as } t \to t_f = \frac{1}{c+k} \ln \left\{ \frac{\Delta(0)x(0)}{\Delta(0)x(0) - (c+k)} \right\}. \tag{3.46}$$

Note that this finite escape can not be eliminated by making c larger: for any values of c and k and for any nonzero value of $\Delta(0)$ there exist initial conditions $x(0)$ which satisfy the inequality $\Delta(0)x(0) > c + k$. This example shows that in a nonlinear system, neglecting the effects of exponentially decaying disturbances or nonzero initial conditions can be catastrophic: assuming that $x(0) = 0$ or $\Delta(0) = 0$ artificially eliminates the finite escape time phenomenon.

Nonlinear damping with κ-terms To overcome this problem and guarantee that the solutions $x(t)$ starting from any initial condition $x(0)$ will remain bounded, we augment the control law $u = -cx$ with a *nonlinear damping term* $s(x)$ which is to be designed:

$$u = -cx + s(x). \tag{3.47}$$

Returning to the system (3.39), we use the quadratic Lyapunov function $V(x) = \frac{1}{2}x^2$ whose derivative is:

$$\begin{aligned} \dot{V} &= x\,u + x\,\varphi(x)\Delta(t) \\ &= -cx^2 + xs(x) + x\,\varphi(x)\Delta(t). \end{aligned} \tag{3.48}$$

The objective of guaranteeing global boundedness of solutions can be equivalently expressed as rendering the derivative of the Lyapunov function negative outside a compact region. This is achieved with the choice

$$s(x) = -\kappa\,x\,\varphi^2(x), \quad \kappa > 0, \tag{3.49}$$

which yields the control

$$u = -cx - \kappa\,x\,\varphi^2(x) \tag{3.50}$$

and the Lyapunov function derivative

$$\begin{aligned} \dot{V} &= -cx^2 - \kappa\,x^2\varphi^2(x) + x\,\varphi(x)\Delta(t) \\ &= -cx^2 - \kappa\left[x\,\varphi(x) - \frac{\Delta(t)}{2\kappa}\right]^2 + \frac{\Delta^2(t)}{4\kappa} \\ &\le -cx^2 + \frac{\Delta^2(t)}{4\kappa}. \end{aligned} \tag{3.51}$$

It is clear that the choice of the nonlinear damping term in (3.49) allows the completion of squares in (3.51). This simple completion of squares can be viewed as a special case of *Young's Inequality*, which, in a simplified form, states that if the constants $p > 1$ and $q > 1$ are such that $(p - 1)(q - 1) = 1$, then for all $\varepsilon > 0$ and all $(x, y) \in \mathbb{R}^2$ we have

$$xy \le \frac{\varepsilon^p}{p}|x|^p + \frac{1}{q\varepsilon^q}|y|^q. \tag{3.52}$$

Choosing $p = q = 2$ and $\varepsilon^2 = 2\kappa$, (3.52) becomes

$$xy \leq \kappa x^2 + \frac{1}{4\kappa} y^2 , \tag{3.53}$$

which is the form used in (3.51):

$$x\,\varphi(x)\Delta(t) \leq \kappa x^2 \varphi^2(x) + \frac{\Delta^2(t)}{4\kappa} . \tag{3.54}$$

Global boundedness and convergence Returning to (3.51), we see that the Lyapunov function derivative is negative whenever $|x(t)| \geq \frac{\Delta(t)}{2\sqrt{\kappa c}}$. Since $\Delta(t)$ is a bounded disturbance, we conclude that $\dot{V}$ is negative outside the compact residual set

$$\mathcal{R} = \left\{ x : |x| \leq \frac{\|\Delta\|_\infty}{2\sqrt{\kappa c}} \right\}. \tag{3.55}$$

Recalling that $V(x) = \frac{1}{2}x^2$, we conclude that $|x(t)|$ decreases whenever $x(t)$ is outside the set $\mathcal{R}$, and hence $x(t)$ is bounded:

$$\|x\|_\infty \leq \max\left\{ |x(0)|, \frac{\|\Delta\|_\infty}{2\sqrt{\kappa c}} \right\}. \tag{3.56}$$

Moreover, we can draw some conclusions about the asymptotic behavior of $x(t)$. Let us rewrite (3.51) as:

$$\frac{d}{dt}\left(x^2\right) \leq -2cx^2 + \frac{\Delta^2(t)}{2\kappa} . \tag{3.57}$$

To obtain explicit bounds on $x(t)$, we consider the signal $x(t)e^{ct}$. Using (3.57) we get

$$\frac{d}{dt}\left(x^2 e^{2ct}\right) = \frac{d}{dt}\left(x^2\right)e^{2ct} + 2cx^2 e^{2ct} \leq \frac{\Delta^2(t)}{2\kappa}e^{2ct} . \tag{3.58}$$

Integrating both sides over the interval $[0, t]$ yields

$$x^2(t)e^{2ct} \leq x^2(0) + \int_0^t \frac{1}{2\kappa}\Delta^2(\tau)e^{2c\tau}\,d\tau$$

$$\leq x^2(0) + \frac{1}{4\kappa c}\left[\sup_{0\leq\tau\leq t} \Delta^2(\tau)\right]\left(e^{2ct} - 1\right). \tag{3.59}$$

Multiplying both sides with e^{-2ct} and using the fact that $a^2 \leq b^2 + c^2 \Rightarrow |a| \leq |b| + |c|$, we obtain an explicit bound for $x(t)$:

$$|x(t)| \leq |x(0)|e^{-ct} + \frac{1}{2\sqrt{\kappa c}}\left[\sup_{0\leq\tau\leq t} |\Delta(\tau)|\right]\sqrt{1 - e^{-2ct}}. \tag{3.60}$$

Since $\sup_{0\leq\tau\leq t} |\Delta(\tau)| \leq \sup_{0\leq\tau<\infty} |\Delta(\tau)| \stackrel{\Delta}{=} \|\Delta\|_\infty$, (3.60) leads to

$$|x(t)| \leq |x(0)|e^{-ct} + \frac{\|\Delta\|_\infty}{2\sqrt{\kappa c}}\sqrt{1 - e^{-2ct}}, \tag{3.61}$$

which shows that $x(t)$ converges to the compact set $\mathcal{R}$ defined in (3.55):

$$\lim_{t \to \infty} \text{dist}\,\{x(t), \mathcal{R}\} = 0. \tag{3.62}$$

It is important to note that these properties of boundedness (cf. (3.56)) and convergence (cf. (3.62)) are guaranteed for any bounded disturbance $\Delta(t)$ and for any smooth nonlinearity $\varphi(x)$ (including $\varphi(x) = x^2$). Furthermore, the nonlinear control law (3.50) does not assume knowledge of a bound on the disturbance, nor does it have to use large values for the gains κ and c. Indeed, the residual set $\mathcal{R}$ defined in (3.55) is compact for any finite value of $\|\Delta\|_\infty$ and for any positive value of κ and c. Hence, *global boundedness is guaranteed in the presence of bounded disturbances with possibly unknown bounds, regardless of how small the gains κ and c are chosen.* Of course the size of $\mathcal{R}$ can not be estimated *a priori* if no bound for $\|\Delta\|_\infty$ is given, but it can be reduced *a posteriori* by increasing the values of κ and c.

This property is achieved by the "κ-term" $-\kappa x \varphi^2(x)$ in (3.50), which renders the effective gain of (3.50) "selectively high": when κ and c are chosen to be small, the gain is low around the origin, but it becomes high when x is in a region where $\varphi(x)$ is large enough to make the term $\kappa\varphi^2(x)$ large. If we interpret the nonlinearity $\varphi(x)$ as the "disturbance gain", since it multiplies the disturbance $\Delta(t)$ in (3.39), we see that the term $-\kappa\varphi^2(x)$ in (3.50) guarantees that when the disturbance gain becomes large, the control gain becomes large enough to keep the state bounded.

Finally, we should note that if the disturbance $\Delta(t)$ converges to zero in addition to being bounded, then the control (3.50) guarantees convergence of $x(t)$ to zero in addition to global boundedness. To show this, let $\bar{\Delta}(t)$ be a continuous nonnegative *monotonically decreasing* function such that $|\Delta(t)| \leq \bar{\Delta}(t)$ and $\lim_{t \to \infty} \bar{\Delta}(t) = 0$. Then, starting with the first inequality from (3.59), we obtain

$$|x(t)|^2 \leq |x(0)|^2 e^{-2ct} + \frac{1}{2\kappa} \int_0^t e^{-2c(t-\tau)} \bar{\Delta}^2(\tau) d\tau$$

$$\leq |x(0)|^2 e^{-2ct} + \frac{1}{2\kappa} \int_0^{t/2} e^{-2c(t-\tau)} \sup_{0 \leq \tau \leq t/2} \{\bar{\Delta}^2(\tau)\} d\tau$$

$$+ \frac{1}{2\kappa} \int_{t/2}^t e^{-2c(t-\tau)} \sup_{t/2 \leq \tau \leq t} \{\bar{\Delta}^2(\tau)\} d\tau$$

$$= |x(0)|^2 e^{-2ct} + \frac{1}{2\kappa} \bar{\Delta}^2(0) \int_0^{t/2} e^{-2c(t-\tau)} d\tau$$

$$+ \frac{1}{2\kappa} \bar{\Delta}^2(t/2) \int_{t/2}^t e^{-2c(t-\tau)} d\tau$$

$$= |x(0)|^2 e^{-2ct} + \frac{1}{4\kappa c} \bar{\Delta}^2(0) e^{-ct} \left(1 - e^{-ct}\right)$$

$$+ \frac{1}{4\kappa c} \bar{\Delta}^2(t/2) \left(1 - e^{-ct}\right), \tag{3.63}$$

which leads to

$$|x(t)| \leq |x(0)|e^{-ct} + \frac{1}{2\sqrt{\kappa c}} \left(\bar{\Delta}(0)e^{-\frac{c}{2}t} + \bar{\Delta}(t/2) \right) . \tag{3.64}$$

Since $\lim_{t \to \infty} \bar{\Delta}(t/2) = 0$, we see that $\lim_{t \to \infty} x(t) = 0$.

Nonlinear operator interpretation The effect of the term $-\kappa x \varphi^2(x)$ in (3.50) can also be interpreted from an operator point of view on the basis of (3.56), which is repeated here for convenience:

$$\|x\|_\infty \leq \max \left\{ |x(0)|, \frac{\|\Delta\|_\infty}{2\sqrt{\kappa c}} \right\} . \tag{3.65}$$

Assuming that the initial condition $|x(0)|$ is small enough, we obtain

$$\|x\|_\infty \leq \frac{1}{2\sqrt{\kappa c}} \|\Delta\|_\infty , \tag{3.66}$$

which shows that the nonlinear operator K mapping the disturbance $\Delta(t)$ to the output $x(t)$, depicted in Figure 14, is bounded, and its $\mathcal{L}_\infty$-induced gain is

$$\|K\|_{\infty \, \text{ind}} \leq \frac{1}{2\sqrt{\kappa c}} . \tag{3.67}$$

Note that the nonlinear damping term renders the operator K bounded for *any* positive values of c and κ. Note also that (3.65) provides a more complete description of this operator than (3.66), because it explicitly shows the effect of initial conditions. In contrast to the linear operator case, neglecting the effects of initial conditions can be quite dangerous for nonlinear systems, as the finite escape time example (3.41)–(3.46) demonstrates.

ISS interpretation Perhaps the most appropriate interpretation of the effect of the nonlinear damping term $-\kappa x \varphi^2(x)$ in (3.50) is that it renders the closed-loop system ISS (input-to-state stable [25, 26]) with respect to the disturbance input $\Delta(t)$. Let us recall from [25, 26] that the system $\dot{x} = f(t, x, u)$ is ISS with respect to u if there exist a class $\mathcal{KL}$ function $\beta_{\mathcal{KL}}$ and a class $\mathcal{K}$ function $\gamma_{\mathcal{K}}$ such that, for any $x(0)$ and for any input $u(\cdot)$ which is continuous and bounded on $[0, \infty)$, the solution $x(t)$ exists for all $t \geq 0$ and satisfies

$$|x(t)| \leq \beta_{\mathcal{KL}}(|x(t_0)|, t - t_0) + \gamma_{\mathcal{K}} \left(\sup_{t_0 \leq \tau \leq t} |u(\tau)| \right) \tag{3.68}$$

for all t_0 and t such that $0 \leq t_0 \leq t$. To show that this is true for our closed-loop system, we repeat the argument that led from (3.58) to (3.60), this time integrating over the interval $[t_0, t]$. The result is

$$|x(t)| \leq |x(t_0)|e^{-c(t-t_0)} + \frac{1}{2\sqrt{\kappa c}} \left[\sup_{t_0 \leq \tau \leq t} |\Delta(\tau)| \right], \tag{3.69}$$

which is identical to (3.68) with $\beta_{\mathcal{KL}}(r, s) = re^{-cs}$, $\gamma_{\mathcal{K}}(r) = \frac{1}{2\sqrt{\kappa c}} r$ and $u(\tau)$ replaced by the disturbance $\Delta(\tau)$.

Plant

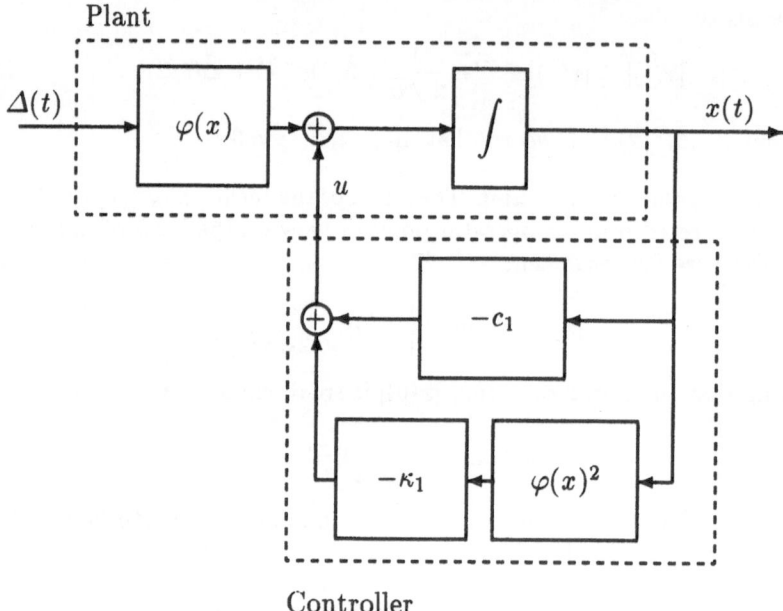

Controller

Fig. 14. The bounded nonlinear operator $K : \Delta(t) \to x(t)$.

Adaptive design The nonlinear damping with κ-terms can easily be incorporated into the adaptive design procedures we have discussed so far. The resulting adaptive controllers guarantee boundedness even when the adaptation is switched off, and their transient performance can be improved in a systematic way through *trajectory initialization* and the choice of design parameters.

To illustrate the design with κ-terms and the process of trajectory initialization, we consider again the system (2.25) with the output $y = x_1$:

$$\begin{aligned} \dot{x}_1 &= x_2 + \theta\varphi(x_1) \\ \dot{x}_2 &= u \\ y &= x_1 \,. \end{aligned} \tag{3.70}$$

The control objective is to asymptotically track a reference output $y_r(t)$ with the output y of the system (3.70). We assume that not only y_r, but also its first two derivatives $\dot{y}_r$, $\ddot{y}_r$ are known and uniformly bounded, and, in addition, $\dddot{y}_r$ is piecewise continuous.

Step 1. The first error variable is now the *tracking error*

$$z_1 = y - y_r = x_1 - y_r \,, \tag{3.71}$$

whose derivative is

$$\dot{z}_1 = x_2 + \theta^{\mathrm{T}}\varphi_1(x_1) - \dot{y}_r \,. \tag{3.72}$$

Viewing x_2 as the virtual control we define the stabilizing function

$$\alpha_1 = -c_1 z_1 - \kappa_1 z_1 \varphi^2 - \hat{\theta}\varphi + \dot{y}_r . \tag{3.73}$$

Comparing (3.73) with (2.26) we note two new terms in (3.73). The term $\dot{y}_r$, which is intended to cancel the corresponding term in (3.72), is due to the tracking objective. The nonlinear damping term $-\kappa_1 z_1 \varphi^2$ is motivated the previous discussion. It contains the square of the term (φ) which multiplies the parametric uncertainty in the error equation obtained by substituting $z_2 = x_2 - \alpha_1$ and (3.73) into (3.72):

$$\dot{z}_1 = -c_1 z_1 - \kappa_1 z_1 \varphi^2 + z_2 + \tilde{\theta}\varphi . \tag{3.74}$$

The derivative of the Lyapunov function $V_1 = \frac{1}{2}z_1^2 + \frac{1}{2\gamma}\tilde{\theta}^2$ becomes

$$\dot{V}_1 = z_1 z_2 - c_1 z_1^2 - \kappa_1 z_1^2 \varphi^2 + \tilde{\theta}\left(\varphi z_1 - \frac{1}{\gamma}\dot{\hat{\theta}}\right) . \tag{3.75}$$

Step 2. As in (2.28b), the derivative of $z_2 = x_2 - \alpha_1$ is

$$
\begin{aligned}
\dot{z}_2 &= u - \frac{\partial \alpha_1}{\partial x_1}(x_2 + \theta\varphi) - \frac{\partial \alpha_1}{\partial y_r}\dot{y}_r - \frac{\partial \alpha_1}{\partial \dot{y}_r}\ddot{y}_r - \frac{\partial \alpha_1}{\partial \hat{\theta}}\dot{\hat{\theta}} \\
&= u - \frac{\partial \alpha_1}{\partial x_1}x_2 - \hat{\theta}\frac{\partial \alpha_1}{\partial x_1}\varphi - \tilde{\theta}\frac{\partial \alpha_1}{\partial x_1}\varphi - \frac{\partial \alpha_1}{\partial y_r}\dot{y}_r - \ddot{y}_r - \frac{\partial \alpha_1}{\partial \hat{\theta}}\dot{\hat{\theta}} , \tag{3.76}
\end{aligned}
$$

where in the last equality we have used the identity $\frac{\partial \alpha_1}{\partial \dot{y}_r} = 1$. Using (3.75) and (3.76), the derivative of the Lyapunov function

$$V_2 = V_1 + \frac{1}{2}z_2^2 = \frac{1}{2}z_1^2 + \frac{1}{2}z_2^2 + \frac{1}{2\gamma}\tilde{\theta}^2 \tag{3.77}$$

is expressed as

$$
\begin{aligned}
\dot{V}_2 &= z_1 z_2 - c_1 z_1^2 - \kappa_1 z_1^2 \varphi^2 + \tilde{\theta}\left(\varphi z_1 - \frac{1}{\gamma}\dot{\hat{\theta}}\right) \\
&\quad + z_2\left[u - \frac{\partial \alpha_1}{\partial x_1}x_2 - \hat{\theta}\frac{\partial \alpha_1}{\partial x_1}\varphi - \tilde{\theta}\frac{\partial \alpha_1}{\partial x_1}\varphi - \frac{\partial \alpha_1}{\partial y_r}\dot{y}_r - \ddot{y}_r - \frac{\partial \alpha_1}{\partial \hat{\theta}}\dot{\hat{\theta}}\right] \\
&= -c_1 z_1^2 - \kappa_1 z_1^2 \varphi^2 + \tilde{\theta}\left[\varphi z_1 - z_2\frac{\partial \alpha_1}{\partial x_1}\varphi - \frac{1}{\gamma}\dot{\hat{\theta}}\right] \\
&\quad + z_2\left[z_1 + u - \frac{\partial \alpha_1}{\partial x_1}x_2 - \hat{\theta}\frac{\partial \alpha_1}{\partial x_1}\varphi - \frac{\partial \alpha_1}{\partial y_r}\dot{y}_r - \ddot{y}_r - \frac{\partial \alpha_1}{\partial \hat{\theta}}\dot{\hat{\theta}}\right] . \tag{3.78}
\end{aligned}
$$

As in (2.32), all the terms containing $\tilde{\theta}$ have been grouped together. To eliminate them, the update law is chosen as

$$\dot{\hat{\theta}} = \gamma\left(\varphi z_1 - \frac{\partial \alpha_1}{\partial x_1}\varphi z_2\right) . \tag{3.79}$$

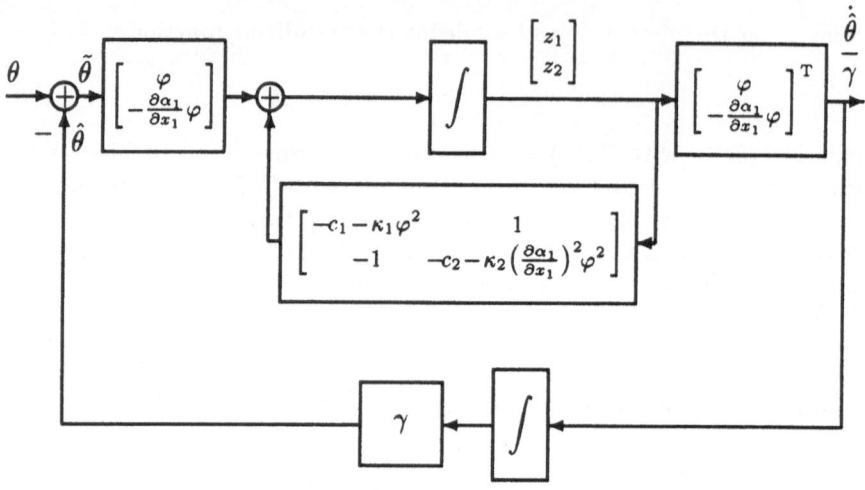

Fig. 15. The closed-loop adaptive system (3.82).

The control law is now chosen to render the last bracketed term in (3.78) equal to $-c_2 z_2^2 - \kappa_2 z_2^2 \left(\frac{\partial \alpha_1}{\partial x_1} \varphi\right)^2$, instead of just equal to $-c_2 z_2^2$ as in (2.34):

$$u = -z_1 - c_2 z_2 - \kappa_2 z_2 \left(\frac{\partial \alpha_1}{\partial x_1} \varphi\right)^2 + \frac{\partial \alpha_1}{\partial x_1} x_2 + \hat{\theta} \frac{\partial \alpha_1}{\partial x_1} \varphi + \frac{\partial \alpha_1}{\partial y_r} \dot{y}_r + \ddot{y}_r + \frac{\partial \alpha_1}{\partial \hat{\theta}} \dot{\hat{\theta}} \quad (3.80)$$

where we replace $\dot{\hat{\theta}}$ with the analytical expression of the update law (3.79).

Substituting the expressions (3.79) and (3.80) into (3.78) we obtain:

$$\dot{V}_2 = -c_1 z_1^2 - \kappa_1 z_1^2 \varphi^2 - c_2 z_2^2 - \kappa_2 z_2 \left(\frac{\partial \alpha_1}{\partial x_1} \varphi\right)^2 \leq 0, \quad (3.81)$$

while the complete error system becomes (see block diagram in Figure 10):

$$\frac{d}{dt}\begin{bmatrix} z_1 \\ z_2 \end{bmatrix} = \begin{bmatrix} -c_1 - \kappa_1 \varphi^2 & 1 \\ -1 & -c_2 - \kappa_2 \left(\frac{\partial \alpha_1}{\partial x_1} \varphi\right)^2 \end{bmatrix} \begin{bmatrix} z_1 \\ z_2 \end{bmatrix} + \begin{bmatrix} \varphi \\ -\frac{\partial \alpha_1}{\partial x_1} \varphi \end{bmatrix} \tilde{\theta}$$

$$\dot{\hat{\theta}} = \gamma \left[\varphi - \frac{\partial \alpha_1}{\partial x_1} \varphi \right] \begin{bmatrix} z_1 \\ z_2 \end{bmatrix}. \quad (3.82)$$

Comparing (3.82) with (2.37), we see that the system matrix in (3.82) is not constant: its diagonal terms have been "fortified" with additional nonlinear damping terms. These terms contain the squares of the elements of the vector that multiplies the parameter error $\tilde{\theta}$.

Let us now study the properties of the error system (3.82):

Global stability and asymptotic tracking. Using (3.77) and (3.81) we conclude that the $(z, \tilde{\theta})$-system has a globally uniformly stable equilibrium at the origin, and

$$\lim_{t \to \infty} z(t) = 0. \tag{3.83}$$

In particular, this implies that the state of the system (3.70) is globally uniformly bounded (since y_r, $\dot{y}_r$, $\ddot{y}_r$ are bounded), and that the tracking error $z_1 = y - y_r$ converges to zero asymptotically.

Boundedness without adaptation. It is also straightforward to see that the designed controller guarantees global uniform boundedness even when the adaptation is turned off, i.e., even with $\gamma = 0$. In that case, the closed-loop system (3.82) becomes

$$\frac{d}{dt} \begin{bmatrix} z_1 \\ z_2 \end{bmatrix} = \begin{bmatrix} -c_1 - \kappa_1 \varphi^2 & 1 \\ -1 & -c_2 - \kappa_2 \left(\frac{\partial \alpha_1}{\partial x_1} \varphi \right)^2 \end{bmatrix} \begin{bmatrix} z_1 \\ z_2 \end{bmatrix} + \begin{bmatrix} \varphi \\ -\frac{\partial \alpha_1}{\partial x_1} \varphi \end{bmatrix} \tilde{\theta}. \tag{3.84}$$

A candidate Lyapunov function for this system is given by

$$V(z) = \frac{1}{2} |z|^2 = \frac{1}{2} \left(z_1^2 + z_2^2 \right). \tag{3.85}$$

Its derivative along the solutions of (3.84) satisfies:

$$\begin{aligned}
\dot{V}_{(3.84)} &= -c_1 z_1^2 - c_2 z_2^2 - \kappa_1 z_1^2 \varphi^2 - \kappa_2 z_2^2 \left(\frac{\partial \alpha_1}{\partial x_1} \varphi \right)^2 + z_1 \varphi \tilde{\theta} - z_2 \frac{\partial \alpha_1}{\partial x_1} \varphi \tilde{\theta} \\
&\leq -c_1 z_1^2 - c_2 z_2^2 - \kappa_1 \left(z_1 \varphi - \frac{\tilde{\theta}}{2\kappa_1} \right)^2 + \frac{\tilde{\theta}^2}{4\kappa_1} - \kappa_2 \left(z_2 \frac{\partial \alpha_1}{\partial x_1} \varphi + \frac{\tilde{\theta}}{2\kappa_2} \right)^2 + \frac{\tilde{\theta}^2}{4\kappa_2} \\
&\leq -c_1 z_1^2 - c_2 z_2^2 + \frac{\tilde{\theta}^2}{4\kappa_1} + \frac{\tilde{\theta}^2}{4\kappa_2} \\
&\leq -c_0 |z|^2 + \frac{\tilde{\theta}^2}{4\kappa_0},
\end{aligned} \tag{3.86}$$

where the constants c_0 and κ_0 are defined as

$$c_0 = \min\{c_1, c_2\}, \quad \frac{1}{\kappa_0} = \frac{1}{\kappa_1} + \frac{1}{\kappa_2}. \tag{3.87}$$

It is clear from (3.86) that, for any positive values of c_0 and κ_0, the state of the error system (and hence the state of the plant) is uniformly bounded, since $\dot{V} < 0$ whenever $|z|^2 > |\tilde{\theta}|^2 / 4\kappa_0 c_0$, where $\tilde{\theta} = \theta - \hat{\theta}(0)$ is constant since adaptation is turned off.

Transient performance improvement with trajectory initialization. Let us now investigate the transient performance of the adaptive closed-loop system (3.82). The derivative of the nonnegative function $V(z)$ defined in (3.85) along the solutions of (3.82) satisfies the same inequality as in (3.86):

$$\frac{d}{dt}\left(\frac{1}{2}|z|^2\right) \leq -c_0|z|^2 + \frac{\tilde{\theta}^2}{4\kappa_0}. \tag{3.88}$$

Since the boundedness of $\tilde{\theta}$ has already been established from (3.77) and (3.81), we can strengthen the inequality in (3.88) by replacing $\tilde{\theta}^2$ with its bound $\|\tilde{\theta}\|_\infty^2$. This bound is estimated from (3.77) using the fact that V_2 is nonincreasing:

$$\frac{1}{2\gamma}|\tilde{\theta}(t)|^2 \leq \frac{1}{2}|z(t)|^2 + \frac{1}{2\gamma}\tilde{\theta}(t)^2 = V_2(t)$$

$$\leq V_2(0) = \frac{1}{2}|z(0)|^2 + \frac{1}{2\gamma}\tilde{\theta}(0)^2, \tag{3.89}$$

which implies[2]

$$\|\tilde{\theta}\|_\infty^2 \leq \gamma|z(0)|^2 + \tilde{\theta}(0)^2. \tag{3.90}$$

Combining (3.88) and (3.90) we obtain

$$\frac{d}{dt}\left(|z|^2\right) \leq -2c_0|z|^2 + \frac{1}{2\kappa_0}\left[\gamma|z(0)|^2 + \tilde{\theta}(0)^2\right]. \tag{3.91}$$

Multiplying both sides of (3.91) by $e^{2c_0 t}$ and integrating over the interval $[0, t]$ results in

$$|z(t)|^2 \leq |z(0)|^2 e^{-2c_0 t} + \frac{1}{4\kappa_0 c_0}\left[\gamma|z(0)|^2 + \tilde{\theta}(0)^2\right]. \tag{3.92}$$

The bound (3.92) suggests that the transient behavior of the error system can be influenced through the choice of design constants c_0, κ_0 and γ. What is not clear, however, is that an increase of $\kappa_0 c_0$ alone may not reduce the maximum value of $|z(t)|$ and will certaintly not reduce the computable $\mathcal{L}_\infty$-bound of z. In fact, it may even *increase* this bound by increasing the initial value $|z(0)|$. To clarify this point, let us recall the definitions of z_1 and z_2:

$$z_1 = x_1 - y_r$$
$$z_2 = x_2 - \alpha_1 = x_2 + c_1 z_1 + \kappa_1 z_1 \varphi^2 + \hat{\theta}\varphi - \dot{y}_r.$$

Suppose now that $z_1(0)$ is different than zero. In that case, an increase of c_1 and κ_1 may increase the value of $z_2(0)$ and thus also the value of $|z(0)|$. Moreover, this increase may more than offset the decreasing effect of the term $1/4\kappa_0 c_0$ in (3.92), since $|z(0)|^2$ will increase in proportion to c_1^2 and κ_1^2.

---

[2] The bound (3.90) may seem too conservative, because it bounds the parameter error by the initial value of the Lyapunov function. However, since we are dealing with Lyapunov-based adaptation and we know that $z(t) \to 0$ as $t \to \infty$, this is the tightest bound that one can obtain in this case.

It would seem that the dependence of $z(0)$ on the design constants $c_1, c_2, \kappa_1, \kappa_2$ eliminates any possibility of systematically improving the transient performance of the error system through the choice of c_0 and κ_0. Fortunately, it is not so. The remedy for this problem is to use *trajectory initialization* to render $z(0) = 0$ *independently* of the choice of these design constants. The initialization procedure, presented for the general case in Section 4.3.2, is straightforward and is dictated by the definitions of the z-variables:

- Starting with z_1, set $z_1(0) = 0$ by choosing

$$y_r(0) = x_1(0). \tag{3.93}$$

- Since $z_1(0) = 0$, (3.73) shows that

$$\alpha_1(0) = \dot{y}_r(0) - \hat{\theta}(0)\varphi(0), \tag{3.94}$$

where we use the notation $\varphi(0) = \varphi(x_1(0))$. From (3.94) it is clear that we can set $z_2(0) = 0$ with the choice

$$\dot{y}_r(0) = x_2(0) + \hat{\theta}(0)\varphi(0). \tag{3.95}$$

With the trajectory initialization defined by (3.93) and (3.95), we have set $z(0) = 0$. In the case of model reference control, this is achieved by adjusting the initial conditions of the reference model. If, on the other hand, the reference trajectory is given as a precomputed function of time, then it can be initialized through the addition of exponentially decaying terms which define the *reference transients*.

We note that (3.93) and (3.95) are independent of the design constants $c_1, c_2, \kappa_1, \kappa_2$. This means that different choices of c_0 and κ_0 will still result in $z(0) = 0$ with the same values of $y_r(0)$ and $\dot{y}_r(0)$. Returning to (3.92), we substitute $z(0) = 0$ to obtain

$$|z(t)|^2 \leq \frac{1}{4\kappa_0 c_0}\tilde{\theta}(0)^2, \tag{3.96}$$

which implies

$$\|z\|_\infty \leq \frac{1}{2\sqrt{\kappa_0 c_0}}|\tilde{\theta}(0)|. \tag{3.97}$$

Hence, the $\mathcal{L}_\infty$-bound on the transient performance of the error system is directly proportional to the initial parametric uncertainty and can be reduced arbitrarily by increasing the values of c_0 and κ_0. In particular, this implies that the transients of the tracking error $z_1 = y - y_r$ are directly influenced by the design constants c_i and κ_i. This possibility of arbitrary reduction may seem peculiar, since it can be achieved for all initial conditions. We must remember, however, that this error is defined with respect to the reference signals which have in turn been initialized to set $z(0) = 0$. Hence, the effect of the plant initial conditions has been "absorbed" into the reference transients.

To provide some further insight into the process of trajectory initialization, let us return to the Lyapunov function (3.77). When $z(0) = 0$, the initial value of this function is reduced to the initial value of the parametric uncertainty. If we interpret the value of this function as a distance between the actual system trajectory and the reference trajectory, we see that *trajectory initialization places the initial point of the reference trajectory as close as possible to the initial point of the system trajectory.* If the parametric uncertainty were zero, trajectory initialization would have placed the reference output and its derivatives at the true values of the plant output and its derivatives. This is easily seen if $\hat{\theta}(0)$ is replaced by θ in (3.93) and (3.95):

$$y_r(0) = x_1(0) = y(0)$$
$$\dot{y}_r(0) = x_2(0) + \theta(0)\varphi(0) = \dot{y}(0).$$

Since the parameter θ is unknown, however, trajectory initialization placed the reference output at the true value of the plant output and the derivatives of the reference output at the *estimated* values of the plant output derivatives. Thus, the guidelines for this initialization are not only dictated by the design procedure itself, but they also correspond to the intuitive idea of matching the initial values of the reference and plant outputs and their derivatives as closely as possible. Through this process, the initial value of the Lyapunov function becomes as small as possible in the presence of parametric uncertainty:

$$V_2(0) = \frac{1}{2\gamma}\tilde{\theta}(0)^2. \tag{3.98}$$

4   Output-feedback schemes

The design procedures illustrated so far required full-state feedback. As we mentioned in Section 2, however, these procedures can be modified to handle the case where only the output of a system is measured, provided that the system can be transformed into the output-feedback canonical form (5.4), in which the nonlinearities depend only on the measured output.

To illustrate the output-feedback schemes, let us first consider the nonlinear system in output-feedback canonical form

$$\dot{x}_1 = x_2 + \theta\varphi_1(y)$$
$$\dot{x}_2 = x_3 + \theta\varphi_2(y) + u \tag{4.1}$$
$$\dot{x}_3 = u$$
$$y = x_1,$$

where now we assume that only the output $y = x_1$ is measured, i.e., the states x_2 and x_3 are not available for feedback. We want to design an adaptive nonlinear output-feedback controller that guarantees asymptotic tracking of the reference signal $y_r(t)$ by the output y while keeping all the states of the closed-loop system bounded.

Since x_2 and x_3 are not measured and θ is an unknown parameter, we must attempt to reconstruct the full state of the system through the use of filters. In this case, we need only two filters: one for the known part of the system (i.e., the part that does not contain θ), and one for the unknown part (since we only have one unknown parameter). That is, we are trying to reconstruct the state through two filters ξ_0 and ξ_1 as:

$$x = \xi_0 + \theta\xi_1 + \varepsilon. \tag{4.2}$$

It is clear that the estimate $\xi_0 + \theta\xi_1$ is only a "virtual estimate", since it depends on the unknown parameter θ. Nevertheless, its components ξ_0 and ξ_1 are known and this is good enough for our control purposes, provided that the error ε tends to zero asymptotically. To ensure this, the two filters are defined as follows:

$$\begin{aligned}
\dot{\xi}_{01} &= -k_1(\xi_{01} - y) + \xi_{02} & \dot{\xi}_{11} &= -k_1\xi_{11} + \xi_{12} + \varphi_1(y) \\
\dot{\xi}_{02} &= -k_2(\xi_{01} - y) + \xi_{03} + u & \dot{\xi}_{12} &= -k_2\xi_{11} + \xi_{13} + \varphi_2(y) \\
\dot{\xi}_{03} &= -k_3(\xi_{01} - y) + u & \dot{\xi}_{13} &= -k_3\xi_{11},
\end{aligned} \tag{4.3}$$

where k_1, k_2, k_3 are chosen so that the matrix

$$A_0 = \begin{bmatrix} -k_1 & 1 & 0 \\ -k_2 & 0 & 1 \\ -k_3 & 0 & 0 \end{bmatrix} \tag{4.4}$$

is Hurwitz. To justify these definitions and to show why they achieve our goal, let us rewrite the system in the x-coordinates in the following form:

$$\dot{x} = A_0 x + ky + bu + \theta\varphi(y), \quad k = \begin{bmatrix} -k_1 \\ -k_2 \\ -k_3 \end{bmatrix}, b = \begin{bmatrix} 0 \\ 1 \\ 1 \end{bmatrix}, \varphi(y) = \begin{bmatrix} \varphi_1(y) \\ \varphi_2(y) \\ 0 \end{bmatrix}. \tag{4.5}$$

Now the filters ξ_0 and ξ_1 are rewritten in the more transparent form:

$$\dot{\xi}_0 = A_0\xi_0 + ky + bu, \quad \dot{\xi}_1 = A_0\xi_1 + \varphi(y). \tag{4.6}$$

Combining these equations, we obtain for the error $\varepsilon = x - \xi_0 - \theta\xi_1$:

$$\begin{aligned}
\dot{\varepsilon} &= \dot{x} - \dot{\xi}_0 - \theta\dot{\xi}_1 \\
&= A_0 x + ky + bu + \theta\varphi(y) - A_0\xi_0 - ky - bu - A_0\theta\xi_1 - \theta\varphi(y) \\
&= A_0(x - \xi_0 - \theta\xi_1) \\
&= A_0\varepsilon. \tag{4.7}
\end{aligned}$$

Since A_0 is a Hurwitz matrix, we can find a positive definite symmetric matrix P_0 which satisfies the Lyapunov matrix equation:

$$P_0 A_0 + A_0^T P_0 = -I. \tag{4.8}$$

This implies that

$$\frac{d}{dt}\left(\varepsilon^T P_0\varepsilon\right) = -\varepsilon^T\varepsilon. \tag{4.9}$$

We are now ready to design our adaptive nonlinear output-feedback controller.

Step 1. The control objective is to track the reference signal $y_r(t)$ with the output y, so the first error variable is the tracking error:

$$z_1 = y - y_r. \tag{4.10}$$

The derivative of z_1 is:

$$\dot{z}_1 = \dot{y} - \dot{y}_r = x_2 + \theta\varphi_1(y) - \dot{y}_r. \tag{4.11}$$

If x_2 were measured, it would be our virtual control. Since it is not measured, we replace it with the sum of its "virtual estimate" and the corresponding error:

$$x_2 = \xi_{02} + \theta\xi_{12} + \varepsilon_2, \tag{4.12}$$

to obtain

$$\dot{z}_1 = \xi_{02} + \theta\underbrace{[\varphi_1(y) + \xi_{12}]}_{\omega} - \dot{y}_r + \varepsilon_2. \tag{4.13}$$

Now we must pick one of the known variables appearing in the above equation to play the role of the virtual control. Looking at the filter equations, we see that only ξ_{02} contains the control u in its first derivative, so it is the only candidate. Defining the second error variable as

$$z_2 = \xi_{02} - \alpha_1, \tag{4.14}$$

we get

$$\dot{z}_1 = z_2 + \alpha_1 + \theta\omega - \dot{y}_r + \varepsilon_2. \tag{4.15}$$

The first stabilizing function α_1 is chosen as

$$\alpha_1 = -c_1 z_1 - d_1 z_1 + \dot{y}_r - \hat{\theta}\omega. \tag{4.16}$$

With this choice, the $\dot{z}_1$-equation becomes

$$\dot{z}_1 = z_2 - c_1 z_1 - d_1 z_1 + \tilde{\theta}\omega + \varepsilon_2. \tag{4.17}$$

The first partial Lyapunov function is chosen as

$$V_1 = \frac{1}{2}z_1^2 + \frac{1}{2\gamma}\tilde{\theta}^2 + \frac{1}{d_1}\varepsilon^T P_0\varepsilon, \tag{4.18}$$

where $\gamma > 0$ is the adaptation gain. Its derivative satisfies:

$$\dot{V}_1 = z_1\dot{z}_1 - \frac{1}{\gamma}\tilde{\theta}\dot{\hat{\theta}} + \frac{1}{d_1}\frac{d}{dt}\left(\varepsilon^T P_0\varepsilon\right)$$

$$= z_1 z_2 - c_1 z_1^2 - d_1 z_1^2 + \tilde{\theta}\left(z_1\omega - \frac{1}{\gamma}\dot{\hat{\theta}}\right) + z_1\varepsilon_2 - \frac{1}{d_1}\varepsilon^T\varepsilon$$

$$= z_1 z_2 - c_1 z_1^2 + \tilde{\theta}\left(z_1\omega - \frac{1}{\gamma}\dot{\hat{\theta}}\right)$$

$$-d_1\left[z_1 - \frac{1}{2d_1}\varepsilon_2\right]^2 + \frac{1}{4d_1}\varepsilon_2^2 - \frac{1}{d_1}\varepsilon^T\varepsilon$$

$$\leq z_1 z_2 - c_1 z_1^2 + \tilde{\theta}\left(z_1\omega - \frac{1}{\gamma}\dot{\hat{\theta}}\right) - \frac{3}{4d_1}\varepsilon^T\varepsilon. \tag{4.19}$$

The reason for including the term $-d_1 z_1^2$ in α_1 should now be clear: that term was used to complete squares with the cross-term $z_1 \varepsilon_2$. But the question that is left unanswered is why this completion of squares is necessary. After all, the term ε_2 is exponentially decaying, so why do we need to explicitly account for its presence? The answer is that in this first equation, where ε_2 is multiplied only with z_1, the presence of the d_1-term is not necessary. However, in subsequent steps we will see that this exponentially decaying error ε_2 is multiplied with nonlinear functions of y. In that case, we do indeed need to explicitly counteract the effect of ε_2, since its presence may even cause finite escape times. To prevent such phenomena, our design procedure incorporates *nonlinear damping* terms such as the term $-d_1 z_1^2$.

If this were the final step of the design procedure, we would eliminate the $\tilde{\theta}$-term from the derivative of V_1 by choosing the update law $\dot{\hat{\theta}} = \gamma \tau_1$, where τ_1 is the first tuning function

$$\tau_1 = z_1 \omega . \tag{4.20}$$

Since the control u has not appeared yet, the design procedure must continue. Therefore, the choice of update law is postponed. Using the tuning function τ_1, the V_1-inequality is rewritten as:

$$\dot{V}_1 \le z_1 z_2 - c_1 z_1^2 + \tilde{\theta} \left(\tau_1 - \frac{1}{\gamma} \dot{\hat{\theta}} \right) - \frac{3}{4 d_1} \varepsilon^T \varepsilon . \tag{4.21}$$

The first error system is:

$$\begin{aligned}
\dot{z}_1 &= -c_1 z_1 - d_1 z_1 + z_2 + \tilde{\theta} \omega + \varepsilon_2 \\
\dot{\varepsilon} &= A_0 \varepsilon \\
\tau_1 &= \omega z_1 .
\end{aligned} \tag{4.22}$$

Step 2. The control u appears in the derivative of z_2, and hence this is the last design step:

$$\begin{aligned}
\dot{z}_2 &= \dot{\xi}_{02} - \dot{\alpha}_1 \\
&= u - k_2(\xi_{01} - y) + \xi_{03} - \frac{\partial \alpha_1}{\partial y} \dot{y} \\
&\quad - \frac{\partial \alpha_1}{\partial \xi_{12}} \underbrace{(-k_2 \xi_{11} + \xi_{13} + \varphi_2(y))}_{\dot{\xi}_{12}} - \frac{\partial \alpha_1}{\partial y_r} \dot{y}_r - \underbrace{\frac{\partial \alpha_1}{\partial \dot{y}_r} \ddot{y}_r}_{1} - \frac{\partial \alpha_1}{\partial \hat{\theta}} \dot{\hat{\theta}} \\
&= u - k_2(\xi_{01} - y) + \xi_{03} - \frac{\partial \alpha_1}{\partial y} \underbrace{(\xi_{02} + \theta \omega + \varepsilon_2)}_{\dot{y}} \\
&\quad - \frac{\partial \alpha_1}{\partial \xi_{12}}(-k_2 \xi_{11} + \xi_{13} + \varphi_2(y)) - \frac{\partial \alpha_1}{\partial y_r} \dot{y}_r - \ddot{y}_r - \frac{\partial \alpha_1}{\partial \hat{\theta}} \dot{\hat{\theta}} \\
&= u + \beta_2 - \frac{\partial \alpha_1}{\partial y} \tilde{\theta} \omega - \frac{\partial \alpha_1}{\partial y} \varepsilon_2 - \ddot{y}_r - \frac{\partial \alpha_1}{\partial \hat{\theta}} \dot{\hat{\theta}} ,
\end{aligned} \tag{4.23}$$

where β_2 encompasses all the known terms except u and $\ddot{y}_r$:

$$\beta_2 = -k_2(\xi_{01} - y) + \xi_{03} - \frac{\partial \alpha_1}{\partial y}(\xi_{02} + \hat{\theta}\omega) - \frac{\partial \alpha_1}{\partial \xi_{12}}(-k_2\xi_{11} + \xi_{13} + \varphi_2(y)) - \frac{\partial \alpha_1}{\partial y_r}\dot{y}_r.$$
$$(4.24)$$

To design the update law $\dot{\hat{\theta}}$ and the control u, we consider the augmented partial Lyapunov function

$$V_2 = V_1 + \frac{1}{2}z_2^2 + \frac{1}{d_2}\varepsilon^T P_0 \varepsilon,$$
$$(4.25)$$

whose derivative is

$$\dot{V}_2 = \dot{V}_1 + z_2\dot{z}_2 - \frac{1}{d_2}\varepsilon^T \varepsilon$$

$$\leq z_1 z_2 - c_1 z_1^2 + \tilde{\theta}\left(\tau_1 - \frac{1}{\gamma}\dot{\hat{\theta}}\right) - \frac{3}{4d_1}\varepsilon^T \varepsilon$$

$$+ z_2\left[u + \beta_2 - \tilde{\theta}\frac{\partial \alpha_1}{\partial y}\omega - \frac{\partial \alpha_1}{\partial y}\varepsilon_2 - \ddot{y}_r - \frac{\partial \alpha_1}{\partial \hat{\theta}}\dot{\hat{\theta}}\right] - \frac{1}{d_2}\varepsilon^T \varepsilon$$

$$= -c_1 z_1^2 + \tilde{\theta}\left(\tau_1 - z_2\frac{\partial \alpha_1}{\partial y}\omega - \frac{1}{\gamma}\dot{\hat{\theta}}\right) - \frac{3}{4d_1}\varepsilon^T \varepsilon$$

$$+ z_2\left[u + z_1 + \beta_2 - \ddot{y}_r - \frac{\partial \alpha_1}{\partial \hat{\theta}}\dot{\hat{\theta}}\right] - z_2\frac{\partial \alpha_1}{\partial y}\varepsilon_2 - \frac{1}{d_2}\varepsilon^T \varepsilon$$

$$= -c_1 z_1^2 + \tilde{\theta}\left(\tau_1 - z_2\frac{\partial \alpha_1}{\partial y}\omega - \frac{1}{\gamma}\dot{\hat{\theta}}\right) - \frac{3}{4d_1}\varepsilon^T \varepsilon$$

$$+ z_2\left[u + z_1 + \beta_2 - \ddot{y}_r - \frac{\partial \alpha_1}{\partial \hat{\theta}}\dot{\hat{\theta}}\right]$$

$$+ d_2 z_2^2\left(\frac{\partial \alpha_1}{\partial y}\right)^2 - d_2\left[z_2\frac{\partial \alpha_1}{\partial y} - \frac{1}{2d_2}\varepsilon_2\right]^2 + \frac{1}{4d_2}\varepsilon_2^2 - \frac{1}{d_2}\varepsilon^T \varepsilon$$

$$\leq -c_1 z_1^2 + \tilde{\theta}\left(\tau_1 - z_2\frac{\partial \alpha_1}{\partial y}\omega - \frac{1}{\gamma}\dot{\hat{\theta}}\right) - \frac{3}{4}\left(\frac{1}{d_1} + \frac{1}{d_2}\right)\varepsilon^T \varepsilon$$

$$+ z_2\left[u + z_1 + d_2 z_2\left(\frac{\partial \alpha_1}{\partial y}\right)^2 + \beta_2 - \ddot{y}_r - \frac{\partial \alpha_1}{\partial \hat{\theta}}\dot{\hat{\theta}}\right].$$
$$(4.26)$$

The $\tilde{\theta}$-term is eliminated with the update law:

$$\dot{\hat{\theta}} = \gamma\tau_2 = \gamma\left(\tau_1 - z_2\frac{\partial \alpha_1}{\partial y}\omega\right),$$
$$(4.27)$$

and the last term in (4.26) is rendered equal to $-c_2 z_2^2$ with the control law:

$$u = -c_2 z_2 - z_1 - d_2\left(\frac{\partial \alpha_1}{\partial y}\right)^2 z_2 - \beta_2 + \ddot{y}_r + \frac{\partial \alpha_1}{\partial \hat{\theta}}\gamma\tau_2.$$
$$(4.28)$$

With these choices, the derivative of $\dot{V}_2$ satisfies the inequality:

$$\dot{V}_2 \leq -c_1 z_1^2 - c_2 z_2^2 - \frac{3}{4}\left(\frac{1}{d_1} + \frac{1}{d_2}\right)\varepsilon^T \varepsilon,$$
$$(4.29)$$

which guarantees the boundedness of $z_1, z_2, \tilde{\theta}, \varepsilon$ and the convergence of z_1, z_2, ε to zero. In particular, this implies that the tracking error $y - y_r$ converges to zero and that ξ_0, ξ_1, x_1, x_2 are bounded. The only thing that remains is to show the boundedness of x_3. To this end, we define the variable $\zeta = x_3 - x_2$. If we can show that ζ is bounded, then $x_3 = \zeta + x_2$ will also be bounded. We have:

$$
\begin{aligned}
\dot{\zeta} &= \dot{x}_3 - \dot{x}_2 \\
&= u - x_3 - \theta\varphi_2(y) - u \\
&= -\zeta - x_2 - \theta\varphi_2(y).
\end{aligned}
\tag{4.30}
$$

This shows that ζ is the output of a strictly proper linear system which has an asymptotically stable pole at $s = -1$ and whose input $(x_2 + \theta\varphi_2(y))$ is bounded. Thus, the output ζ of the system is also bounded.

To demonstrate the effect of the nonlinear damping terms, let us write the error system in its matrix form:

$$
\begin{bmatrix} \dot{z}_1 \\ \dot{z}_2 \end{bmatrix} = \begin{bmatrix} -c_1 - d_1 & 1 \\ -1 & -c_2 - d_2\left(\frac{\partial\alpha_1}{\partial y}\right)^2 \end{bmatrix} \begin{bmatrix} z_1 \\ z_2 \end{bmatrix} + \begin{bmatrix} 1 \\ -\frac{\partial\alpha_1}{\partial y} \end{bmatrix} \left(\omega\tilde{\theta} + \varepsilon_2\right)
$$

$$
\dot{\varepsilon} = A_0\varepsilon
\tag{4.31}
$$

$$
\tau_2 = \begin{bmatrix} 1 & -\frac{\partial\alpha_1}{\partial y} \end{bmatrix} \omega \begin{bmatrix} z_1 \\ z_2 \end{bmatrix},
$$

We see that the nonlinear damping terms strengthen the negativity of the diagonal entries by including the squares of the terms which multiply the state estimation error ε_2.

5 General results

The design procedures illustrated by the examples in this paper have been generalized [7–11,13,14] to classes of nonlinear systems characterized by *canonical forms*: every member of a class can be transformed into the corresponding canonical form via a diffeomorphism. Most of these classes can be characterized in a *coordinate-free* fashion through differential geometric conditions which are necessary and sufficient for the existence of the appropriate diffeomorphism. The discussion of these geometric characterizations is beyond the scope of this tutorial, and thus here we assume that the system at hand has already been transformed into one of the canonical forms.

Once in the canonical form, there are several underlying assumptions which are common to all the results stated in this paper:

- The unknown parameters enter linearly into the system equations, or they can be reparametrized to yield a linear parametrization.
- The nonlinearities satisfy the strict-feedback condition, i.e., they depend only on state-variables which are "fed back", but are otherwise not restricted by any growth conditions.

- The nonlinearities can depend only on measured variables.
- The relative degree ρ of the system is known.
- The dynamic order of the system n is also known.[3]
- The zero dynamics subsystem is bounded-input-bounded-state (BIBS) stable, where any states (other than the subsystem states) entering the equations of the subsystem are treated as inputs. If the states of the zero dynamics are not measured, then the zero dynamics must be linear and exponentially stable.
- The sign of the leading coefficient of the control u is known.[4]

Under these assumptions, the following results can be obtained:

1. Global stability and tracking under full-state feedback for systems in the *strict-feedback canonical form*:

$$
\begin{aligned}
\dot{x}_1 &= x_2 + \theta^T \varphi_1(x_1) \\
\dot{x}_2 &= x_3 + \theta^T \varphi_2(x_1, x_2) \\
&\vdots \\
\dot{x}_{n-1} &= x_n + \theta^T \varphi_{n-1}(x_1, \ldots, x_{n-1}) \\
\dot{x}_n &= \sigma(x)u + \varphi_0(x) + \theta^T \varphi_n(x),
\end{aligned}
\tag{5.1}
$$

where $\sigma(x) \neq 0$ for all $x \in \mathbb{R}$ and θ is the vector of unknown parameters.

2. Global stability and tracking under full-state feedback for systems in the *partially strict-feedback canonical form*:

$$
\begin{aligned}
\dot{x}_1 &= x_2 + \theta^T \varphi_1(x_1) \\
\dot{x}_2 &= x_3 + \theta^T \varphi_2(x_1, x_2) \\
&\vdots \\
\dot{x}_{q-1} &= x_q + \theta^T \varphi_{q-1}(x_1, \ldots, x_{q-1}) \\
\dot{x}_q &= x_{q+1} + \theta^T \varphi_q(x_1, \ldots, x_q, x^r) \\
&\vdots \\
\dot{x}_{\rho-1} &= x_\rho + \theta^T \varphi_{\rho-1}(x_1, \ldots, x_{\rho-1}, x^r) \\
\dot{x}_\rho &= x_{\rho+1} + \varphi_0(x) + \theta^T \varphi_\rho(x) + b_{n-\rho}\sigma(x)u \\
\dot{x}^r &= \Phi_0(x_1, \ldots, x_q, x^r) + \theta^T \Phi(x_1, \ldots, x_q, x^r) \\
y &= x_1,
\end{aligned}
\tag{5.2}
$$

where θ and $b_{n-\rho}$ are unknown, the sign of $b_{n-\rho}$ is known, $\sigma(x) \neq 0$ for all $x \in \mathbb{R}$, and the x^r-subsystem is BIBS stable with respect to $x_1, \ldots, x_q$ as its inputs.

---

[3] If the system is given in the canonical form, then this assumption can be relaxed to require knowledge of only an upper bound on the system order. However, if one has to verify geometric conditions and perform a diffeomorphism, the system order must be known.

[4] This is the nonlinear analog of the known high-frequency gain assumption in adaptive control of linear systems.

3. Regional stability and tracking under full-state feedback for systems in the *pure-feedback canonical form*:

$$\dot{x}_1 = x_2 + \theta^{\mathrm{T}}\varphi_1(x_1, x_2)$$
$$\dot{x}_2 = x_3 + \theta^{\mathrm{T}}\varphi_2(x_1, x_2, x_3)$$
$$\vdots \qquad\qquad\qquad\qquad\qquad (5.3)$$
$$\dot{x}_{n-1} = x_n + \theta^{\mathrm{T}}\varphi_{n-1}(x_1, \ldots, x_n)$$
$$\dot{x}_n = \sigma(x)u + \varphi_0(x) + \theta^{\mathrm{T}}\varphi_n(x).$$

This form is clearly more general than the strict-feedback form (5.1), since the nonlinearities are allowed to depend on one more state variable in each equation. The price paid for this enlargement of the class of nonlinear systems is the loss of globality: stability and tracking can now be guaranteed only in a region around the origin.

4. Global stability and tracking under output feedback for systems in the *output-feedback canonical form*:

$$\dot{x}_1 = x_2 + \varphi_{0,1}(y) + \theta^{\mathrm{T}}\varphi_1(y)$$
$$\vdots$$
$$\dot{x}_{\rho-1} = x_\rho + \varphi_{0,\rho-1}(y) + \theta^{\mathrm{T}}\varphi_{\rho-1}(y) \qquad\qquad (5.4)$$
$$\dot{x}_\rho = x_{\rho+1} + \varphi_{0,\rho}(y) + \theta^{\mathrm{T}}\varphi_\rho(y) + b_{n-\rho}\sigma(y)u$$
$$\vdots$$
$$\dot{x}_n = \varphi_{0,n}(y) + \theta^{\mathrm{T}}\varphi_n(y) + b_0\sigma(y)u$$
$$y = x_1,$$

where θ and $b_0, \ldots, b_{n-\rho}$ are unknown, $\sigma(y) \neq 0$ for all $y \in \mathbb{R}$, the sign of $b_{n-\rho}$ is known, the polynomial $B(s) = b_{n-\rho}s^{n-\rho} + \cdots + b_1 s + b_0$ is Hurwitz, and the only measured variable is the output y.

5. Global stability and tracking under partial-state feedback for systems in the *partial-feedback canonical form*:

$$\dot{x}_1 = x_2 + \theta^{\mathrm{T}}\varphi_1(x_1)$$
$$\dot{x}_2 = x_3 + \theta^{\mathrm{T}}\varphi_2(x_1, x_2)$$
$$\vdots$$
$$\dot{x}_{q-1} = x_q + \theta^{\mathrm{T}}\varphi_{\rho-1}(x_1, \ldots, x_{q-1})$$
$$\dot{x}_q = x_{q+1} + \varphi_{0,q}(x_1, \ldots, x_q) + \theta^{\mathrm{T}}\varphi_q(x_1, \ldots, x_q)$$
$$\vdots$$
$$\dot{x}_{\rho-1} = x_\rho + \varphi_{0,\rho-1}(x_1, \ldots, x_q) + \theta^{\mathrm{T}}\varphi_{\rho-1}(x_1, \ldots, x_q) \qquad (5.5)$$
$$\dot{x}_\rho = x_{\rho+1} + \varphi_{0,\rho}(x_1, \ldots, x_q)$$
$$\qquad + \theta^{\mathrm{T}}\varphi_\rho(x_1, \ldots, x_q) + b_{n-\rho}\sigma(x_1, \ldots, x_q)u$$
$$\vdots$$
$$\dot{x}_n = \varphi_{0,n}(x_1, \ldots, x_q) + \theta^{\mathrm{T}}\varphi_n(x_1, \ldots, x_q) + b_0\sigma(x_1, \ldots, x_q)u,$$

where θ and $b_0, \ldots, b_{n-\rho}$ are unknown, $\sigma(x_1, \ldots, x_q) \neq 0$ for all $(x_1, \ldots, x_q)$ $\in \mathbb{R}^q$, the sign of $b_{n-\rho}$ is known, the polynomial $B(s) = b_{n-\rho}s^{n-\rho} + \cdots + b_1 s + b_0$ is Hurwitz, and the only measured variables are $x_1, \ldots, x_q$.

6 Application to linear systems

The output-feedback design procedure illustrated in Section 4 is applicable to nonlinear systems that can be transformed into the output-feedback canonical form, which is repeated here for convenience:

$$\dot{x}_1 = x_2 + \varphi_{0,1}(y) + \theta^T \varphi_1(y)$$
$$\dot{x}_2 = x_3 + \varphi_{0,2}(y) + \theta^T \varphi_2(y)$$

$$\vdots$$

$$\dot{x}_{\rho-1} = x_\rho + \varphi_{0,\rho-1}(y) + \theta^T \varphi_{\rho-1}(y) \qquad (6.1)$$
$$\dot{x}_\rho = x_{\rho+1} + \varphi_{0,\rho}(y) + \theta^T \varphi_\rho(y) + b_{n-\rho}\sigma(y)u$$

$$\vdots$$

$$\dot{x}_n = \varphi_{0,n}(y) + \theta^T \varphi_n(y) + b_0\sigma(y)u$$
$$y = x_1 \, ,$$

where θ and $b_0, \ldots, b_{n-\rho}$ are unknown, $\sigma(y) \neq 0$ for all $y \in \mathbb{R}$, the sign of $b_{n-\rho}$ is known, and the polynomial $B(s) = b_{n-\rho}s^{n-\rho} + \cdots + b_1 s + b_0$ is Hurwitz.

Clearly, any minimum-phase linear system can be transformed into this form, which in that case becomes the well-known observer canonical form. Indeed, any linear system

$$y(s) = \frac{b_m s^m + \cdots + b_1 s + b_0}{s^n + a_{n-1}s^{n-1} + \cdots + a_1 s + a_0} u(s) \qquad (6.2)$$

can be expressed in the state-space representation

$$\dot{x}_1 = x_2 - a_{n-1}y$$

$$\vdots$$

$$\dot{x}_{\rho-1} = x_\rho - a_{m+1}y$$
$$\dot{x}_\rho = x_{\rho+1} - a_m y + b_m u \qquad (6.3)$$

$$\vdots$$

$$\dot{x}_n = -a_0 y + b_0 u$$
$$y = x_1 \, ,$$

which is identical to (6.1) with $\theta^T = [-a_{n-1}, \ldots, -a_0]$, $\rho = n - m$, and $\sigma(y) = 1$.

To briefly illustrate the design procedure for linear systems, let us consider the unstable relative-degree-three plant

$$y(s) = \frac{1}{s^2(s - \theta)} u(s) \, , \qquad (6.4)$$

where $\theta > 0$ is considered to be unknown. The control objective is to asymptotically track the output of the reference model

$$y_{\mathrm{r}}(s) = \frac{1}{(s+1)^3} r(s). \tag{6.5}$$

To derive the adaptive controller resulting from our nonlinear design, the plant (6.4) is first rewritten in the state-space form (6.3):

$$\begin{aligned}
\dot{x}_1 &= x_2 + \theta x_1 \\
\dot{x}_2 &= x_3 \\
\dot{x}_3 &= u \\
y &= x_1.
\end{aligned} \tag{6.6}$$

The filters required for the "virtual estimates" of x_2 and x_3 are

$$\dot{\eta} = A_0 \eta + e_3 y, \quad \xi_2 = A_0^2 \eta, \quad \xi_3 = -A_0^3 \eta \tag{6.7}$$

$$\dot{\lambda} = A_0 \lambda + e_3 u, \quad v = \lambda, \quad A_0 = \begin{bmatrix} -k_1 & 1 & 0 \\ -k_2 & 0 & 1 \\ -k_3 & 0 & 0 \end{bmatrix}. \tag{6.8}$$

The signals $y_{\mathrm{r}}, \dot{y}_{\mathrm{r}}, \ddot{y}_{\mathrm{r}}, y_{\mathrm{r}}^{(3)}$ are implemented from the reference model (6.5) as follows:

$$y_{\mathrm{r}} = r_1, \quad \dot{y}_{\mathrm{r}} = r_2, \quad \ddot{y}_{\mathrm{r}} = r_3, \quad y_{\mathrm{r}}^{(3)} = -3r_3 - 3r_2 - r_1 + r, \tag{6.9}$$

where $\dot{r}_1 = r_2, \dot{r}_2 = r_3, \dot{r}_3 = -3r_3 - 3r_2 - r_1 + r$.

The virtual estimate of x is $\xi_3 + \theta \xi_2 + v$, and by defining $\omega = \xi_{2,2} + y$ the results of the three steps of our design procedure are:

Step 1.

$$z_1 = y - y_{\mathrm{r}} \tag{6.10}$$

$$\tau_1 = \omega z_1 \tag{6.11}$$

$$\alpha_1 = -c_1 z_1 - d_1 z_1 - \xi_{3,2} + \dot{y}_{\mathrm{r}} - \omega \hat{\theta}. \tag{6.12}$$

Step 2.

$$z_2 = v_2 - \alpha_1 \tag{6.13}$$

$$\tau_2 = \tau_1 - \frac{\partial \alpha_1}{\partial y} \omega z_2 \tag{6.14}$$

$$\begin{aligned}
\alpha_2 = &-c_2 z_2 - d_2 \left(\frac{\partial \alpha_1}{\partial y} \right)^2 z_2 - z_1 + k_2 v_1 + \frac{\partial \alpha_1}{\partial y} (v_2 + \xi_{3,2}) \quad (6.15) \\
&+ \frac{\partial \alpha_1}{\partial y_{\mathrm{r}}} \dot{y}_{\mathrm{r}} + \frac{\partial \alpha_1}{\partial \dot{y}_{\mathrm{r}}} \ddot{y}_{\mathrm{r}} + \frac{\partial \alpha_1}{\partial \xi_3} (A_0 \xi_3 + ky) + \frac{\partial \alpha_1}{\partial \xi_2} (A_0 \xi_2 + e_1 y) \\
&+ \frac{\partial \alpha_1}{\partial y} \omega \hat{\theta} + \frac{\partial \alpha_1}{\partial \hat{\theta}} \gamma \tau_2.
\end{aligned}$$

Step 3.

$$z_3 = v_3 - \alpha_2 \tag{6.16}$$

$$\tau_3 = \tau_2 - \frac{\partial \alpha_2}{\partial y} \omega z_3 \tag{6.17}$$

$$u = -c_3 z_3 - d_3 \left(\frac{\partial \alpha_2}{\partial y} \right)^2 z_3 - z_2 + k_3 v_1 + \frac{\partial \alpha_2}{\partial y} (v_2 + \xi_{3,2}) \tag{6.18}$$

$$+ \frac{\partial \alpha_2}{\partial y_r} \dot{y}_r + \frac{\partial \alpha_2}{\partial \dot{y}_r} \ddot{y}_r + \frac{\partial \alpha_2}{\partial \ddot{y}_r} y_r^{(3)} + \frac{\partial \alpha_2}{\partial \xi_3} (A_0 \xi_3 + ky)$$

$$+ \frac{\partial \alpha_2}{\partial \xi_2} (A_0 \xi_2 + e_1 y) + \frac{\partial \alpha_2}{\partial v_1} (v_2 - k_1 v_1) + \frac{\partial \alpha_2}{\partial v_2} (v_3 - k_2 v_1)$$

$$+ \frac{\partial \alpha_2}{\partial y} \omega \hat{\theta} + \frac{\partial \alpha_2}{\partial \hat{\theta}} \gamma \tau_3 - \gamma z_2 \frac{\partial \alpha_1}{\partial \hat{\theta}} \frac{\partial \alpha_2}{\partial y} \omega .$$

The matrix form of the error system is:

$$\begin{bmatrix} \dot{z}_1 \\ \dot{z}_2 \\ \dot{z}_3 \end{bmatrix} = \begin{bmatrix} -c_1 - d_1 & 1 & 0 \\ -1 & -c_2 - d_2 \left(\frac{\partial \alpha_1}{\partial y} \right)^2 & 1 + \gamma \sigma \omega \\ 0 & -1 - \gamma \sigma \omega & -c_3 - d_3 \left(\frac{\partial \alpha_2}{\partial y} \right)^2 \end{bmatrix} \begin{bmatrix} z_1 \\ z_2 \\ z_3 \end{bmatrix}$$

$$+ \begin{bmatrix} 1 \\ -\frac{\partial \alpha_1}{\partial y} \\ -\frac{\partial \alpha_2}{\partial y} \end{bmatrix} \omega \tilde{\theta} + \begin{bmatrix} 1 \\ -\frac{\partial \alpha_1}{\partial y} \\ -\frac{\partial \alpha_2}{\partial y} \end{bmatrix} \varepsilon_2 \tag{6.19}$$

$$\dot{\varepsilon} = A_0 \varepsilon$$

$$\tau_3 = \begin{bmatrix} 1 & -\frac{\partial \alpha_1}{\partial y} & -\frac{\partial \alpha_2}{\partial y} \end{bmatrix} \omega \begin{bmatrix} z_1 \\ z_2 \\ z_3 \end{bmatrix} ,$$

where $\sigma = \frac{\partial \alpha_1}{\partial \hat{\theta}} \frac{\partial \alpha_2}{\partial y}$. Note again the skew-symmetry of the off-diagonal entries and the stabilizing role of the diagonal entries, whose negativity has been strengthened by the nonlinear damping terms.

This adaptive scheme, developed in [14], was compared in the same paper with a standard indirect certainty-equivalence scheme [2, 20] on the basis of transient performance and control effort. In the indirect scheme, the plant equation $s^2(s - \theta)y(s) = u(s)$ is filtered by a Hurwitz observer polynomial $s^3 + k_1 s^2 + k_2 s + k_3$ to obtain the estimation equation:

$$\phi = \psi \theta$$

$$\phi = \frac{s^3}{s^3 + k_1 s^2 + k_2 s + k_3} y(s) - \frac{1}{s^3 + k_1 s^2 + k_2 s + k_3} u(s) \tag{6.20}$$

$$\psi = \frac{s^2}{s^3 + k_1 s^2 + k_2 s + k_3} y(s) ,$$

and the parameter update law is a normalized gradient (the simulation results with a least-squares update law were virtually identical):

$$\dot{\theta} = \gamma \frac{\psi e}{1 + \psi^2} \ , \quad e = \phi - \psi\hat{\theta} \ . \tag{6.21}$$

The control law is

$$u = r + \left[\frac{\delta_0 s^2 + \delta_1 s + \delta_2}{s^2 + m_1 s + m_2} \right] y + \left[\frac{\delta_3 s + \delta_4}{s^2 + m_1 s + m_2} \right] u \ , \tag{6.22}$$

where $s^2 + m_1 s + m_2$ is a Hurwitz polynomial and the controller parameters $\delta_0, \ldots, \delta_4$ are computed from the Bezout identity

$$s^5 + s^4[m_1 - \delta_3 - \theta] + s^3[m_2 - \delta_4 - a(m_1 - \delta_3)] - s^2[\delta_0 + (m_2 - \delta_4)\theta] =$$
$$= (s + 1)^3(s^2 + m_1 s + m_2) + \delta_1 s + \delta_2 \tag{6.23}$$

which gives $\delta_3 = -(3 + \hat{\theta})$, $\delta_4 = -[3 + 3m_1 + \hat{\theta}(m_1 - \delta_3)]$, $\delta_0 = -[1 + 3m_1 + 3m_2 + (m_2 - \delta_4)\hat{\theta}]$, $\delta_1 = -m_1 - 3m_2$, $\delta_2 = -m_2$.

The above indirect adaptive linear scheme and the adaptive nonlinear scheme were applied to the plant (6.3) with the true parameter $\theta = 3$. In all tests the initial parameter estimate was $\hat{\theta}(0) = 0$, so that, with the adaptation switched off, both closed-loop systems were unstable. The reference input was $r(t) = \sin t$. As is evident from the simulation results in Figure 16, the nonlinear scheme consistently demonstrated much better transient performance without an increase in control effort.[5]

This improvement can be attributed to the different construction of the control law in the two schemes, whose corresponding block diagrams are given in Figures 17 and 18. Figure 17 shows the standard structure of the input and output filters feeding into an estimator/controller block which consists of a parameter estimator and a certainty-equivalence "linear" controller. As seen in Figure 18, the adaptive nonlinear controller retains much of the familiar filter structure. The fundamental difference is that the estimator/controller block now becomes a nonlinear controller designed via the three-step procedure outlined above: in the control law (6.18) both parameter estimates and filter signals enter nonlinearly.

---

[5] For the control plots as well as for further details, the reader is referred to the original paper [14].

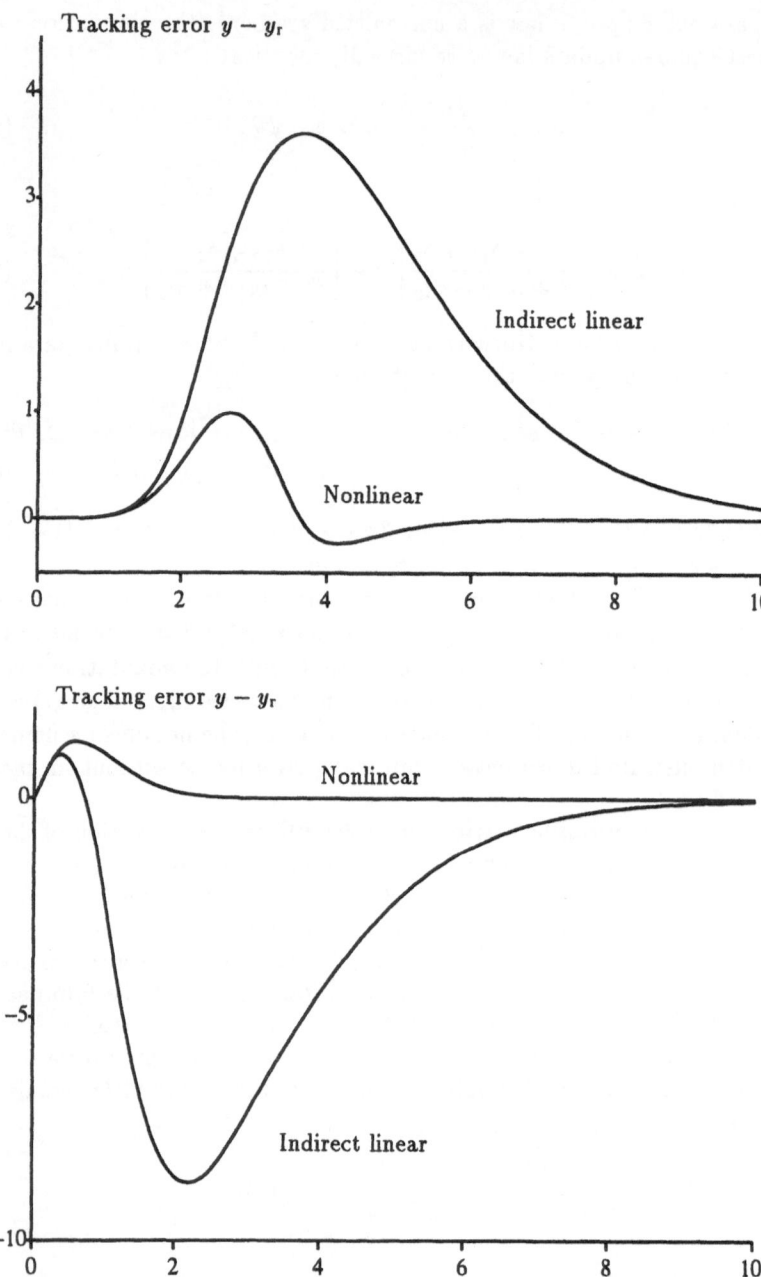

Fig. 16. Transient performance comparison of indirect linear scheme with new nonlinear scheme when $y(0) = 0$ (top) and when $y(0) = 1$ (bottom). In both cases, the nonlinear scheme achieves a dramatic performance improvement without any increase in control effort.

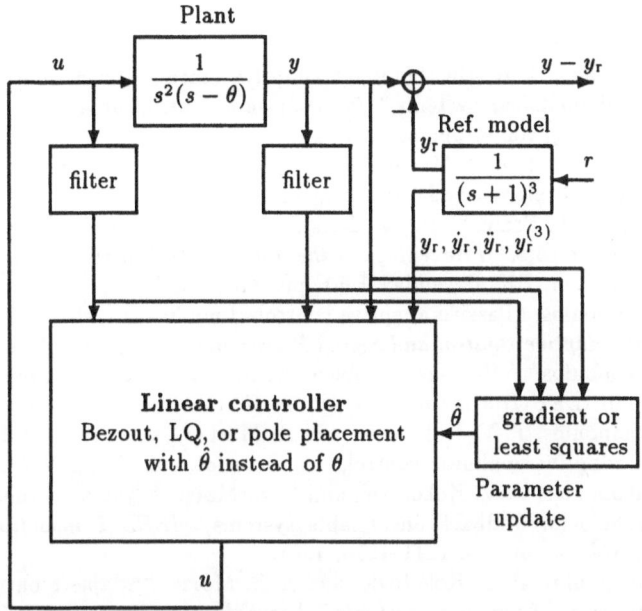

Fig. 17. The indirect scheme is designed according to the traditional structure of adaptive "linear" control which is based on the certainty-equivalence principle.

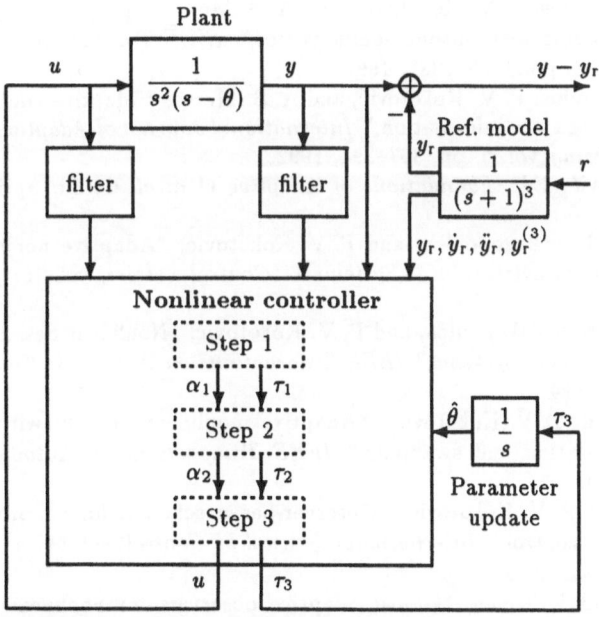

Fig. 18. The distinguishing feature of the new adaptive system is the "Nonlinear controller" block. In contrast to the certainty-equivalence design of Fig. 17, the three-step nonlinear procedure produces a control law in which both parameter estimates and filter signals enter nonlinearly.

References

1. G. Campion and G. Bastin, "Indirect adaptive state-feedback control of linearly parametrized nonlinear systems," *International Journal of Adaptive Control and Signal Processing*, vol. 4, pp. 345–358, 1990.

2. G. C. Goodwin and D. Q. Mayne, "A parameter estimation perspective of continuous time model reference adaptive control," *Automatica*, vol. 23, pp. 57-70, 1987.

3. Z. P. Jiang and L. Praly, "Iterative designs of adaptive controllers for systems with nonlinear integrators," *Proceedings of the 30th IEEE Conference on Decision and Control*, Brighton, UK, December 1991, pp. 2482–2487.

4. I. Kanellakopoulos, "Passive adaptive control of nonlinear systems," *International Journal of Adaptive Control and Signal Processing*, vol. 7, pp. 339–352, 1993.

5. I. Kanellakopoulos, " 'Low-gain' robust control of uncertain nonlinear systems, submitted to *IEEE Transactions on Automatic Control*.

6. I. Kanellakopoulos, P. V. Kokotović, and R. Marino, "An extended direct scheme for robust adaptive nonlinear control," *Automatica*, vol. 27, pp. 247–255, 1991.

7. I. Kanellakopoulos, P. V. Kokotović, and A. S. Morse, "Systematic design of adaptive controllers for feedback linearizable systems," *IEEE Transactions on Automatic Control*, vol. 36, pp. 1241–1253, 1991.

8. I. Kanellakopoulos, P. V. Kokotović, and A. S. Morse, "Adaptive output-feedback control of a class of nonlinear systems," *Proceedings of the 30th IEEE Conference on Decision and Control*, Brighton, UK, December 1991, pp. 1082–1087.

9. I. Kanellakopoulos, P. V. Kokotović, and A. S. Morse, "A toolkit for nonlinear feedback design," *Systems & Control Letters*, vol. 18, pp. 83–92, 1992.

10. I. Kanellakopoulos, P. V. Kokotović, and A. S. Morse, "Adaptive output-feedback control of systems with output nonlinearities," *IEEE Transactions on Automatic Control*, vol. 37, pp. 1266–1282, 1992.

11. I. Kanellakopoulos, P. V. Kokotović, and A. S. Morse, "Adaptive nonlinear control with incomplete state information," *International Journal of Adaptive Control and Signal Processing*, vol. 6, pp. 367–394, 1992.

12. P. V. Kokotović, Ed., *Foundations of Adaptive Control*, Berlin: Springer-Verlag, 1991.

13. M. Krstić, I. Kanellakopoulos, and P. V. Kokotović, "Adaptive nonlinear control without overparametrization," *Systems & Control Letters*, vol. 19, pp. 177–185, 1992.

14. M. Krstić, I. Kanellakopoulos, and P. V. Kokotović, "Nonlinear design of adaptive controllers for linear systems," *IEEE Transactions on Automatic Control*, vol. 39, pp. 783–752, 1994.

15. M. Krstić and P. V. Kokotović, "Adaptive nonlinear design with controller-identifier separation and swapping," *IEEE Transactions on Automatic Control*, to appear, 1995.

16. M. Krstić and P. V. Kokotović, "Observer-based schemes for adaptive nonlinear state-feedback control," *International Journal of Control*, vol. 59, pp. 1373–1381, 1994.

17. R. Marino and P. Tomei, "Global adaptive observers for nonlinear systems via filtered transformations," *IEEE Transactions on Automatic Control*, vol. 37, pp. 1239–1245, 1992.

18. R. Marino and P. Tomei, "Global adaptive output-feedback control of nonlinear systems, Part I: linear parameterization," *IEEE Transactions on Automatic Control*, vol. 38, pp. 17–32, 1993.

19. R. Marino and P. Tomei, "Global adaptive output-feedback control of nonlinear systems, Part II: nonlinear parameterization," *IEEE Transactions on Automatic Control*, vol. 38, pp. 33–49, 1993.

20. R. H. Middleton, "Indirect continuous time adaptive control," *Automatica*, vol. 23, pp. 793–795, 1987.

21. K. Nam and A. Arapostathis, "A model-reference adaptive control scheme for pure-feedback nonlinear systems," *IEEE Transactions on Automatic Control*, vol. 33, pp. 803–811, 1988.

22. J. B. Pomet and L. Praly, "Adaptive nonlinear regulation: estimation from the Lyapunov equation," *IEEE Transactions on Automatic Control*, vol. 37, pp. 729–740, 1992.

23. S. S. Sastry and A. Isidori, "Adaptive control of linearizable systems," *IEEE Transactions on Automatic Control*, vol. 34, pp. 1123–1131, 1989.

24. D. Seto, A. M. Annaswamy and J. Baillieul, "Adaptive control of nonlinear systems with a triangular structure," *IEEE Transactions on Automatic Control*, vol. 39, pp. 1411–1428, 1994.

25. E. D. Sontag, "Smooth stabilization implies coprime factorization," *IEEE Transactions on Automatic Control*, vol. 34, pp. 435–443, 1989.

26. E. D. Sontag, "Input/output and state-space stability," in *New Trends in System Theory*, G. Conte et al., Eds., Boston: Birkhäuser, 1991.

27. D. G. Taylor, P. V. Kokotović, R. Marino and I. Kanellakopoulos, "Adaptive regulation of nonlinear systems with unmodeled dynamics," *IEEE Transactions on Automatic Control*, vol. 34, pp. 405–412, 1989.

28. A. R. Teel, R. R. Kadiyala, P. V. Kokotović and S. S. Sastry, "Indirect techniques for adaptive input-output linearization of non-linear systems," *International Journal of Control*, vol. 53, pp. 193–222, 1991.

29. A. R. Teel, "Error-based adaptive non-linear control and regions of feasibility," *International Journal of Adaptive Control and Signal Processing*, vol. 6, pp. 319–327, 1992.

This article was processed using the LaTeX macro package with LLNCS style

Nonlinear Output Feedback Control

Hassan K. Khalil

Department of Electrical Engineering
Michigan State University
East Lansing, MI 48824-1226, USA

Abstract. It is shown that the performance of a globally bounded (possibly dynamic) partial state feedback control of an input-output linearizable system can be recovered by a sufficiently fast high-gain observer. The performance recovery includes recovery of boundedness of trajectories, recovery of the property that the trajectories reach a certain set in finite time, and recovery of convergence to a positively invariant set.

1   Introduction

A few years ago, we introduced a new technique in the design of robust output feedback control for input-output linearizable system [1]. The basic ingredients of this technique are

(1) A high-gain observer that robustly estimates the derivatives of the output;
(2) A globally bounded state feedback control, usually obtained by saturating a continuous state feedback function outside a compact region of interest, that meets the design objectives using nonlinear robust control techniques. The global boundedness of the control protects the state of the plant from peaking when the high-gain observer estimates are used instead of the true states.

This technique has been the impetus for several results we have obtained over the past few years. It was used in [1] and [6] to achieve stabilization and semiglobal stabilization of fully-linearizable systems, in [5] to design robust servomechanisms for fully linearizable systems, in [10] and [11] to extend the results of [5] to system having nontrivial zero dynamics. It was used also in adaptive control [4], variable structure control [12], and speed control of induction motors [7].

As the results of [1] became known, other researchers adopted our technique in their work. Teel and Praly [13, 14] used it in a few papers to achieve semiglobal stabilization. Lin and Saberi [8] used it also to achieve semiglobal stabilization. Jankovic [3] used it in an adaptive control problem.

In all these papers, except [3, 12], the controller is designed in two steps. First, a globally bounded state feedback control is designed to meet the design objective. Second, a high-gain observer, designed to be fast enough, recovers the performance achieved under state feedback. This recovery is shown using asymptotic analysis of a singularly perturbed closed-loop system. Despite the similarity of the argument used in all these papers, the technical argument had to be developed independently in each case due to different technical assumptions. Our goal in the current paper is to develop this recovery property in a generic form that can be applied in different problems; thus avoiding the duplication of effort in proving recovery, and allowing us to concentrate attention on the performance that can be achieved under state feedback.

2   Input-Output Linearizable Systems

We consider a single-input single-output nonlinear system represented by

$$\dot{e} = Ae + B[f_1(e, z, d(t)) + g_1(e, z, d(t))u] \ , \tag{1}$$

$$\dot{z} = f_2(e, z, u, d(t)) \ , \tag{2}$$

$$y = Ce \ , \tag{3}$$

$$\zeta = h(e, z) \ , \tag{4}$$

where u is the control input, y is the controlled output, assumed to be measured, $e \in R^r$ and $z \in R^m$ constitute the state vector, $d(t)$ is a vector of piecewise continuous and bounded functions of time, and ζ is a vector of additional measured variables. The functions f_1, f_2, and g_1 are locally Lipschitz in their arguments, and $g_1(\cdot) \neq 0$ over the domain of interest. The $r \times r$ matrix A, the $r \times 1$ matrix B, and the $1 \times r$ matrix C are given by

$$A = \begin{bmatrix} 0 & 1 & 0 & \cdots & 0 \\ 0 & 0 & 1 & \cdots & 0 \\ \vdots & & & & \vdots \\ 0 & 0 & \cdots & 1 & 0 \\ 0 & 0 & \cdots & 0 & 1 \\ 0 & 0 & \cdots & 0 & 0 \end{bmatrix}, \quad B = \begin{bmatrix} 0 \\ 0 \\ \vdots \\ 0 \\ 0 \\ 1 \end{bmatrix},$$

$$C = \begin{bmatrix} 1 & 0 & \cdots & 0 & 0 \end{bmatrix} \ .$$

The main goal is to design an output feedback controller, using only measurement of y and ζ, to regulate y to zero while ensuring boundedness of all state variables. Depending on the problem there may be other objectives like stabilization of an equilibrium point at which $y = 0$. The system is already represented in error coordinates where the output y is the tracking error. In such representation the vector $d(t)$ comprises exogenous disturbances and reference signals.

The model (1)–(4) is input-output linearizable because the state feedback control

$$u = \frac{1}{g_1(e, z, d)}[v - f_1(e, z, d)]$$

yields a linear map from the input v to the output y. The main source of this model is the normal form of a nonlinear system having relative degree r. It is well known [2] that if the nonlinear system

$$\dot{\chi} = f(\chi) + g(\chi)u \ ,$$

$$\sigma = h(\chi)$$

has relative degree r, then it can be transformed into the normal form

$$\dot{\xi} = A\xi + B[\bar{f}_1(\xi, z) + \bar{g}_1(\xi, z)u] ,$$
$$\dot{z} = \bar{f}_2(\xi, z) ,$$
$$\sigma = C\xi .$$

Taking $y = e_1 = \xi_1 - r(t)$, $e_2 = \xi_2 - \dot{r}(t)$, and so on, brings the equation into form (1)–(3). In this case the only measured variable is y and equation (4) is dropped.

Another source of the model (1)–(4), where equation (4) is relevant, arises when an nth-order system of relative degree ρ is extended by augmenting a series of $(n - \rho)$ integrators at the input side [15, 14, 4]. Suppose that the system is modeled by the nth-order differential equation

$$y_o^{(n)} = f_0(\cdot) + g_0(\cdot)\mu^{(n-\rho)}$$

where μ is the input, y_o is the output, f_0 and g_0 are functions of y_o, $y_o^{(1)}$, ... , $\sigma^{(n-1)}$, μ, ..., $\mu^{(n-\rho-1)}$, and $w(t)$ (a time-varying disturbance). Augmenting $(n-\rho)$ integrators at the input side, denoting their states by $z_i = \mu^{(i-1)}$, setting $u = \mu^{(n-\rho)}$ as the control input of the augmented system, and taking $e_1 = y_o - r(t)$, $e_2 = y_o^{(1)} - r^{(1)}$, up to $e_n = y_o^{(n-1)} - r^{(n-1)}$, results in a system of the form (1)–(4) with $r = n$ and $m = n - \rho$. In this case all the components of z are measured; hence $h(e, z) = z$ in (4).

The model (1)–(4) may arise also from the physical description of the system. For example, in a recent study of the control of induction motors [7], the motor was represented, after some transformations and inclusion of a flux estimator, by

$$\dot{e} = Ae + B[f_1(e, \lambda, \nu, d(t)) + g_1(\lambda, \nu)v_q] ,$$
$$\dot{\lambda} = A_2\lambda + B_2[f_3(e, \lambda, \nu) + g_3v_d] ,$$
$$\dot{\nu} = A_3(x, \lambda, \nu, d(t))\nu + f_4(x, \lambda, \nu, d(t)) ,$$

where $e = [\delta, \dot{\delta}, \ddot{\delta}]^T$, with $\delta = \theta - \theta_{\mathrm{ref}}(t)$ as the rotor position error, $\lambda = [\psi_d, i_d]^T$ comprises a flux estimate ψ_d and a measured current i_d, $\nu = [e_d, e_q]^T$ contains errors in flux estimation, and v_d and v_q are the control inputs. The model takes the form (1)–(4) with $z = [\lambda^T, \nu^T]^T$ and $\zeta = [\lambda^T, i_q]^T$, where i_q is a measured current that can be expressed as a function of (x, λ, ν, d).

3 Dynamic Partial State Feedback Control

Assuming that e and ζ are available for feedback, we design a state feedback control to meet the design objectives. We allow the state feedback control to be dynamic, which is the case for example in the servomechanism design of [5, 9] and the adaptive control of [4]. The state feedback control is assumed in the form

$$\dot{\vartheta} = \phi(\vartheta, e, \zeta, d(t)) \ , \tag{5}$$

$$u = \psi(\vartheta, e, \zeta, d(t)) \ , \tag{6}$$

where $\psi(\cdot)$ and $\phi(\cdot)$ are locally Lipschitz functions in their arguments and globally-bounded functions of e. The global boundedness of ψ and ϕ is a key property of the control, which can be always achieved by saturating ψ and ϕ outside a compact region of interest. The control is allowed to depend on $d(t)$ since it may depend on the reference signal and its derivatives. The closed-loop system under state feedback is described by

$$\dot{e} = Ae + B[f_1(e, z, d(t)) + g_1(e, z, d(t))\psi(\vartheta, e, \zeta, d(t))] \ ,$$

$$\dot{z} = f_2(e, z, \psi(\vartheta, e, \zeta, d(t)), d(t)) \ ,$$

$$\dot{\vartheta} = \phi(\vartheta, e, \zeta, d(t)) \ .$$

We rewrite these equations in the compact form

$$\dot{x} = F_s(x, d(t)) \ , \tag{7}$$

where

$$x = \begin{bmatrix} e \\ z \\ \vartheta \end{bmatrix} \ .$$

We do not concern ourselves here with the design of the state feedback control. Instead, we describe certain properties of the closed-loop system (7) which are achieved under the state feedback design. Then in Section 5 we will establish the recovery of these properties using a high-gain observer. We describe three properties: boundedness of trajectories, reaching a positively invariant set in finite time, and exponential convergence to a positively invariant set. In each case we characterize the property by means of Lyapunov functions, which will be used to show recovery in Section 5.

Property I: Boundedness. There exist locally-Lipschitz nonnegative scalar functions $U_i(x)$, and positive constants c_i for $i = 1, \ldots, p$, $p \geq 1$, such that

$$\Omega = \bigcap_{i=1,\ldots,p} \{U_i(x) \leq c_i\}$$

is bounded and either

$$\dot{U}_i\Big|_{(7)} \leq -\beta_i < 0, \quad \text{on } \Omega \cap \{U_i(x) = c_i\} \ , \tag{8}$$

for all $i = 1, \ldots, p$, or (8) is satisfied for $i = 1, \ldots, p_1$ while

$$\dot{U}_i\Big|_{(7)} = 0, \quad \text{on } \Omega \cap \{U_i(x) = c_i\} \ , \tag{9}$$

for $i = p_1 + 1, \ldots, p$, independent of the variable e in $\psi(\cdot)$ and $\phi(\cdot)$; that is, (9) is satisfied even when e in $\psi(\cdot)$ and $\phi(\cdot)$ is replaced by any vector in R^n.

The notation $\dot{U}_i\big|_{(7)}$ denotes the derivative of U_i along the trajectories of (7). Either one of the foregoing conditions ensures that the set Ω is positively invariant. Hence, all trajectories starting in Ω remain in it for all future time; consequently they are bounded. The simplest form of this property happens when there is a positive definite function $U(x)$ whose derivative is negative on the boundary $\{U(x) = c\}$. We allow several functions, which are not necessarily positive definite, to include cases where boundedness is established using vector Lyapunov functions. Allowing one or more of the functions to satisfy the condition (9) accommodates cases that appear in the servomechanism problem [5] and adaptive control [4], as it will be illustrated later on.

Property II: Finite-Time Arrival at a Positively Invariant Set. Suppose that Property I is satisfied, and there exists locally-Lipschitz nonnegative functions $V_i(x)$ and positive constants δ_i, for $i = 1, \ldots, q, q \geq 1$, such that the sets

$$\Delta_i = \Omega \cap \{V_1(x) \leq \delta_1\} \cap \cdots \cap \{V_{i-1}(x) \leq \delta_{i-1}\} \cap \{V_i(x) > \delta_i\}$$

are nonempty and

$$\dot{V}_i\Big|_{(7)} \leq -\bar{\beta}_i < 0, \quad \forall\, x \in \Delta_i \;, \tag{10}$$

for $i = 1, \ldots, q$, where $\Delta_0 \overset{\text{def}}{=} \Omega$. Inequality (10) ensures that

$$\Delta \overset{\text{def}}{=} \Omega \cap \{V_1(x) \leq \delta_1\} \cap \cdots \cap \{V_q(x) \leq \delta_q\}$$

is positively invariant and $x(t)$ reaches Δ in finite time.

Property III: Exponential Convergence to a Positively Invariant Set. Suppose that either Property I or Property II is satisfied, and set $\mho = \Omega$ in the first case and $\mho = \Delta$ in the second one. Suppose there is a set $\Gamma \subset \mho$ such that

(1) Γ is a positively invariant set with respect to (7).
(2) $\forall\, x(0) \in \mho$, the trajectories of (7) satisfy

$$|x|_\Gamma(t) \leq k_s e^{-\alpha_s t} |x|_\Gamma(0) \;, \tag{11}$$

for some positive constants α_s and k_s, where $|x|_\Gamma$ is the distance from x to Γ. The following lemma states a converse Lyapunov theorem for the exponential convergence property. The proof of this lemma is a straight-forward extension of the proofs of Theorem 19.2 and 22.5 of [16].

Lemma 1. *There exists a locally-Lipschitz function $W(t, x)$, defined for all $x \in \mho$ and all $t \geq 0$, such that*

$$|x|_\Gamma \leq W(t, x) \leq k_u |x|_\Gamma \;, \tag{12}$$

$$\dot{W}\Big|_{(7)} \leq -\gamma W, \quad \gamma > 0 \;. \tag{13}$$

We recall three examples to illustrate these properties.

Example 1 [6]: Robust Semiglobal Stabilization

Consider the nonlinear system

$$\dot{\mathcal{X}} = f(\mathcal{X}, \theta) + g(\mathcal{X}, \theta)u \ , \tag{14}$$

$$y = h(\mathcal{X}, \theta) \ , \tag{15}$$

where $\mathcal{X} \in R^n$ is the state, $u \in R$ is the control input, $y \in R$ is the measured output, and θ is a vector of constant unknown parameters which belongs to a compact set Θ. We assume that f, g, and h are sufficiently smooth in $\mathcal{X}$, $f(0, \theta) = 0$, and $h(0, \theta) = 0$, $\forall \ \mathcal{X} \in R^n$, $\forall \ \theta \in \Theta$. Hence, the origin $\mathcal{X} = 0$ is an equilibrium point of the open-loop system (14)–(15) when $u = 0$. We also assume that f, g, h, and their partial derivatives with respect to $\mathcal{X}$, are continuous in θ, $\forall \ \mathcal{X} \in R^n$, $\forall \ \theta \in \Theta$. The class of nonlinear systems is characterized by the following assumption.

Assumption 1: For all $\theta \in \Theta$ and all $\mathcal{X} \in R^n$,

- The system (14)–(15) has a uniform relative degree n, i.e.,

$$L_g h(\mathcal{X}, \theta) = \cdots = L_g L_f^{n-2} h(\mathcal{X}, \theta) = 0$$

and

$$L_g L_f^{n-1} h(\mathcal{X}, \theta) \neq 0 \ .$$

- The mapping $e = T(\mathcal{X}, \theta)$, defined by

$$e_j = L_f^{j-1} h(\mathcal{X}, \theta), \quad 1 \leq j \leq n \ , \tag{16}$$

is proper, i.e.,

$$\lim_{\|\mathcal{X}\| \to \infty} \|T(\mathcal{X}, \theta)\| = \infty \ . \tag{17}$$

The change of variables $e = T(\mathcal{X}, \theta)$ transforms the system (14)–(15) into the normal form

$$\dot{e}_1 = e_2 \ ,$$

$$\vdots = \vdots$$

$$\dot{e}_{n-1} = e_n \ ,$$

$$\dot{e}_n = F(e, \theta) + G(e, \theta)u \ ,$$

$$y = e_1 \ .$$

The condition (17) ensures that $e = T(\mathcal{X}, \theta)$ is a global diffeomorphism of R^n onto R^n; hence the normal form is defined globally. The global normal form can be rewritten in the form

$$\dot{e} = Ae + B[F(e, \theta) + G(e, \theta)u] \ , \tag{18}$$

$$y = Ce \ . \tag{19}$$

It follows from Assumption 1 that $G(e, \theta) \neq 0$ for all $e \in R^n$ and all $\theta \in \Theta$. The transformation into global normal form is valid for every $\theta \in \Theta$. However, the change of variables (16) depends on the unknown parameters θ. This means that the new state e would not be accessible even if the original state X was available for measurement. This point will not cause a problem in our design since the controller will use only measurements of the output y.

We now show how to achieve semiglobal stabilization using state feedback. Let $F_0(e)$ and $G_0(e)$ be known nominal models of $F(e, \theta)$ and $G(e, \theta)$, respectively. Suppose that $F_0(e)$ and $G_0(e)$ are sufficiently smooth, $F_0(0) = 0$, and $G_0(e) \neq 0$ for all $e \in R^n$.

Assumption 2: For every compact set $\mathcal{U} \subset R^n$, there exist a scalar nonnegative locally Lipschitz function $\tilde{\rho}(e)$ and a positive constant k, both known and possibly dependent on $\mathcal{U}$, such that

$$|F(e, \theta) - G(e, \theta)G_0^{-1}(e)F_0(e)| \leq \tilde{\rho}(e) \tag{20}$$

and

$$\frac{G(e, \theta)}{G_0(e)} \geq k > 0 \tag{21}$$

for all $e \in \mathcal{U}$ and all $\theta \in \Theta$.

The existence of $\tilde{\rho}(e)$ satisfying (20) is guaranteed by the smoothness of the left-hand side function and the compactness of Θ. If there is no parameter uncertainty and F and G are exactly known, we take $F_0 = F$ and $G_0 = G$; then inequalities (20) and (21) are satisfied with $\tilde{\rho} = 0$ and $k = 1$. On the other hand, if no nominal models of F and G are available, we can take $F_0 = 0$ and $G_0 = 1$; then inequalities (20) and (21) take the form

$$|F(e, \theta)| \leq \tilde{\rho}(e), \quad G(e, \theta) \geq k > 0$$

Finally, notice that if (20) and (21) hold globally in e, then they hold on any compact set $\mathcal{U} \subset R^n$ with $\tilde{\rho}(e)$ and k independent of the set $\mathcal{U}$.

The design of a state feedback control starts by choosing a matrix K such that $(A + BK)$ is Hurwitz, which is always possible since (A, B) is controllable. Let $P = P^T > 0$ be the solution of the Lyapunov equation

$$P(A + BK) + (A + BK)^T P = -I \tag{22}$$

and set $V(e) = e^T P e$. Given any compact set $\mathcal{U}_0$, find a positive constant $c > 0$ such that

$$\mathcal{U}_0 \subset \Omega_c \overset{\text{def}}{=} \{e \in R^n \mid V(e) \leq c\} .$$

This is always possible since $V(e)$ is radially unbounded. Find a locally Lipschitz function $\rho(e)$ and a constant k that satisfy (21) and

$$|F(e, \theta) - G(e, \theta)G_0^{-1}(e)F_0(e) - Ke| \leq \rho(e) \tag{23}$$

over the set Ω_c. The existence of $\rho(e)$ and k is guaranteed by Assumption 2. Without loss of generality, choose $\rho(e)$ such that

$$\rho(e) = \rho_0 > 0, \quad \forall \, \|e\| \leq r \, , \tag{24}$$

for some $r > 0$. Consider the state feedback control

$$u = \psi(e) \stackrel{\text{def}}{=} -G_0^{-1}(e)F_0(e) - \frac{1}{k}G_0^{-1}(e)\eta(e) \, , \tag{25}$$

where

$$\eta(e) = \begin{cases} \rho(e) \, \frac{w}{\|w\|} & \text{if } \rho(e)\|w\| \geq \mu \\[2mm] \rho^2(e) \, \frac{w}{\mu} & \text{if } \rho(e)\|w\| < \mu \end{cases} \, , \tag{26}$$

$w = 2B^T P e$, and μ is a positive constant. Using $V(e)$ as a Lyapunov function candidate for the closed-loop system, it can be shown that

$$\dot{V}(e) \leq -\|e\|^2 + \frac{\mu}{4}, \quad \forall \, e \in \Omega_c \, .$$

Taking $\beta = \mu\gamma\lambda_{max}(P)/4$, for some $\gamma > 1$, and choosing $\mu < 4c/\gamma\lambda_{max}(P)$, we can see that Properties I and II are satisfied with $\Omega = \Omega_c$ and $\Delta = \Omega_\beta \stackrel{\text{def}}{=} \{V(e) \leq \beta\} \subset \Omega_c$. Hence, all trajectories starting in Ω_c reach Ω_β in finite time. Since $F(0, \theta) = F_0(0) = 0$ and $F(e, \theta) - G(e, \theta)G_0^{-1}(e)F_0(e)$ is continuously differentiable, it follows that

$$\|F(e, \theta) - G(e, \theta)G_0^{-1}(e)F_0(e) - Ke\| \leq \ell\|e\|, \quad \forall \, \|e\| \leq r \tag{27}$$

with the positive constant ℓ possibly dependent on r. Choose

$$\mu < \frac{4r^2\lambda_{min}(P)}{\gamma\lambda_{max}(P)}$$

so that $\Omega_\beta \subset \{\|e\| < r\}$. Then for all $e \in \Omega_\beta$ and $\rho_0\|w\| < \mu$, we have

$$\dot{V}(e) \leq -\|e\|^2 + \ell\|w\|\|e\| - \frac{\rho_0^2}{\mu}\|w\|^2$$

$$\leq -\tfrac{1}{2}\|e\|^2, \quad \text{for } \mu < \frac{2\rho_0^2}{\ell^2}$$

while for $e \in \Omega_\beta$ and $\rho_0\|w\| \geq \mu$, we have

$$\dot{V}(e) \leq -\|e\|^2$$

Thus, Property III is satisfied with $\mho = \Delta$ and $\Gamma = \{e = 0\}$ (the origin), and we arrive at the conclusion summarized in the following lemma.

Lemma 2. *Suppose $\rho(e)$ and k are chosen such that inequalities (21) and (23) are satisfied over Ω_c, and $\rho(e)$ satisfies (24) for some $r > 0$. Let ℓ be the constant appearing in (27). Suppose that μ is small enough to satisfy*

$$\mu < \min\left\{ \frac{4c}{\gamma\lambda_{max}(P)}, \frac{4r^2\lambda_{min}(P)}{\gamma\lambda_{max}(P)}, \frac{2\rho_0^2}{\ell^2} \right\} , \tag{28}$$

then the origin of the closed-loop system, formed of (18) and (25), is exponentially stable and Ω_c belongs to its region of attraction. This conclusion holds for every $\theta \in \Theta$.

The conclusion of this lemma remains valid when the state feedback control is saturated outside the set Ω_c. In particular, let $S = \max_{e \in \Omega_c} |\psi(e)|$ and define $\psi^s(e)$ by

$$\psi^s(e) = S \operatorname{sat}\left(\frac{\psi(e)}{S}\right) , \tag{29}$$

where sat$(\cdot)$ is the saturation function. Then $\psi^s(e) = \psi(e)$ for all $e \in \Omega_c$, and Lemma 2 holds for the state feedback control $u = \psi^s(e)$.

Example 2 [5]: Servomechanism

Consider the single-input–single-output nonlinear system

$$\dot{z} = f(z, \theta) + g(z, \theta)u + \sum_{i=1}^{\ell} p_i(z, \theta)d_i(t) , \tag{30}$$

$$y_o = h(z, \theta) + Q(\theta)d(t) , \tag{31}$$

where $z \in R^n$ is the state, $u \in R$ is the control input, $y_o \in R$ is the measured output, $d \in R^\ell$ is a time-varying disturbance input, and θ is a vector of constant unknown parameters which belongs to a compact set Θ. The output y_o is to track a time-varying reference input $r(t)$. We assume that both $r(t)$ and $d(t)$ are sufficiently smooth functions of t and they, together with their derivatives up to order n, are bounded functions of t for all $t \geq 0$. We assume also that f, g, p_i, and h are sufficiently smooth in z, $\forall\, \theta \in \Theta$, $\forall\, z \in M_\theta$, where $M_\theta \subset R^n$ is a domain that could be dependent on θ. The class of nonlinear systems we shall deal with is characterized by the following assumption.

Assumption 3: For all $\theta \in \Theta$ and all $z \in M_\theta$,

- The system (30)–(31) has a uniform relative degree n, i.e.,

$$L_g h(z, \theta) = \cdots = L_g L_f^{n-2} h(z, \theta) = 0$$

and

$$L_g L_f^{n-1} h(z, \theta) \neq 0 .$$

- The mapping $\zeta = T(z, \theta)$, defined by

$$\zeta_i = L_f^{i-1} h(z, \theta), \quad 1 \leq i \leq n \tag{32}$$

is a proper map of M_θ into $T(M_\theta, \theta)$.

– The vector fields p_i satisfy the disturbance-strict-feedback condition

$$[X, p_i] \in \mathcal{G}^j, \quad \forall X \in \mathcal{G}^j, \quad 0 \le j \le n-2, \quad 1 \le i \le \ell , \tag{33}$$

where

$$\mathcal{G}^j = span\{g, ad_f g, \ldots, ad_f^j g\} .$$

The uniform relative degree assumption is a necessary and sufficient condition for the mapping $\zeta = T(z, \theta)$ to be a local diffeomorphism which transforms the system (30)–(31), with $d = 0$, into the normal form. By the properness of the mapping $\zeta = T(z, \theta)$, it is a diffeomorphism of M_θ onto $T(M_\theta, \theta)$. Therefore, the normal form is defined for all $\zeta \in T(M_\theta, \theta)$. With $d \ne 0$, the disturbance-strict-feedback condition (33) ensures that the change of variables $\zeta = T(z, \theta)$ brings the system into the strict feedback normal form

$$\dot{\zeta}_i = \zeta_{i+1} + \sum_{j=1}^{\ell} \gamma_{ij}(\zeta_1, \ldots, \zeta_i, \theta) d_j(t), \quad i = 1, \ldots, n-1 , \tag{34}$$

$$\dot{\zeta}_n = \tilde{F}(\zeta, \theta) + \tilde{G}(\zeta, \theta) u + \sum_{j=1}^{\ell} \gamma_{nj}(\zeta, \theta) d_j(t) , \tag{35}$$

$$y_o = \zeta_1 + Q(\theta) d(t) . \tag{36}$$

This form allows us to use a time-varying change of variables to transform the system equation into an error space where the goal of the design will be to drive the state of the system to zero. In particular, let $e_1 = y = y_o - r$ be the tracking error, and set $e_i = d^{i-1} y / dt^{i-1}$ for $2 \le i \le n$. It can be easily seen that

$$e_1 = \zeta_1 + Q(\theta) d - r ,$$
$$e_i = \zeta_i + \varphi_i(\zeta_1, \ldots, \zeta_{i-1}, d, \ldots, d^{(i-2)}, \theta) + Q(\theta) d^{(i-1)} - r^{(i-1)}, \quad 2 \le i \le n , \tag{37}$$

where $d^{(i)}$ and $r^{(i)}$ denote the ith derivative with respect to t, and $\varphi_i(\cdot)$ is a polynomial function of the components of d to $d^{(i-2)}$. Viewing (37) as a change of variables from the ζ variables to the e variables, it is clear that the change of variables is invertible, since, for every e, ζ is uniquely determined by (37). We rewrite (37) as

$$e = \Phi(\zeta, \nu(t), \theta) \tag{38}$$

where ν is a vector of exogenous signals that comprises d, r, and their derivatives up to order n. The function $\nu(t)$ is smooth and bounded by assumption. Let

$$e = \Psi(z, \nu(t), \theta) \tag{39}$$

be the composition of $\zeta = T(z, \theta)$ and $e = \Phi(\zeta, \nu(t), \theta)$; that is, $\Psi(z, \nu, \theta) = \Phi(T(z, \theta), \nu, \theta)$. It can be easily verified that the change of variables $e = \Psi(z, \nu(t), \theta)$ transforms the system (30)–(31) into the normal form

$$\dot{e} = Ae + B[F(e, \nu(t), \theta) + G(e, \nu(t), \theta) u] , \tag{40}$$

$$y = Ce . \tag{41}$$

Let Λ be a bounded set such that $\nu(t) \in \Lambda$ for all $t \geq 0$.

Assumption 4: For every $\theta \in \Theta$ there is a domain $N_\theta \subset R^n$ which contains the origin and $\Psi^{-1}(N_\theta, \nu, \theta) \subset M_\theta, \forall \nu \in \Lambda$.

This is a reasonable assumption in view of the fact that the functions φ_i of (37) are polynomials in the components of ν. In particular, if z_θ^* is in the interior of M_θ and $f(z_\theta^*, \theta) = 0$ and $h(z_\theta^*, \theta) = 0$, then $T(z_\theta^*, \theta) = 0$ and $\Phi(z_\theta^*, 0, \theta) = 0$. Hence, there is a neighborhood N_θ of the origin and a constant $k > 0$ such that $\Psi^{-1}(N_\theta, \nu, \theta) \subset M_\theta$ for all $\|\nu\| < k$. Moreover, if $M_\theta = R^n$ and $T(z, \theta)$ is a proper map of R^n into R^n; i.e., $\lim_{\|z\| \to \infty} \|T(z, \theta)\| = \infty$, then the normal form (40)–(41) is defined globally and we can take $N_\theta = R^n$. Our analysis of the system (40)–(41) will be restricted to the domain N_θ.

We start by designing a servo controller assuming that measurements of the state e are available. Let $F_0(e, \nu)$ and $G_0(e, \nu)$ be nominal models of $F(e, \nu, \theta)$ and $G(e, \nu, \theta)$, respectively. The nominal functions F_0 and G_0 are allowed to depend on ν since in some cases they may depend on the reference signal r and its derivatives. If no such nominal functions are available, take $F_0 = 0$ and $G_0 = 1$. Suppose that for all $e \in N_\theta$ and all $\nu \in \Lambda$, $G_0(e, \nu) \neq 0$ and $G(e, \nu, \theta) \neq 0$; the latter condition follows from the normal form. In the normal form (40)–(41), the uncertain terms F and G satisfy the matching condition. Hence it is possible, under reasonable assumptions, to design a state feedback controller to drive the state x to a residual set in the neighborhood of the origin. The size of the residual set can be made arbitrarily small at the expense of increasing the feedback gain near the origin. Driving the state to the origin requires the use of a discontinuous controller. We are interested in a case where the exogenous signal $\nu(t)$ satisfies a known characteristic equation, referred to as the internal model. Our goal is to include a servo-compensator that duplicates this internal model so that the state is driven to the origin by the presence of the servo-compensator rather than by increasing the feedback gain near the origin.

The first step in the design is to identify the internal model. Toward that end, note that a successful servo controller design would create a zero-error manifold where $y(t) \equiv 0$. From equation (40)–(41) we see that $y(t) \equiv 0$ implies that $e(t) \equiv 0$. Hence, on the zero-error manifold we must have

$$F(0, \nu, \theta) + G(0, \nu, \theta)u = 0 .$$

Using the nominal functions F_0 and G_0, we write u as $u = G_0^{-1}(-F_0 + \tilde{u})$, and use the fact that $G(0, \nu, \theta) \neq 0$ and $G_0(0, \nu) \neq 0$ to arrive at the condition

$$G_0(0, \nu)G^{-1}(0, \nu, \theta)F(0, \nu, \theta) - F_0(0, \nu) + \tilde{u} = 0$$

which must be satisfied on the zero-error manifold. This leads us to the following assumption.

Assumption 5: There exist a $q \times q$ matrix S, with distinct eigenvalues on the imaginary axis, and a $1 \times q$ matrix $\Gamma_1(\theta)$ such that

$$G_0(0, \nu(t))G^{-1}(0, \nu(t), \theta)F(0, \nu(t), \theta) - F_0(0, \nu(t)) \equiv \Gamma_1(\theta)v(t) \qquad (42)$$

for all $t \in R$ and for every $\theta \in \Theta$, where $v(t)$ satisfies the equation

$$\dot{v}(t) = Sv(t) \ . \qquad (43)$$

Assumption 5 identifies equation (43) as the desired internal model. Since the model is linear, we consider the linear servo compensator

$$\dot{\sigma} = S\sigma + Je \ , \qquad (44)$$

where J is a matrix chosen such that the pair (S, J) is controllable. We augment (44) with the system (40)–(41) to obtain the augmented system

$$\dot{\xi} = \mathcal{A}\xi + \mathcal{B}[F(e, \nu(t), \theta) + G(e, \nu(t), \theta)u] \qquad (45)$$

where

$$\mathcal{A} = \begin{bmatrix} S & JC \\ 0 & A \end{bmatrix}, \quad \mathcal{B} = \begin{bmatrix} 0 \\ B \end{bmatrix}, \quad \xi = \begin{bmatrix} \sigma \\ e \end{bmatrix} \ .$$

It is known from linear servomechanism theory that the pair $(\mathcal{A}, \mathcal{B})$ is controllable. In the augmented system (45), the uncertain terms F and G still satisfy the matching condition. We proceed to design a robust state feedback control that drives the state ξ toward the origin. Choose K such that $(\mathcal{A} + \mathcal{B}K)$ is a Hurwitz matrix, and let $P = P^T > 0$ be the solution of the Lyapunov equation

$$P(\mathcal{A} + \mathcal{B}K) + (\mathcal{A} + \mathcal{B}K)^T P = -I \ . \qquad (46)$$

Take $V(\xi) = \xi^T P \xi$ and find a positive constant c such that

$$\Omega_c \stackrel{\text{def}}{=} \{\xi \in R^{n+q} \mid V(\xi) \le c\} \subset \{\xi \in R^{n+q} \mid e \in N_\theta\} \ . \qquad (47)$$

Assumption 6: There exist a scalar nonnegative locally Lipschitz function $\rho(e)$ and a positive constant k, both known, such that

$$\|F(e, \nu, \theta) - G(e, \nu, \theta)G_0^{-1}(e, \nu)F_0(e, \nu) + [G(e, \nu, \theta)G_0^{-1}(e, \nu) - 1]K\xi\| \le \rho(\xi) \qquad (48)$$

and

$$G(e, \nu, \theta)G_0^{-1}(e, \nu) \ge k > 0 \qquad (49)$$

for all $e \in \Omega_c$, all $\nu \in \Lambda$, and all $\theta \in \Theta$.

Consider now the state feedback control

$$u = \psi(\sigma, e, \nu) = G_0^{-1}(e, \nu)\left[-F_0(e, \nu) + K\xi - \frac{1}{k}\phi(\xi)\right] \qquad (50)$$

where

$$\phi(\xi) = \begin{cases} \eta(\xi)\frac{w}{\|w\|}, & \text{if } \eta(\xi)\|w\| \geq \mu > 0 \\ \\ \eta^2(\xi)\frac{w}{\mu}, & \text{if } \eta(\xi)\|w\| < \mu \end{cases}, \qquad (51)$$

$\eta(\xi) \geq \rho(\xi)$, and $w = 2B^T P\xi$. Using $V(\xi)$ as a Lyapunov function candidate for the closed-loop system

$$\dot{\xi} = (A + BK)\xi + B[F - GG_0^{-1}F_0 + (GG_0^{-1} - 1)K\xi] - \frac{1}{k}BGG_0^{-1}\phi(\xi), \quad (52)$$

it can be shown that

$$\dot{V} \leq -\|\xi\|^2 + \frac{\mu}{4}, \quad \forall \, \xi \in \Omega_c .$$

Taking $\beta = \alpha\mu\lambda_{max}(P)/4$, for some $\alpha > 1$, and choosing μ small enough to satisfy $\beta < c$, we conclude that

$$\Omega_\beta = \{V(\xi) \leq \beta\} \subset \Omega_c$$

is a positively invariant set and all trajectories starting in Ω_c enter Ω_β in finite time. Let us note here that Properties I and II are satisfied. If we augment the closed-loop equation (52) with equation (43), we can see that Property I is satisfied with $U_1 = V$ and $U_2 = v^T M^{-T} M^{-1} v$, where M is the modal matrix of S. The derivative of U_1 satisfies inequality of the form (8), while $\dot{U}_2 = 0$; since $M^{-1}SM$ is a skew symmetric matrix. Notice that $\dot{U}_2 = 0$ irrespective of the right-hand side of (52). Property II is satisfied with $\Delta = \Omega_\beta$.

Inside Ω_β, we want $\phi(\xi)$ to behave eventually as a linear function of w. This can be achieved by taking

$$\eta(\xi) = \max\{\eta_0, \rho(\xi)\} \qquad (53)$$

where

$$\eta_0 \geq \mu_1 + \max_{\xi\in\Omega_\beta}\rho(\xi) + \max_{\xi\in\Omega_\beta}\left[\frac{\|B^T P(A + BK)\xi\|}{B^T PB}\right], \quad \mu_1 > 0 .$$

The inequality $\eta_0 > \max_{\xi\in\Omega_\beta}\rho(\xi)$ implies that $\eta(\xi) = \eta_0$ for all $\xi \in \Omega_\beta$. Moreover, it can be shown that there are positive constants c_1 and $\mu_2 < \mu$ such that

$$w^T \dot{w} \leq -c_1 < 0$$

for all $\xi \in \Omega_\beta$ and $\|w\| \geq \mu_2/\eta_0$. Hence, the set $\bar{\Omega}_\mu$, defined by

$$\bar{\Omega}_\mu = \Omega_\beta \cap \left\{\|w\| \leq \frac{\mu_2}{\eta_0}\right\}$$

is a positively invariant set and $\xi(t)$ enters $\bar{\Omega}_\mu$ in finite time. Thus, we revise our earlier statement on Property II and take $\Delta = \bar{\Omega}_\mu$.
Inside $\bar{\Omega}_\mu$ the control u is given by

$$u = \psi(\sigma, e, \nu) = -G_0^{-1}(e, \nu)F_0(e, \nu) + G_0^{-1}(e, \nu)K\xi - \frac{2\eta_0^2}{k\mu}G_0^{-1}(e, \nu)B^T P\xi \quad (54)$$

and the closed-loop system is described by

$$\dot{\sigma} = S\sigma + JCe \ , \tag{55}$$

$$\dot{e} = BK_1\sigma + (A + BK_2)e - \frac{2\eta_0^2}{k\mu}BGG_0^{-1}(B^T P_{12}^T\sigma + B^T P_{22}e)$$

$$+ B[F - GG_0^{-1}F_0 + (GG_0^{-1} - 1)(K_1\sigma + K_2e)] \ , \tag{56}$$

$$\dot{v} = Sv \ , \tag{57}$$

where K and P are partitioned as

$$K = [\, K_1 \ K_2 \,], \quad P = \begin{bmatrix} P_{11} & P_{12} \\ P_{12}^T & P_{22} \end{bmatrix} \ .$$

Our next task is to show that the closed-loop system (55)–(57) has a zero-error invariant manifold of the form $\Gamma = \{\sigma = L(\theta)v, \ e = 0\}$. This can be shown in two steps. First, it can be verified that there is a matrix $L(\theta)$ that satisfies the equations

$$L(\theta)S = SL(\theta); \quad \Gamma_1(\theta) = \frac{2\eta_0^2}{k\mu}B^T P_{12}^T L(\theta) - K_1 L(\theta) \ . \tag{58}$$

Then, it can be easily seen that Γ is an invariant manifold of the closed-loop system (55)–(57).

To study attractivity of this manifold, let

$$\tilde{\xi} = \begin{bmatrix} \sigma - L(\theta)v \\ e \end{bmatrix}$$

and rewrite the closed-loop equations (55)–(56) as

$$\dot{\tilde{\xi}} = (A + BK)\tilde{\xi} - \frac{2\eta_0^2}{k\mu}BG(e, \nu, \theta)G_0^{-1}(e, \nu)B^T P\tilde{\xi}$$

$$+ B[\delta(e, \nu, \theta) - (G(e, \nu, \theta)G_0^{-1}(e, \nu) - 1)K\tilde{\xi}] \tag{59}$$

where

$$\delta(e, \nu, \theta) = G(e, \nu, \theta)G_0^{-1}(e, \nu)[G_0(e, \nu)G^{-1}(e, \nu, \theta)F(e, \nu, \theta) - F_0(e, \nu) - \Gamma_1(\theta)v] \ .$$

In view of Assumption 5, we have $\delta(0, \nu, \theta) = 0$. Thus, there exists a positive constant k_1 such that the inequality

$$\|\delta(e, \nu, \theta) - (G(e, \nu, \theta)G_0^{-1}(e, \nu) - 1)K\tilde{\xi}\| \le k_1\|\tilde{\xi}\| \tag{60}$$

is satisfied for all $\xi \in \bar{\Omega}_\mu$. Using $\tilde{V} = \tilde{\xi}^T P\tilde{\xi}$ as a Lyapunov function candidate, it can be shown that

$$\dot{\tilde{V}} < -k_2\|\tilde{\xi}\|^2, \quad \forall \, \xi \in \bar{\Omega}_\mu \ ,$$

for some $k_2 > 0$, provided $\eta_0 > k_1\sqrt{\mu}/2$. Thus Property III is satisfied with $\mathcal{U} = \bar{\Omega}_\mu$ and $\Gamma = \{\sigma = L(\theta)v, \ e = 0\}$, and we conclude that every trajectory in $\bar{\Omega}_\mu$ approaches the manifold $\{\sigma = L(\theta)v, \ e = 0\}$ as $t \to \infty$.

Example 3 [4]: Adaptive Control

We consider a single-input–single-output nonlinear system represented globally by the nth-order differential equation

$$y_o^{(n)} = f_0(\cdot) + \sum_{i=1}^{p} f_i(\cdot)\theta_i + \left(g_0 + \sum_{i=1}^{p} g_i\theta_i\right)\eta^{(m)}, \quad \eta = \sigma(y_o)u \,, \tag{61}$$

where u is the control input, y_o is the measured output, $y_o^{(i)}$ denotes the ith derivative of y_o, and $m < n$. The nonlinearities f_i and σ are known smooth nonlinearities with $\sigma(y_o) \neq 0$ for all $y_o \in R$, and f_i could depend on $y_o, y_o^{(1)}, \cdots,$ $y_o^{(n-1)}$, $\eta, \eta^{(1)}, \cdots, \eta^{(m-1)}$, e.g.,

$$f_0(\cdot) = f_0(y_o, y_o^{(1)}, \cdots, y_o^{(n-1)}, \eta, \eta^{(1)}, \cdots, \eta^{(m-1)}) \,.$$

The constant parameters g_0 to g_p are known, while the constant parameters θ_1 to θ_p are unknown, but the vector $\theta = [\theta_1, \cdots, \theta_p]^T$ belongs to $\Theta = \{\theta^T\theta \leq k\}$, with a known constant $k > 0$.

Assumption 7: $(g_0 + \theta^T g) \neq 0 \ \forall \ \theta \in \Theta_\delta = \{\theta^T\theta \leq k + \delta\}$, for some $\delta > 0$.

We assume that $y_r(t)$ is bounded, has bounded derivatives up to the $(n+1)$th order. Let

$$\mathcal{Y}(t) = [y(t), \ y^{(1)}(t), \ \cdots, \ y^{(n-1)}(t)]^T \,,$$
$$\mathcal{Y}_r(t) = [y_r(t), \ y_r^{(1)}(t), \ \cdots, \ y_r^{(n-1)}(t)]^T \,,$$
$$\mathcal{Y}_R(t) = [y_r(t), \ y_r^{(1)}(t), \ \cdots, \ y_r^{(n-1)}(t), \ y_r^{(n)}(t)]^T \,,$$

and Y and Y_R be any given compact subsets of R^n and R^{n+1}, respectively. Our objective is to design the adaptive output feedback controller such that for all $\mathcal{Y}(0) \in Y$, for all $\mathcal{Y}_R(t) \in Y_R$, and for all $\theta \in \Theta$, all variables of the closed-loop system are bounded for all $t \geq 0$, and

$$\lim_{t \to \infty} |y_o(t) - y_r(t)| = 0 \,.$$

The compact sets Y, Y_R, and Θ are arbitrary, but we assume that they are known. In other words, our goal is to design a semiglobal adaptive controller.

We represent an extended version of (61) by a state-space model. We augment a series of m integrators at the input side of the system. We denote the states of these integrators by $z_1 = \eta$, $z_2 = \eta^{(1)}$, up to $z_m = \eta^{(m-1)}$, and set $v = \eta^{(m)}$ as the control input of the augmented system. The actual control input u can be calculated from η using $u = \eta/\sigma(y)$. By taking $x_1 = y$, $x_2 = y^{(1)}$, up to $x_n = y^{(n-1)}$, we can represent the augmented system by the state-space model

$$\left.\begin{aligned}
\dot{x}_i &= x_{i+1}, && 1 \leq i \leq n-1 \\
\dot{x}_n &= f_0(x, z) + \theta^T f(x, z) + (g_0 + \theta^T g)v \\
\dot{z}_i &= z_{i+1}, && 1 \leq i \leq m-1 \\
\dot{z}_m &= v \\
y &= x_1
\end{aligned}\right\} \tag{62}$$

where

$$x = [x_1, \cdots, x_n]^T, \quad z = [z_1, \cdots, z_m]^T ,$$
$$f = [f_1, \cdots, f_p]^T, \quad g = [g_1, \cdots, g_p]^T ,$$

Assumption 7 implies that the augmented system (62) is input-output linearizable by full state feedback for every $\theta \in \Theta$. It also guarantees that for every $\theta \in \Theta$, there is a globally defined normal form for (62). In particular, the change of variables

$$\zeta_i = z_i - \frac{x_{n-m+i}}{g_0 + \theta^T g}, \quad 1 \le i \le m \tag{63}$$

transforms the last m state equations of (62) into

$$\left. \begin{array}{l} \dot{\zeta}_i = \zeta_{i+1}, \quad 1 \le i \le m-1 \\[2mm] \dot{\zeta}_m = - \left. \frac{f_0(x,z) + \theta^T f(x,z)}{g_0 + \theta^T g} \right|_{z_i = \zeta_i + x_{n-m+i}/(g_0 + \theta^T g)} \end{array} \right\} \tag{64}$$

which, together with the first n state equations of (62), define the global normal form. Setting $x = 0$ in (64) results in the zero dynamics of (62).

Assumption 8: For every $\theta \in \Theta$, the system (64) has the following properties:

- There exists a positive definite, radially unbounded Lyapunov function $W(\zeta)$ such that

$$\dot{W} \le -\kappa_1(\|\zeta\|), \quad \forall \, \|\zeta\| \ge \kappa_2(\|x\|) ,$$

for all $x \in R^n$ and $\zeta \in R^m$, for some class $\mathcal{K}$ functions $\kappa_1(\cdot)$ and $\kappa_2(\cdot)$. Consequently, for any bounded $x(t)$ and $z(0)$, the state $\zeta(t)$ is bounded.
- Let $\bar{\zeta}(t)$ be the solution when $x(t) = \mathcal{Y}_r(t)$ and $z(0) = 0$. For any bounded $x(t)$ such that $x(t) \to \mathcal{Y}_r(t)$ as $t \to \infty$, and any $z(0)$, the state $\zeta(t) \to \bar{\zeta}(t)$ as $t \to \infty$.

Using (63) to define $\bar{z}(t)$ as

$$\bar{z}_i(t) = \bar{\zeta}_i(t) + \frac{y_r^{(n-m+i-1)}(t)}{g_0 + \theta^T g}, \quad 1 \le i \le m , \tag{65}$$

it can be verified that $\bar{z}_1(t) = \bar{\eta}(t)$ is the solution of the mth-order differential equation that results from (61) when $y_o(t) \equiv y_r(t)$ and $\eta^{(i)}(0) = 0, 0 \le i \le m-1$. It can be also seen that $\bar{z}_i(t) = \bar{\eta}^{(i-1)}(t)$.

Assuming that the full state (x, z) is available for feedback, we proceed to design an adaptive state feedback controller. Taking

$$
\begin{array}{lll}
e_1 = y_o - y_r & = x_1 - y_r , \\
e_2 = \dot{y}_o - \dot{y}_r & = x_2 - \dot{y}_r , \\
\vdots & \\
e_n = y_o^{(n-1)} - y_r^{(n-1)} & = x_n - y_r^{(n-1)} ,
\end{array}
$$

and

$$e = [e_1, \ e_2, \ \cdots, \ e_n]^T \ ,$$

we rewrite the system (62) as

$$\begin{aligned}
\dot{e} &= Ae + b\{f_0(e + \mathcal{Y}_r, z) + \theta^T f(e + \mathcal{Y}_r, z) \\
&\quad + (g_0 + \theta^T g)v - y_r^{(n)}\} \ ,
\end{aligned} \tag{66}$$

$$\dot{z} = A_2 z + b_2 v \ , \tag{67}$$

where (A, b) and (A_2, b_2) are controllable canonical pairs. Choose a matrix K such that $A_m = A - bK$ is Hurwitz, and rewrite equation (66) as

$$\begin{aligned}
\dot{e} &= A_m e + b\{Ke + f_0(e + \mathcal{Y}_r, z) + \theta^T f(e + \mathcal{Y}_r, z) \\
&\quad + (g_0 + \theta^T g)v - y_r^{(n)}\} \ .
\end{aligned} \tag{68}$$

Let $P = P^T > 0$ be the solution of the Lyapunov equation

$$PA_m + A_m^T P = -Q, \quad Q = Q^T > 0 \ ,$$

and consider the Lyapunov function candidate

$$V = e^T Pe + \tfrac{1}{2}\tilde{\theta}^T \Gamma_1^{-1} \tilde{\theta}$$

where $\Gamma_1 = \Gamma_1^T > 0$, $\tilde{\theta} = \hat{\theta} - \theta$, and $\hat{\theta}$ is an estimate of θ to be determined by the parameter adaptation rule. The derivative of V along the trajectories of the system is given by

$$\dot{V} = -e^T Qe + 2e^T Pb[f_0 + \theta^T f + (g_0 + \theta^T g)v + Ke - y_r^{(n)}] + \tilde{\theta}^T \Gamma_1^{-1} \dot{\hat{\theta}} \ .$$

Taking

$$v = \frac{-Ke + y_r^{(n)} - f_0(e + \mathcal{Y}_r, z) - \hat{\theta}^T f(e + \mathcal{Y}_r, z)}{g_0 + \hat{\theta}^T g} \overset{\text{def}}{=} \psi(e, z, \mathcal{Y}_R, \hat{\theta}) \tag{69}$$

and setting

$$\phi = 2e^T Pbw = \phi(e, z, \mathcal{Y}_R, \hat{\theta}) \tag{70}$$

where

$$w(t) = f(e(t) + \mathcal{Y}_r(t), z(t)) + g\psi(e(t), z(t), \mathcal{Y}_R(t), \hat{\theta}(t)) \ , \tag{71}$$

we can rewrite the expression for $\dot{V}$ as

$$\dot{V} = -e^T Qe + \tilde{\theta}^T \Gamma_1^{-1}[\dot{\hat{\theta}} - \Gamma_1 \phi] \ .$$

Taking the parameter adaptation rule as

$$\dot{\hat{\theta}} = \text{Proj}(\hat{\theta}, \phi) \tag{72}$$

where

$$\text{Proj}(\hat{\theta}, \phi) = \begin{cases} \Gamma_1 \phi, & \text{if } \hat{\theta}^T \hat{\theta} < k \text{ or} \\ & \text{if } \hat{\theta}^T \hat{\theta} \geq k \text{ and } \hat{\theta}^T \Gamma_1 \phi \leq 0 \\ \Gamma_1 \phi - \frac{(\hat{\theta}^T \hat{\theta} - k) \hat{\theta}^T \Gamma_1 \phi}{\delta \hat{\theta}^T \Gamma_1 \hat{\theta}} \Gamma_1 \hat{\theta}, & \text{otherwise} \end{cases} \tag{73}$$

ensures that

$$\tilde{\theta}^T \Gamma_1^{-1} [\dot{\hat{\theta}} - \Gamma_1 \phi] \leq 0, \text{ which implies } \dot{V} \leq -e^T Q e \ , \tag{74}$$

and

$$\hat{\theta}^T \text{Proj}(\hat{\theta}, \phi) \leq 0, \quad \text{on } \{\hat{\theta}^T \hat{\theta} = k + \delta\} \ . \tag{75}$$

Inequality (75) shows that $\hat{\theta}(0) \in \Theta \Rightarrow \hat{\theta}(t) \in \Theta_\delta, \ \forall \ t \geq 0$. Note that (75) is satisfied irrespective of the vector ϕ on the right-hand side of (72). To establish boundedness of all state variables, we assume that all the initial conditions are bounded. In particular, $\hat{\theta}(0) \in \Theta$, $e(0) \in E_0$, and $z(0) \in Z_0$, where E_0 and Z_0 are compact sets. The sets E_0 and Z_0 can be chosen large enough to cover any given bounded initial conditions, but once they are chosen we cannot allow initial conditions outside them. From (63) we can find a a constant $c_0 > 0$ such that $\|\zeta(0)\| \leq c_0$. We have already seen that $\hat{\theta}(t) \in \Theta_\delta$ for all $t \geq 0$. Let

$$c_1 = \max_{e \in E_0} e^T P e, \quad c_2 = \max_{\theta \in \Theta, \hat{\theta} \in \Theta_\delta} \frac{1}{2} (\hat{\theta} - \theta)^T \Gamma_1^{-1} (\hat{\theta} - \theta) \ .$$

and $c_3 > c_1 + c_2$. For all $(e, \tilde{\theta}) \in \{V \leq c_3\}$, we have $e \in E \overset{\text{def}}{=} \{e^T P e \leq c_3\}$. By the boundedness of $\mathcal{Y}_r(t)$, there exists a constant c_4 such that $e \in E \Rightarrow \|x\| \leq c_4$. Let $c_5 = \max\{c_0, \kappa_2(c_4)\}$, and take $c_6 > \max_{\|\zeta\| \leq c_5} W(\zeta)$. Define Ω as

$$\Omega = \{V \leq c_3\} \cap \{\hat{\theta} \in \Theta_\delta\} \cap \{W \leq c_6\} \ . \tag{76}$$

It can be seen that

$$\dot{V} \leq -c_7 < 0, \quad \text{on } \Omega \cap \{V = c_3\} \tag{77}$$

and

$$\dot{W} \leq -c_8 < 0, \quad \text{on } \Omega \cap \{W = c_6\} \tag{78}$$

for some positive constants c_7 and c_8. Inequalities (75), (77), and (78) show that Property I is satisfied and Ω is positively invariant. Thus $e(t)$, $\hat{\theta}(t)$, and $z(t)$ are bounded for all $t \geq 0$. With all signals bounded, we conclude from $\dot{V} \leq -e^T Q e$ and the invariance theorem that

$$e(t) \to 0 \quad \text{as } t \to \infty \ . \tag{79}$$

It follows from Assumption 8 that

$$z(t) \to \bar{z}(t) \quad \text{as } t \to \infty \ . \tag{80}$$

Furthermore, since $y_r^{(n+1)}$ is bounded, it can be shown that

$$\dot{e}(t) \to 0 \quad \text{as} \quad t \to \infty \ . \tag{81}$$

Using (79), (80), and (81) in (68) we obtain

$$\psi \to \frac{1}{g_0 + \theta^T g} \left[y_r^{(n)}(t) - f_0(\mathcal{Y}_r(t), \bar{z}(t)) - \theta^T f(\mathcal{Y}_r(t), \bar{z}(t)) \right] \quad \text{as} \quad t \to \infty \ .$$

Therefore

$$w(t) \to w_r(t) \quad \text{as} \quad t \to \infty \tag{82}$$

where

$$w_r(t) = f(\mathcal{Y}_r(t), \bar{z}(t)) + \frac{1}{g_0 + \theta^T g} g \left[y_r^{(n)}(t) - f_0(\mathcal{Y}_r(t), \bar{z}(t)) - \theta^T f(\mathcal{Y}_r(t), \bar{z}(t)) \right] \ . \tag{83}$$

We show next that under an additional persistence of excitation condition, the parameter estimate $\hat{\theta}(t)$ approaches the true unknown parameter vector θ. The closed-loop system under state feedback can be written as

$$\dot{e} = A_m e - b\tilde{\theta}^T w(t) \ , \tag{84}$$

$$\dot{\tilde{\theta}} = 2\hat{\Gamma}_1(t)w(t)b^T Pe \ , \tag{85}$$

$$\dot{z} = A_2 z + b_2 \psi \ , \tag{86}$$

where $\hat{\Gamma}_1(t)$ is a continuous function of $\hat{\theta}(t)$ determined by (72). We rewrite equations (84) and (85) as the homogeneous linear time-varying system

$$\begin{bmatrix} \dot{e} \\ \dot{\tilde{\theta}} \end{bmatrix} = \begin{bmatrix} A_m & -bw^T(t) \\ 2\hat{\Gamma}_1(t)w(t)b^T P & 0 \end{bmatrix} \begin{bmatrix} e \\ \tilde{\theta} \end{bmatrix} \ . \tag{87}$$

The derivative of $V = e^T Pe + \frac{1}{2}\tilde{\theta}^T \Gamma_1 \tilde{\theta}$ along the trajectories of (87) satisfies $\dot{V} \leq 0$. By an argument similar to the one used in the traditional adaptive control theory it can be shown that if $w(t)$ is persistently exciting then

$$\left\| \begin{bmatrix} e(t) \\ \tilde{\theta}(t) \end{bmatrix} \right\| \leq ke^{-\gamma t} \left\| \begin{bmatrix} e(0) \\ \tilde{\theta}(0) \end{bmatrix} \right\|, \quad \forall \, t \geq 0 \ , \tag{88}$$

for some $k > 0$ and $\gamma > 0$. The persistence of excitation condition on $w(t)$ is not a verifiable condition since the signal $w(t)$ is generated during the operation of the adaptive control system and is not known a priori. In view of the limit relationships (82), we make the following assumption.

Assumption 9: $\forall \, \theta \in \Theta$, the signal $w_r(t)$ is persistently exciting.

Assumption 9 is independent of the operation of the adaptive control system. It can be verified directly from the given data. Assumption 9 and the limit relationship (82) guarantee that $w(t)$ is persistently exciting. Inequality (88) shows that Property III is satisfied with $\mho = \Omega$ and $\Gamma = \{e = 0, \tilde{\theta} = 0\}$.

In preparation for the output feedback controller, we saturate the control v and the vector ϕ outside the domain of interest. We know that $e(t) \in E$, $\zeta(t) \in \{W(\zeta) \leq c_6\}$, and $\theta(t) \in \Theta_\delta$, for all $t \geq 0$. Using (63), we can find a compact set Z such that $z(t) \in Z$ for all $t \geq 0$. The control input $v = \psi(e, z, \mathcal{Y}_R, \hat{\theta})$ defined by (69) and the vector $\phi(e, z, \mathcal{Y}_R, \hat{\theta})$ defined by (70) are smooth functions of $e, \mathcal{Y}_R$, z, and $\hat{\theta}$. Hence they are bounded on compact sets of these variables. Let

$$S \geq \max|\psi(e, z, \mathcal{Y}_R, \hat{\theta})|, \quad S_i \geq \max|\phi_i(e, z, \mathcal{Y}_R, \hat{\theta})| \ ,$$

where the maximization is taken over all $e \in E$, $z \in Z$, $\mathcal{Y}_R \in Y_R$, $\hat{\theta} \in \Theta_\delta$. Define the saturated functions ψ^s and ϕ^s by

$$\psi^s(e, z, \mathcal{Y}_R, \hat{\theta}) = S \operatorname{sat}\left(\frac{\psi(e, z, \mathcal{Y}_R, \hat{\theta})}{S}\right) \ , \tag{89}$$

$$\phi_i^s(e, z, \mathcal{Y}_R, \hat{\theta}) = S_i \operatorname{sat}\left(\frac{\phi_i(e, z, \mathcal{Y}_R, \hat{\theta})}{S_i}\right), \quad 1 \leq i \leq p \ . \tag{90}$$

For all $e(0) \in E_0$, $z(0) \in Z_0$, and $\hat{\theta}(0) \in \Theta$, we have $|\psi| \leq S$ and $|\phi_i| \leq S_i$ for all $t \geq 0$. Hence the saturation functions will not be effective and the state feedback adaptive controller with ψ and ϕ replaced by ψ^s and ϕ^s will result in the same performance.

The foregoing three examples illustrate Properties I to III. Example 1 is a simple case where both Properties I and II are satisfied with a single Lyapunov function. Example 2 shows a more complex form of Property II, where Δ is defined using two functions: $V_1 = V$ and $V_2 = w^T w$. Example 3 shows a more complex form of Property I, where Ω is defined using three functions: $U_1 = V$, $U_2 = \hat{\theta}^T \hat{\theta}$, and $U_3 = W$. As for Property III, Example 1 shows a case where the positively invariant set Γ is an equilibrium point and $\mho = \Delta$, Example 2 shows a case where Γ is the manifold of zero tracking error and $\mho = \Delta$, and in Example 3, Γ is the manifold of zero tracking error and zero parameter error and $\mho = \Delta$.

4 High-Gain Observer

To implement the state feedback control using output feedback, we need to estimate e. With the goal of recovering the performance achieved under state feedback, we use the high-gain observer

$$\dot{\hat{e}}_i = \hat{e}_{i+1} + \frac{\alpha_i}{\epsilon^i}(y - \hat{e}_1), \quad 1 \leq i \leq r - 1 \ ,$$

$$\dot{\hat{e}}_r = \frac{\alpha_r}{\epsilon^r}(y - \hat{e}_1) + \hat{f}_1(\vartheta, \hat{e}, y, \zeta, d(t)) + \hat{g}_1(\vartheta, \hat{e}, y, \zeta, d(t))\psi(\vartheta, \hat{e}, \zeta, d(t)) \ , \tag{91}$$

where ϵ is a small positive parameter to be specified. The positive constants α_i are chosen such that the roots of

$$s^r + \alpha_1 s^{r-1} + \cdots + \alpha_{r-1} s + \alpha_r = 0 \tag{92}$$

have negative real parts. The functions $\hat{f}_1$ and $\hat{g}_1$ are globally bounded functions of $\hat{e}$. They are allowed to depend on $d(t)$ so that they may be functions of the reference signal and its derivatives. We think of $\hat{f}_1$ and $\hat{g}_1$ as nominal models of f_1 and g_1, respectively. If no such nominal models are available, we can take one or both to be zero.

To eliminate peaking in the implementation of the observer, let

$$\hat{e}_i = \frac{q_i}{\epsilon^{i-1}}, \quad 1 \leq i \leq r \tag{93}$$

Then

$$\epsilon \dot{q}_i = q_{i+1} + \alpha_i (y - q_1), \quad 1 \leq i \leq r-1 \ ,$$

$$\tag{94}$$

$$\epsilon \dot{q}_r = \alpha_r (y - q_1) + \epsilon^r [\hat{f}_1(\vartheta, \hat{e}, y, \zeta, d(t)) + \hat{g}_1(\vartheta, \hat{e}, y, \zeta, d(t)) \psi(\vartheta, \hat{e}, \zeta, d(t))] \ .$$

The system (94) is a standard singularly perturbed model and will not exhibit peaking if the input y and the initial conditions are bounded functions of ϵ. In summary, the output feedback controller is given by

$$\dot{\vartheta} = \phi(\vartheta, \hat{e}, \zeta, d(t)) \ , \tag{95}$$

$$u = \psi(\vartheta, \hat{e}, \zeta, d(t)) \ , \tag{96}$$

where $\hat{e}$ is given by (93)–(94). We note that in (95) and (96) we may replace $\hat{e}_1$ on the right-hand side by the measured signal $e_1 = y$. By taking

$$\eta_i = \frac{e_i - \hat{e}_i}{\epsilon^{r-i}}, \quad 1 \leq i \leq r \ , \tag{97}$$

and $\eta = [\eta_1, \ldots, \eta_r]^T$, we represent the closed-loop system in the standard singularly perturbed form

$$\dot{e} = Ae + B[f_1(e, z, d(t)) + g_1(e, z, d(t)) \psi(\vartheta, e - D(\epsilon)\eta, \zeta, d(t))] \ ,$$
$$\dot{z} = f_2(e, z, \psi(\vartheta, e - D(\epsilon)\eta, \zeta, d(t)), d(t)) \ ,$$
$$\dot{\vartheta} = \phi(\vartheta, e - D(\epsilon)\eta, \zeta, d(t)) \ ,$$
$$\epsilon \dot{\eta} = (A - HC)\eta + \epsilon B\{f_1(\cdot) - \hat{f}_1(\cdot) + [g_1(\cdot) - \hat{g}_1(\cdot)]\psi(\cdot)\} \ ,$$

where $H = [\alpha_1, \ldots, \alpha_r]^T$ and $D(\epsilon)$ is a diagonal matrix with ϵ^{r-i} as the i-th diagonal element. The characteristic equation of $(A - HC)$ is (92); hence it is Hurwitz. We rewrite the closed-loop state equations in the compact form

$$\dot{x} = F(x, D(\epsilon)\eta, d(t)) \ ,$$

$$\tag{98}$$

$$\epsilon \dot{\eta} = A_o \eta + \epsilon G(x, D(\epsilon)\eta, d(t)) \ ,$$

where $A_o = A - HC$, and $F(x, 0, d(t)) = F_s(x, d(t))$. The singularly perturbed system (98) has an exponentially stable boundary-layer model, and its reduced model is the closed-loop system under state feedback (7). The ϵ-dependent scaling (97) causes an impulsive-like behavior in η as $\epsilon \to 0$, but since η enters the slow equation through bounded functions, the slow variables do not exhibit a similar impulsive like behavior.

5 Performance Recovery

We will now show that the output-feedback closed-loop system (98) recovers Properties I to III of the state-feedback closed-loop system (7), for sufficiently small ϵ. We state the recovery of these three properties in three theorems. Before we state the theorem we define a set Ω_o by

$$\Omega_o = \bigcap_{i=1,\ldots,p} \{U_i(x) \le b_i\}$$

where $b_i < c_i$ for all $i = 1, \ldots, p$, or if U_i satisfies (9) for some i we can take $b_i = c_i$. Let Q be any compact subset of R^r.

Theorem 3. *Suppose that System (7) possesses Property I. Then there exists $\epsilon_1^* > 0$, dependent on Ω_o and Q, such that for all $\epsilon \in (0, \epsilon_1^*]$, and all $(x(0), q(0)) \in \Omega_o \times Q$, the solution of (98) satisfies*

$$x(t, \epsilon) \in \Omega, \quad \forall\, t \ge 0 , \tag{99}$$

and

$$\eta(t, \epsilon) = O(\epsilon), \quad \forall\, t \ge T(\epsilon) > 0 , \tag{100}$$

where $T(\epsilon) \to 0$ as $\epsilon \to 0$.

Proof. For all $x \in \Omega$ we have

$$\|F(x, D(\epsilon)\eta, d(t)) - F(x, 0, d(t))\| \le L_1 \|\eta\| , \tag{101}$$

$$\|F(x, D(\epsilon)\eta, d(t))\| \le k_1 , \tag{102}$$

$$\|G(x, D(\epsilon)\eta, d(t))\| \le k_2 , \tag{103}$$

for some nonnegative constants L_1, k_1, and k_2. Let P be the positive definite solution of the Lyapunov equation

$$PA_o + A_o^T P = -I .$$

We start by showing that there exist positive constants ρ and ϵ_1 (dependent on ρ) such that the set $\Lambda \overset{\text{def}}{=} \Omega \times \{\eta^T P\eta \le \rho\epsilon^2\}$ is positively invariant for all $\epsilon \le \epsilon_1$. This can be shown by taking $\nu = \eta^T P\eta$ and verifying that

$$\dot{U}_i\Big|_{(98)} \le \dot{U}_i\Big|_{(7)} + \epsilon k_3, \quad \text{for } (x, \eta) \in \Omega \cap \{U_i(x) = c_i\} \times \{\nu \le \rho\epsilon^2\} ,$$

$$\dot{\nu}\Big|_{(98)} \le -\frac{1}{\epsilon}\|\eta\|^2 + 2\|\eta\|\|P\|k_2, \quad \text{for } (x, \eta) \in \Omega \times \{\nu \le \rho\epsilon^2\} ,$$

where $k_3 = L_1 L_2 \sqrt{\rho/\lambda_{min}(P)}$ and L_2 is a Lipschitz constant for U_i over Ω. Taking $\rho = 16 k_2^2 \|P\|^3$, and $\epsilon_1 = \min_{i=1,\ldots,p} \{\beta_i\}/k_3$, it can be seen that

$$\dot{U}_i\Big|_{(98)} \leq 0, \quad \text{for } (x, \eta) \in \Omega \cap \{U_i(x) = c_i\} \times \{v \leq \rho\epsilon^2\} , \tag{104}$$

$$\dot{v}|_{(98)} \leq 0, \quad \text{for } (x, \eta) \in \Omega \times \{v = \rho\epsilon^2\} , \tag{105}$$

for all $\epsilon \leq \epsilon_1$. From (104) and (105) we conclude that Λ is positively invariant. In the case that U_i satisfies (9) instead of (8), we arrive at the same conclusion by noting that (9) is still satisfied for (98) since it is independent of the variable e in $\psi(\cdot)$ and $\phi(\cdot)$.

Now we consider the initial state $(x(0), \eta(0))$ where $x(0) \in \Omega_o$ and $\eta(0)$ satisfies $\|\eta(0)\| \leq k/\epsilon^{(r-1)}$, for some nonnegative constant k dependent on Ω_o and Q. Using (102), it can be shown that $\|x(t, \epsilon)\| \leq \|x(0)\| + k_1 t$, for all $x(t, \epsilon) \in \Omega$. Since $x(0) \in \Omega_o$, there exists a finite time T_1, independent of ϵ, such that $x(t, \epsilon) \in \Omega$, for all $t \in [0, T_1]$. During this time interval we have

$$\dot{v}|_{(98)} \leq -\frac{1}{2\epsilon} \|\eta\|^2, \quad \text{for } v \geq \rho\epsilon^2$$

Therefore

$$v(t, \epsilon) \leq \frac{\gamma_2}{\epsilon^{2(r-1)}} e^{-\gamma_1 t/\epsilon}$$

where $\gamma_1 = 1/2\|P\|$ and $\gamma_2 = k^2\|P\|$. Choose ϵ_2 small enough that

$$T(\epsilon) \stackrel{def}{=} \frac{\epsilon}{\gamma_1} \ln\left(\frac{\gamma_2}{\rho\epsilon^{2r}}\right) \leq \tfrac{1}{2} T_1$$

for all $0 < \epsilon \leq \epsilon_2$. We note that ϵ_2 exists since the left-hand side of the preceding inequality tends to zero as ϵ tends to zero. It follows that $v(T, \epsilon) \leq \rho\epsilon^2$, for all $0 < \epsilon \leq \epsilon_2$. Thus the trajectory is guaranteed to enter the set Λ during the interval $[0, T(\epsilon)]$ and will remain there for all $t \geq T(\epsilon)$.

Theorem 4. *Suppose that System (7) possesses Property II. Then there exists $\epsilon_2^* > 0$, dependent on Ω_o and Q, such that for all $\epsilon \in (0, \epsilon_2^*]$, and all $(x(0), q(0)) \in \Omega_o \times Q$, the solution of (98) satisfies (99)–(100) and there exists a finite time T_2, independent of ϵ, such that $x(t, \epsilon)$ reaches Δ before T_2.*

Proof. For $t \geq T(\epsilon)$, we have

$$\dot{V}_i\Big|_{(98)} \leq -\dot{V}_i\Big|_{(7)} + \epsilon k_4 \leq -\bar{\beta}_i + \epsilon k_4, \quad \forall \, x \in \Delta_i ,$$

for $i = 1, \ldots, q$, where $k_4 = L_1 L_3 \sqrt{\rho/\lambda_{min}(P)}$ and L_3 is a Lipschitz constant for V_i over Ω. Taking $\epsilon_2 = \min_{i=1,\ldots,q} \{\bar{\beta}_i\}/2k_4$ shows that

$$\dot{V}_i\Big|_{(98)} \leq -\frac{\bar{\beta}_i}{2}, \quad \forall \, x \in \Delta_i ,$$

for all $\epsilon < \epsilon_2$. Hence x reaches Δ in a finite time less than $T_2 \stackrel{def}{=} T_1/2 + 2 \sum_{i=1}^{q} (M_i/\bar{\beta}_i)$, where $M_i = \max_{x \in \Delta_i} \{V_i(x)\}$.

Theorem 5. *Suppose that System (7) possesses Property III and*

$$G(x, 0, d(t))|_{x \in \mathcal{U} \cap \Gamma} = 0 . \tag{106}$$

Then there exists $\epsilon_3^* > 0$, dependent on Ω_o and Q, such that for all $\epsilon \in (0, \epsilon_3^*]$, and all $(x(0), q(0)) \in \Omega_o \times Q$, the solution of (98) satisfies (99)–(100) and reaches the set $\mathcal{U} \times \{\nu \leq \rho\epsilon^2\}$ in finite time $\tau(\epsilon)$. From that time on, the solution satisfies the exponentially decaying bound

$$[|x|_\Gamma(t, \epsilon) + \|\eta(t, \epsilon)\|] \leq k_o e^{-\alpha_o(t-\tau)}[|x|_\Gamma(\tau) + \|\eta(\tau)\|] , \tag{107}$$

for some positive constants k_o and α_o, independent of ϵ.

Proof. We only need to prove (107) since the other claims follow from Theorems 3 and 4. The condition (106) ensures that

$$\|G(x, D(\epsilon)\eta, d(t))\| \leq k_5 |x|_\Gamma + k_6 \|\eta\|$$

for all $(x, \eta) \in \mathcal{U} \times \{\nu \leq \rho\epsilon^2\}$, with nonnegative constants k_5 and k_6 independent of ϵ. Taking $\nu_1 = W + a\sqrt{\nu}$, $a > 0$, and calculating the derivative of ν_1 along the trajectories of (98), it can be verified that there exist $a > 0$ and ϵ_3, dependent on a, such that for all $(x, \eta) \in \mathcal{U} \times \{\nu \leq \rho\epsilon^2\}$ and all $0 < \epsilon \leq \epsilon_3$,

$$k_7(|x|_\Gamma + \|\eta\|) \leq \nu_1 \leq k_8(|x|_\Gamma + \|\eta\|), \quad \dot{\nu}_1 \leq -\frac{\gamma}{2}\nu_1 ,$$

with some positive constants k_7 and k_8 independent of ϵ, which yields (107).

We note that if

$$f_1(e, z, d(t))|_{x \in \mathcal{U} \cap \Gamma} = 0, \text{ and } \psi(\vartheta, e, \zeta, d(t))|_{x \in \mathcal{U} \cap \Gamma} = 0 , \tag{108}$$

then the condition (106) can be always achieved by choosing $\hat{f}_1$ to satisfy

$$\hat{f}_1(\vartheta, \hat{e}, y, \zeta, d(t))\Big|_{x \in \mathcal{U} \cap \Gamma, \eta = 0} = 0 .$$

It can be easily checked that (108) is satisfied for Examples 1 to 3. In [1], an observer for Example 1 meets the condition (106) with $\hat{f}_1$ that vanishes at the origin. In [5] and [4], the observers for Examples 2 and 3 meet the condition (106) by taking $\hat{f}_1 = \hat{g}_1 = 0$, which reduces (91) to a linear observer.

Acknowledgment: This work was supported by the National Science Foundation under Grant no. ECS-9121501 and Grant no. ECS-9402187.

References

1. F. Esfandiari and H.K. Khalil. Output feedback stabilizaton of fully linearizable systems. *Int. J. Contr.*, 56:1007–1037, 1992.
2. A. Isidori. *Nonlinear Control Systems.* Springer-Verlag, New York, second edition, 1989.
3. M. Jankovic. Adaptive nonlinear output feedback tracking with a reduced order high-gain observer. 1993. Submitted for publication.
4. H.K. Khalil. Adaptive output feedback control of nonlinear systems represented by input-output models. In *Proc. IEEE Conf. on Decision and Control*, Orlando, FL, December 1994.
5. H.K. Khalil. Robust servomechanism output feedback controllers for a class of feedback linearizable systems. *AUTOMATICA*, 30(10):1587–1599, 1994.
6. H.K. Khalil and F. Esfandiari. Semiglobal stabilizaton of a class of nonlinear systems using output feedback. *IEEE Trans. Automat. Contr.*, 38(9):1412–1415, 1993.
7. H.K. Khalil and E.G. Strangas. Robust speed control of induction motors using position and current measurements. 1994. Submitted for publication.
8. Z. Lin and A. Saberi. Robust semi-global stabilization of minimum-phase input-output linearizable systems via partial state and output feedback. 1993. Submitted for publication.
9. N.A. Mahmoud. *Robust tracking control for nonlinear systems using output feedback.* PhD thesis, Michigan State University, East Lansing, MI, 1994.
10. N.A. Mahmoud and H.K. Khalil. Asymptotic regulation of minimum phase nonlinear systems using output feedback. In *Proc. American Control Conf.*, pages 1490–1494, San Francisco, CA, June 1993.
11. N.A. Mahmoud and H.K. Khalil. Robust tracking control of nonlinear systems represented by input-output models. In *Proc. IEEE Conf. on Decision and Control*, pages 1960–1965, San Antonio, TX, December 1993.
12. S. Oh and H.K. Khalil. Output feedback stabilization using variable structure control. In *Proc. IEEE Conf. on Decision and Control*, pages 401–406, San Antonio, TX, December 1993.
13. A. Teel and L. Praly. Tools for semiglobal stabilization by partial state and output feedback. 1992. To appear in SIAM.
14. A. Teel and L. Praly. Global stabilizability and observability imply semi-global stabilizability by output feedback. *Systems Contr. Lett.*, 22:313–325, 1994.
15. A. Tornambe'. Output feedback stabilization of a class of non-minimum phase nonlinear systems. *Systems Contr. Lett.*, 19:193–204, 1992.
16. T. Yoshizawa. *Stability Theory By Liapunov's Second Method.* The Mathematical Society of Japan, Tokyo, 1966.

This article was processed using the LaTeX macro package with LLNCS style

Fuzzy Logic Control: Hardware Implementations[1]

M. Jamshidi
University of New Mexico
Albuquerque, New Mexico, USA

Abstract. Fuzzy logic and fuzzy expert control systems are currently among the most active research and development areas of artificial intelligence. Thanks to tremendous technical advances and many industrial applications developed by the Japanese, fuzzy logic enjoys an unprecedented popularity. The object of this paper is to describe a number of hardware implementations of fuzzy control systems. The paper is organized in five sections: Section 1 provides a brief introduction to fuzzy control. Section 2 describes a real-time fuzzy control and tracker for a power system generation unit. Section 3 illustrates the experimental experiences for real-time fuzzy control of a model train on a circular path. A fuzzy PI controller for a thermoelectric cell used for non-chlorofluorocarbon (CFC) air conditioning and refrigeration systems is given in Section 4. Finally, Section 5 presents and discusses some conclusions.

1 Introduction To Fuzzy Control

Most of the current applications of fuzzy logic are fuzzy expert control systems. Fuzzy controllers are expert control systems that smoothly interpolate between otherwise crisp (or predicate logic-based) rules. Rules fire to continuous degrees and the multiple resultant actions are combined into an interpolated result. The processing of uncertain information and energy savings through the use of common-sense rules and natural language statements provide the basis for fuzzy logic control (or "fuzzy control," for short). The use of sensor data in practical control systems involves several tasks that are usually performed by a human in the decision loop, e.g., an astronaut adjusting the position of a satellite or putting it in the proper orbit, a driver adjusting an air conditioning unit, etc. The performance of all such tasks must be based on evaluation of pertinent data according to a set of rules that the human expert has learned from experience or training. Often, if not most of the time, these rules are not crisp (based on binary logic), i.e., they involve common sense and human judgment in the decision-making process. Such problems (or "judgment calls") can be addressed by a set of fuzzy variables and rules which, if calculated and executed properly, can make expert decisions.

The basic structure of a fuzzy controller takes the form of a set of IF-THEN rules whose *antecedents* (IF parts) and *consequents* (THEN parts) are membership functions. Consequents from different rules are numerically combined (typically union via MAX) and are then collapsed (typically taking the centroid or center of gravity of the combined distribution) to yield a single real number

---

[1]This paper is dedicated to Professor Peter V. Kokotovic on the occasion of his 60th birth year.

(binary) output. Within the framework of a fuzzy expert system, as with regular expert systems, typical rules can be the result of a human operator's knowledge, e.g.:

"If the *Temperature* is <u>Hot</u>, and Pressure is <u>High</u> then increase the *Current* to a <u>Medium</u> level.

In this rule, <u>Hot</u>, <u>High</u> and <u>Medium</u> are fuzzy (linguistic) variables. Such natural language rules can then be translated into typical computer language type statements such as:

IF (A is A1 and B is B1 and C is C1 and D is D1) THEN (E is E1 and F is F1)

Using a set of rules such as these, an entire finite number of rules can be derived in the form of natural language statements, as if a human operator were performing the controlling task. In any practical system, such as an air-conditioning unit, the user or operator often fine-tunes, tweaks, or adjusts the knobs until the desired cool (or hot) air can be felt with the desired speed. Such operator knowledge can be utilized in the design of a fuzzy controller for an air-conditioning unit system. One of the most common ways of designing fuzzy controller is through "fuzzy rule-based systems." The controller shows the processes of *fuzzification*, i.e., binary-to-fuzzy transformation and *defuzzification*, i.e., fuzzy-to-binary transformation.

There is an alternative method of implementing a fuzzy control regime which is similar to a conventional adaptive control law. A standard 1-, 2- or 3-term controller such as P, PI, PD, or PID can be used and then, through some adaptation loop, the conventional controller's gains K_p and K_d (for example, in a PD control law, $u = K_p y + K_d (dy/dt)$), can be adjusted through fuzzy rules[1].

Fuzzy logic corresponds to the logic of approximate reasoning. Like classical (crisp) logic, it can handle symbolic manipulation, but more importantly, it handles numerical data as well. Figure 1 shows a block diagram for a fuzzy control system. As shown, the analog plant's output is recognized by sensory elements and the corresponding voltage from the sensor is converted into digital words by an analog-to-digital (A/D) converter, often installed on the digital computer control board. Once the system's output is digitized, it can be changed into a finite number of fuzzy (linguistic) variables through the FUZZIFIER by assigning fuzzy membership functions to various ranges of the output's universe of discourse. Once the digital (crisp) values have been fuzzified, they can be used in the premises (IF parts) of the fuzzy inference engine within the fuzzy expert system, as shown. The resulting fuzzy decision (control) variables in linguistic form will be converted back to crisp values through the process of defuzzification (DE-FUZZ in Figure 1). The crisp control signal, still in digital form, is finally converted back into analog signal to be used to actuate the analog plant. Currently, several possible real-time control architectures can be implemented for real-time self-tuning fuzzy control systems. The implementations can be made by five different methods. These are 1) A loop-up table implementation, 2) fuzzy control software using standard processors, 3) using fuzzy coprocessors, 4) a standard MPU or MCU with a hardware fuzzy engine (chip), and 5) applications specific with a hardware fuzzy engine. The first three implementations can be realized on a PC. The last two implementations are embedded on a microcontroller card [2].

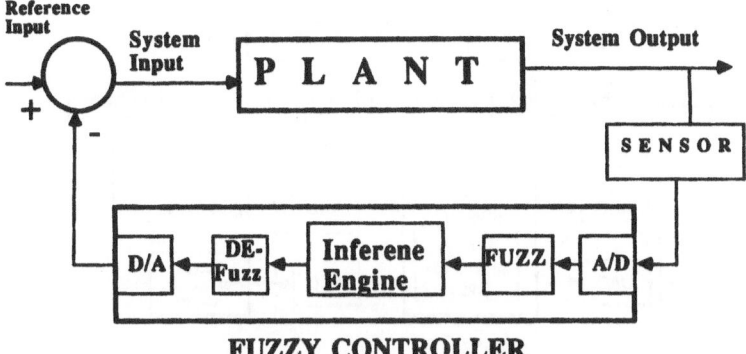

Figure 1. Block diagram for a typical fuzzy control system showing fuzzifier (FUZZ), defuzzifier (DE-FUZZ), and inference engine.

2 Fuzzy Control of A Power Generating System

A real-time fuzzy logic controller was designed for a synchronous generator driven by a DC-motor without knowledge of the systems transfer function. The purpose of the controller was to keep the output frequency and voltage steady at 60 Hz and 120 V_{rms}, respectively, with only the knowledge of the system input/output behavior. From this information the fuzzy logic rule base was formed and implemented into software with the Togai InfraLogic Shell. The controller was tested in real time for different loads on the generator. Comparisons with adaptive crisp controllers, underway at this point, have shown fuzzy control to be remarkable in dealing with load and line disturbances.

2.1 Introduction

In the last few years only a few papers[2-6] have been published on the use of fuzzy logic in power systems. For this reason, and because of the strong nonlinearity of the components in power systems, this project was an interesting challenge to the authors.

This section describes the use of fuzzy logic to control an electric power generator and its prime mover in real-time. The system is stand-alone and is therefore not dependent on the behavior of any other generator. Since a stiff power network is not present in the system, any load change will have a sizable effect on both the output frequency and the output voltage.

The purpose of the fuzzy logic controller is to keep the frequency at 60 Hz and the voltage at 120 V_{rms} for any changes in load, and to remove all setpoint deviations as fast as possible.

2.2 System Set-Up

Figure 2 shows a block diagram of the system set-up. The system consists of the following units:

1. DC-motor, which is the prime mover;
2. Synchronous machine; and
3. Fuzzy logic controller (computer, control circuits, and measurement circuits).

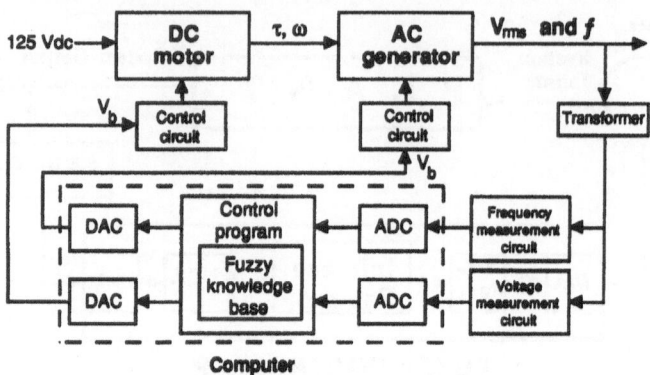

Figure 2. Block diagram of system set-up.

The DC-motor is connected in compound fashion, which provides a simple means of speed control by changes in the rheostat resistant. Because of the need to control the speed with a computer, the rheostat is replaced by a power transistor, which functions as a current controller. The base voltage of the transistor is supplied by the DAC (see Figure 3) in the computer.

The generator is driven by the DC-motor and supplies an output of three-phase 208 V_{rms} at 60 Hz. The change in voltage output is controlled by an excitation field, provided by an external DC source and, to some extent, by the speed of the rotor. The control of the excitation field is implemented with a transistor in a fashion similar to the DC-motor (Figure 3). The transistor controls the magnitude of the current passing through the field windings and, as before, the base voltage of the transistor is supplied by the DAC in the computer.

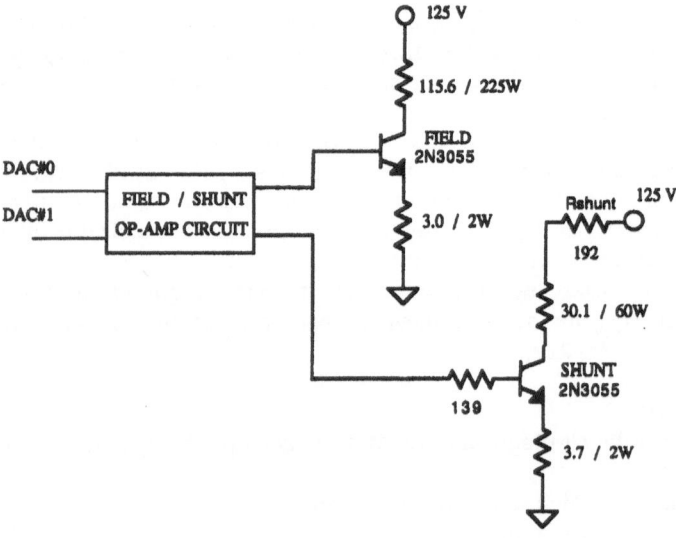

Figure 3. Control circuitry.

To be able to sense the behavior of the generator and its prime mover, the output voltage and output frequency are measured by measurement circuits (see Figure 4). These circuits give an output voltage in the range of 0-5 V_{DC}, proportional to the measured values. These DC-voltage signals are fed in to the computer through a pair of 8-bit ADCs.

The fuzzy logic part is software-based and the ADC readings are sent to the fuzzy logic subroutine to evaluate their membership values; those values are used to execute appropriate rules, which consequently produce the control values. The control values are then sent to the 8/12-bit DACs, which convert them to DC-voltage signals. These signals are fed to the base of the two control transistors, as mentioned above, which in fact closes the control loop.

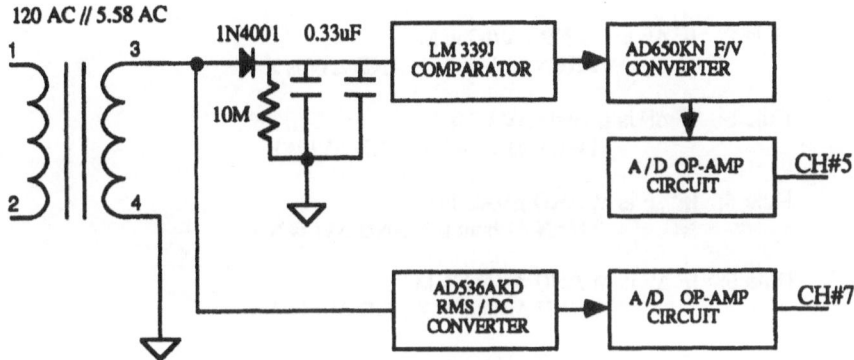

Figure 4. Measurement circuitry.

2.3 Fuzzy Knowledge Base

To be able to form a knowledge base, the input and output signals had to be given a working range. It was decided that the voltage could fluctuate in a range between 110 V_{rms} and 130 V_{rms}, and the frequency in a range between 55 Hz and 65 Hz. These ranges were then divided into sets, and each set given membership functions. The next step was to form a knowledge base. But before that was possible, measurement of the input/output behavior of the system was needed. From these measurements the fuzzy rules were created. Here is an example of the making of a fuzzy rule:

Changing load from infinity to 1000W:

Frequency: 60 Hz -> 57 Hz => $\Delta F = 3$ Hz
Voltage: 120 VAC -> 113 VAC => $\Delta V_{AC} = 7V_{AC}$

    Control actions to get frequency back to 60 Hz and voltage to 120 VAC:

       I_{shunt}: 0.310 VDC -> 0.261 VDC => $\Delta I_{shunt} = -0.049$ ADC
       V_f : 54.35 VDC -> 56.8 VDC => $\Delta V_f = 2.45$ VDC

Then the following rule is generated from the fuzzy sets:

IF ΔF is L **AND** ΔVAC is M
THEN DIshunt is N **AND** ΔVf is P

where L stands for the membership function Low, M for Medium, N for Negative, and P for Positive.

In this way, the entire set of rules was formed, based on the change in input and output. The final rule base is as follows. (L=Low, M=Medium, H=High, N=Negative, Z=Zero, P=Positive)

Rule 1: **IF** ΔF is L **AND** ΔVAC is L
          THEN ΔIshunt is N **AND** ΔVf is N

Rule 2: **IF** ΔF is L **AND** ΔVAC is M
          THEN ΔIshunt is N **AND** ΔVf is Z

Rule 3: **IF** ΔF is L **AND** ΔVAC is H
          THEN ΔIshunt is N **AND** ΔVf is P

Rule 4: **IF** ΔF is M **AND** ΔVAC is L
          THEN ΔIshunt is Z **AND** ΔVf is N

Rule 5: **IF** ΔF is M **AND** ΔVAC is M
          THEN ΔIshunt is Z **AND** ΔVf is Z

Rule 6: **IF** ΔF is M **AND** ΔVAC is H
          THEN ΔIshunt is Z **AND** ΔVf is P

Rule 7: **IF** ΔF is H **AND** DVAC is L
          THEN ΔIshunt is P **AND** ΔVf is N

Rule 8: **IF** ΔF is H **AND** ΔVAC is M
          THEN ΔIshunt is P **AND** ΔVf is Z

Rule 9: **IF** ΔF is H **AND** ΔVAC is H
          THEN ΔIshunt is P **AND** ΔVf is P

Membership functions are shown in Figures 5-7.

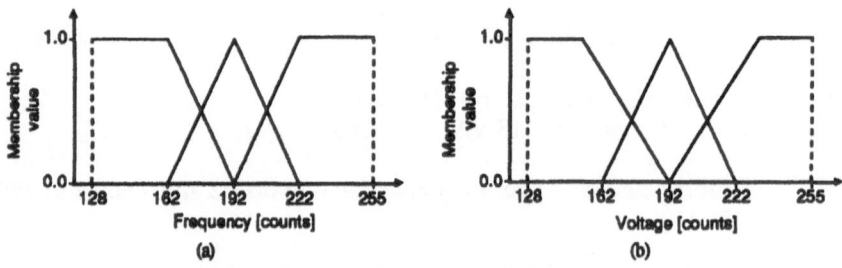

Figure 5. Membership functions for input values of controller.

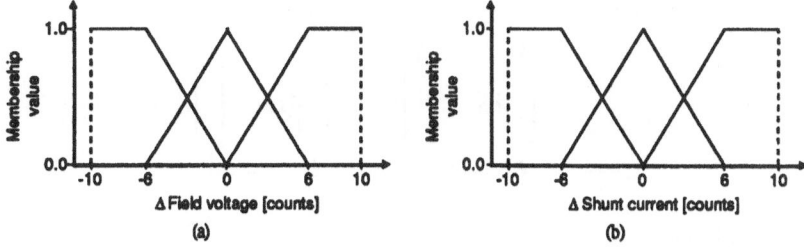

Figure 6. Membership functions for output values of controller using an 8-bit DAC.

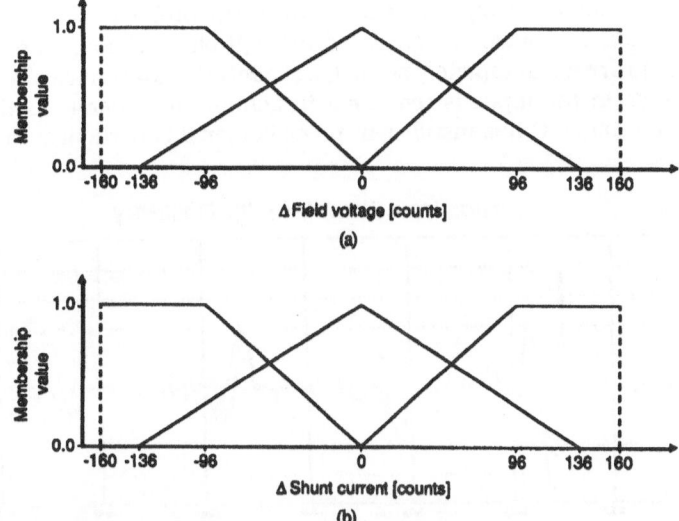

Figure 7. Membership functions for output values of controller using a 12-bit DAC.

The fuzzy sets, their membership functions, and the fuzzy rules were then assembled into a fuzzy knowledge base by using the TIL Shell in MS Windows. The TIL Shell output file was compiled to generate a C-subroutine, which was then linked to the main control program.

2.4 Experiment

To test the system, a load unit was connected to the output of the generator. This load unit is a box of parallel connected resistors, with each resistor activated by a switch. Figure 8 shows the load unit diagram.

Note that each additional load does not add resistance on the output, but does decrease the resistance and in that way demands power. Therefore the generator must increase its power output, which means more power to the DC-motor. Results from different load changes are shown in the following section.

140

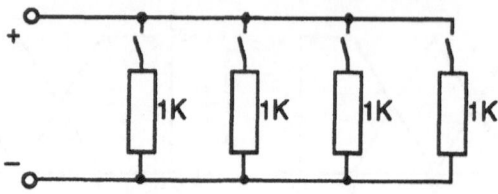

Figure 8. Load unit.

2.5 Results

The frequency and voltage system responses are shown in Figures 9 and 10, respectively. In Figure 9 (8-bit case), signals settle around their optimal values but their fluctuation is too big to be satisfactory. The problem is in the sensitivity of the control signals, shown in Figures 9 and 10 (8-bit case); i.e., even though the system outputs are not completely settled, the control signals have already settled because change in the input is too small to achieve any change in the output. Therefore, the 8-bit DAC has insufficient precision for this particular circuit.

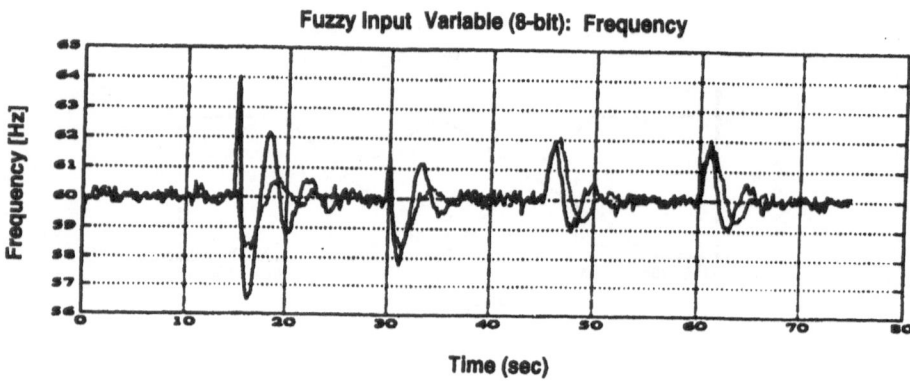

Figure 9. Frequency response for 8- and 12-bit DAC.

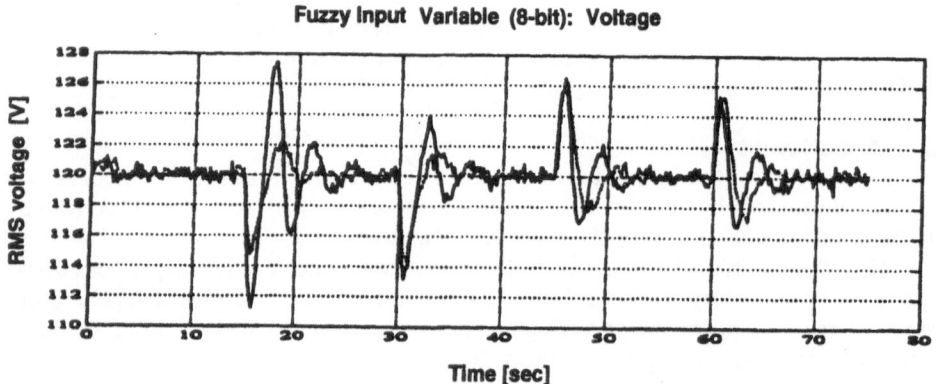

Figure 10. Voltage response for 8- and 12-bit DAC.

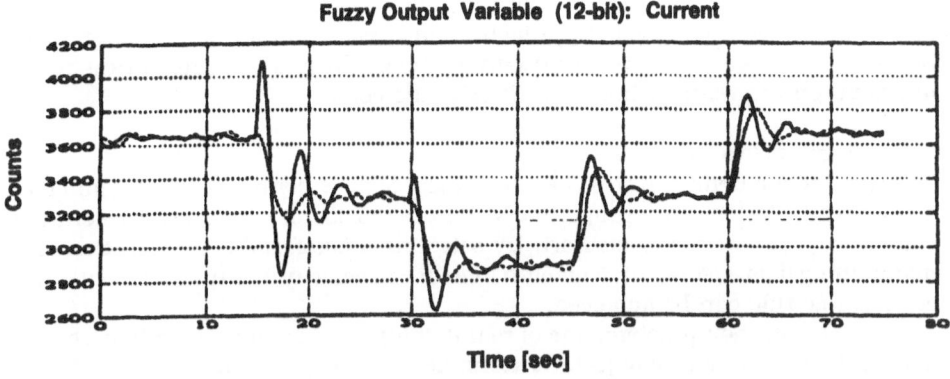

Figure 11. Control signal from 8- and 12-bit DAC for control of field voltage.

To solve this problem the number of DAC bits is increased to 12, which gives a much wider control resolution. Figures 11 and 12 (12-bit case) show the results of the 12 bit control signals.

As shown in Figure 9, the fluctuations around optimal values were minimized with the 12-bit DAC, which supports the use of a more expensive DAC.

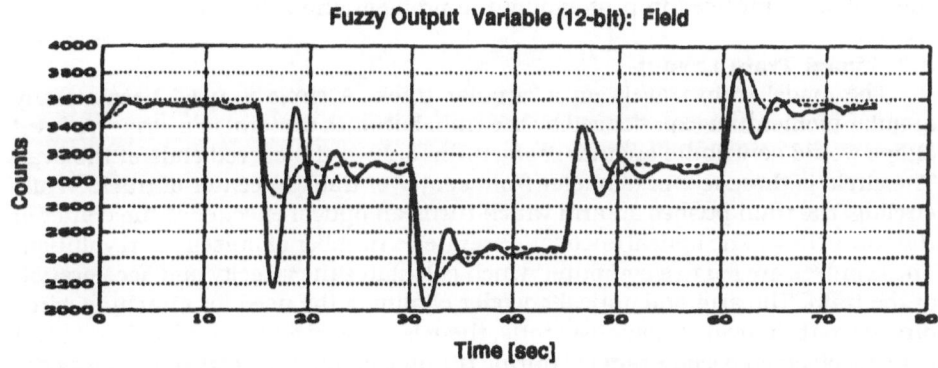

Figure 12. Control signal from 8- and 12-bit DAC for control of shunt current.

3 Fuzzy Acceleration Control of a Model Train

In this section fuzzy control and decision making are used for acceleration control of a model train on a circular path. The fuzzy acceleration control system showed a marked improvement over crisp control in smoothness of ride.

Previous software experiences with the applications of fuzzy control to traffic has shown very promising results. In over 625 different simulation runs, it was shown that nearly one-half of the waiting time behind red lights at a single intersection can be saved with fuzzy control over standard techniques such as sensor-based or timed cycle approaches. In a cost (waiting time) comparison between fuzzy control and two standard methods, the fuzzy controller drastically improved on the waiting time, and thereby the cost.

3.1 Introduction

This section demonstrates the use of fuzzy logic to control the acceleration of a model train. Velocity is a property which humans do not notice, yet acceleration can cause a passenger discomfort. By selecting a comfortable acceleration and maintaining it as a constant, a smooth increase in velocity and thus a smooth and comfortable ride can be achieved.

The acceleration problem, one of maintaining a constant acceleration and/or deceleration for a moving object (for example, a train), has a fuzzy nature. Fuzzy logic was chosen as a means of control for this problem, the idea being that a constant acceleration needs to be achieved as quickly as possible and then maintained. This is also true of deceleration when bringing the train to a stop. The question was: Can fuzzy logic control do better than a human operator in this area?

The first part of the problem involved the selection of a comfortable acceleration constant, which is obviously different for a model train than a real train. For this work a comfortable acceleration was selected by visual inspection of a water cell being pulled by the model train. The most desirable constant acceleration was indicated by the water level in the water cell maintaining an even surface during increase in the train's velocity. From this visual state the acceleration can be calculated with known time and distance. Once the acceleration was selected, the more interesting problem of control was explored. The following sections describe the hardware involved in prototyping the problem and solution.

3.2 Model Train Layout

The model train travels on a circular track. A circle is used because any angular momentum experienced by the train becomes uniform in a circle. An oval shape would contribute angular momentum only at the curved ends of the oval. The circle is three feet in diameter from center of track to center of track. While circling, the train pushes an arm which turns an optical encoder at the center of the track circle. The optical encoder generates a number of pulses per revolution. These pulses are fed to a computer which calculates the velocity and acceleration of the train. The arm and optical encoder eliminate the need for attaching wires directly to the moving locomotive; thus, there is no need for an electrical umbilical cord which could become tangled during running of the train. Figure 13 shows the layout of the train track, train, arm, and optical encoder. The circle represents the train track; the dot at the center is the optical encoder; the gray rectangle is the train locomotive; and the line between the locomotive and optical encoder is the arm. The model train locomotive pulls a gondola car which has a water cell mounted inside. The gondola car has a lead weight mounted to the bottom of it between the wheel bases. This weight helps balance the weight of the water and water cell.

The computer takes the difference of the selected and current accelerations and uses it as an input to the fuzzy controller. The output of the fuzzy controller is an amount of change in acceleration, which might be an increase, a decrease, or nothing. This actually results in the computer increasing, decreasing, or not changing the power to the train. Deceleration follows the same scenario, except that the selected value is negative.

3.3 Computer Input

The movement of the train pushing the arm turns the optical encoder. The rotation of the optical encoder generates digital pulses at the rate of 128 pulses per revolution, or one pulse every 2.8 degrees. The pulses are accumulated in a 6-bit counter which eliminates the need for the computer to be constantly reading pulses. Also, since the pulses are digital, there is no need for an analog-to-digital conversion. The 6-bit counter is constructed of two cascading 4-bit counters.

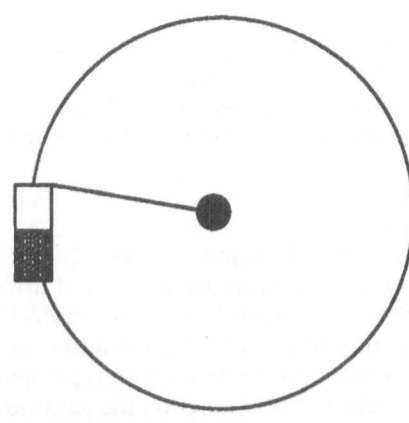

Through experimentation, it was determined that one 4-bit counter was insufficient for counting pulses when the train was moving at full velocity. The train's top velocity generates more than 16 pulses per read (strobe of the game and serial ports) by the computer. The 6-bit counter allows for a maximum of 64 pulses to be accumulated.

The 6-bit counter is connected to the computer through the game port (four lines used) and the serial port (two lines used). This provides six lines for reading the 6-bit value from the counter. Use of the game and serial ports simplifies the interface between the computer and the input from the train. No communications protocol is necessary.

Figure 13. A layout of the model train on a circular path.

3.4 Fuzzy Controller

The fuzzy controller consists of one input and one output. The input to the controller is the difference in the selected acceleration and the train's current acceleration. The membership functions for diff (the input) are Very Negative, Negative, Zero, Positive, and Very Positive (VN, N, Z, P, and VP, respectively). The output from the controller is an amount of change in power. The membership functions for this change (the output) are Large Decrease, Small Decrease, Zero, Small Increase, and Large Increase (LD, SD, Z, SI, and LI, respectively), were chosen in standard fashion as other applications in this chapter. The fuzzy rule-base mapping the input to the output is shown in Figure 14. The number of rules is determined by the number of membership functions in the input, which is a total of five. These membership functions and rules will work for deceleration as well because the selected deceleration value will be the negative of the selected acceleration value.

> IF diff is Z THEN change is Z
>
> IF diff is P THEN change is SI
>
> IF diff is VP THEN change is LI
>
> IF diff is N THEN change is SD
>
> IF diff is VN THEN change is LD

Figure 14. Fuzzy rules for the train acceleration control system.

3.5 Computer Output

The output from the fuzzy controller (change in power) is added to the current power being output to the train. This range of power (in volts) is represented by the computer as a numerical range between zero and 255. The computer outputs a value in this range via the parallel port to a digital-to-analog (D/A) chip. Use of the parallel port for output from the computer again simplifies the interface and avoids the need for communications protocols. The D/A chip converts the eight input lines to one analog output line. The output from the chip is a voltage in the range of zero to 5 volts and is connected to a power regulator circuit. A 15-volt transformer serves as the power supply for the train. This supply is varied by the power regulator circuit as directed by the D/A chip output. The power regulator circuit has an output (to the train track) in the range of 4 to 10 volts, along with an increased current to drive the train. Experimentation showed that 4 volts is just below the amount needed to keep the train moving.

3.6 Software Interface and Results

The software interface consists of three parts: the input module, the fuzzy controller module, and the overall driver program. The input module and driver program are written in C compiled with the Turbo C compiler. The fuzzy controller module is written in Togai InfraLogic fuzzy source code and compiled into C with the Togai Fuzzy-C Compiler [14]. All these programs were executed on a personal computer. The input module read the data lines from the counter on the game and serial ports and built the 6-bit value representing the number of pulses. A driver program was written in C to go with the Togai Fuzzy-C language code. The driver program initialized everything, computed the acceleration, called the fuzzy controller, and wrote the output to the parallel port.

The results are heavily based upon visual inspection of the water cell and observation of the train itself. The hope was that the surface of the water in the water cell would angle "uphill" as the train increased acceleration. When a constant acceleration is achieved the water should level itself, and when deceleration begins the water level should angle "downhill." If this state actually occurred, however, the angle was so small it could not be observed. Perhaps the maximum velocity was not large enough. Because of this it was also difficult to select an acceleration for the model train which represented a "comfortable" (for humans) acceleration in real trains. The time it takes the model train to accelerate to maximum is so short that there is not much to observe.

Two methods of acceleration increase were observed, fuzzy controlled and crisp proportional control. The crisp control method is analogous to setting the "throttle" wide open: the train was observed making a "jackrabbit" start and the water in the water cell showed erratic movement. This is analogous to driving a car while holding a cup of coffee: the coffee sloshes around or even spills when the car is abruptly stopped or started. When acceleration was controlled with the fuzzy controller, the train's behavior was much less abrupt. Some movement in the water cell was observed, but the water surface was smoother than that observed with crisp control.

3.7 Problems and Limitations

The main problem with this experiment is one of granularity. The computer measures its own internal time in clicks, with 18.2 clicks per second. This calculates to be approximately 0.05 seconds per click. In 0.05 seconds at full speed the train travels approximately two pulses. More accurate accelerations could be calculated with more accurate timing of the pulses, which would require a computer whose internal clock generates more clicks per second.

Although the lack of communications protocol simplifies the design of the system, it also causes a potential problem. This problem occurs with reading the values of the counter through the game and serial ports. It is possible for the counter to receive a pulse during the time the computer is reading from those ports. This could cause an erroneous value to be input to the computer. Hence, a check was added to identify and discard any calculated velocities which appeared out of range (faster than the observed maximum velocity of the train).

Another limitation involved the power regulator circuit. Although the power supply offered a maximum of 15 volts, the power regulator circuit was able to vary a range of only 4 to 10 volts. An improved power regulator circuit could obtain smoother control of the train and allow the train to be accelerated over a longer period of time, allowing the fuzzy control to have a greater effect.

4 Fuzzy Control of a Thermoelectric Device-Based Refrigeration System

4.1 Introduction

This section deals with two important issues in manufacturing: (1) Environmentally Conscious Manufacturing (ECM); and (2) fuzzy logic as an alternative method of controlling manufacturing processes, thereby creating a new type of so-called *smart manufacturing products*. The specific application presented here is a refrigeration system which is not based on Chlorofluorocarbon (CFC) or any other chemical substance, but rather on electronic refrigeration using thermo-electric devices (TEDs)[8,9].

Moreover, this section describes the potential of fuzzy logic for system control on the lowest level. Since the chamber temperature is a first-order linear system, its control is easily achieved via a PI linear controller. However, knowing that a TED is a dynamic device, and lacking the exact model of the system, an FLC (fuzzy logic controller) is proposed as a possible solution for temperature control [10,11].

4.2 Thermoelectric Devices

In simple terms, a thermoelectric device (or TED) is a heat pump that transports heat from one location to another. This phenomenon is traceable to the early work of Seebeck and Peltier as reported in [9], whose work led to the contemporary thermocouple and thermoelectric devices. Peltier's original junction would generate only small temperature changes, while contemporary devices can achieve drops as large as 150°C below ambient temperature. The reason for this performance improvement in the contemporary thermoelectric device lies in the high-technology materials used in its fabrication. The selection of the material used in such a cell is based mainly on operational requirements, and can range from antimony, lead, or bismuth all the way to telluride. Figure 15 illustrates the construction of a simple single-junction TED. Each TED contains anywhere from three to 127 such junctions, electrically connected in series, but thermally in parallel. When the electric current is reversed, so is the cool/heat function of the device.

Thermoelectric cooling can be accomplished only if the generated heat is rapidly transported from the junction; otherwise, a transmigration of thermal energy back to the cold side occurs, reversing the previous cooling effect. There are additional drawbacks to using TEDs. Most vital among these is low-energy efficiency–when electrical input energy is substantially higher than thermal output derived. This problem is of significant concern when applying thermoelectric modules to cooling applications. In fact, many companies stopped using TEDs in this application years ago because of their low overall efficiency.

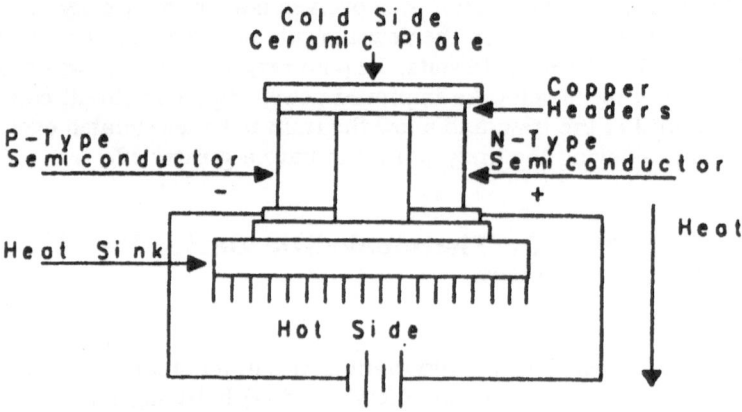

Figure 15. Simple single-junction TED.

Part of the problem in using TEDs is that during normal operation, heat builds up between the hot and cold plates, affecting the module's performance and efficiency. Figures 16-18 show the current response and inter-ceramic response of such a device to the step input of 12 volts, as well as the device's resistance.

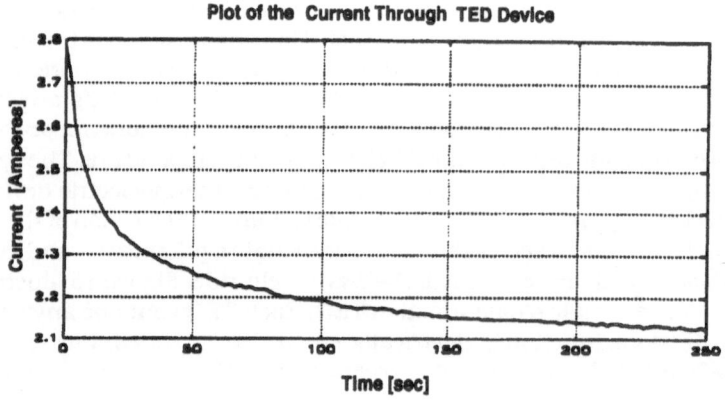

Figure 16. Current through a TED.

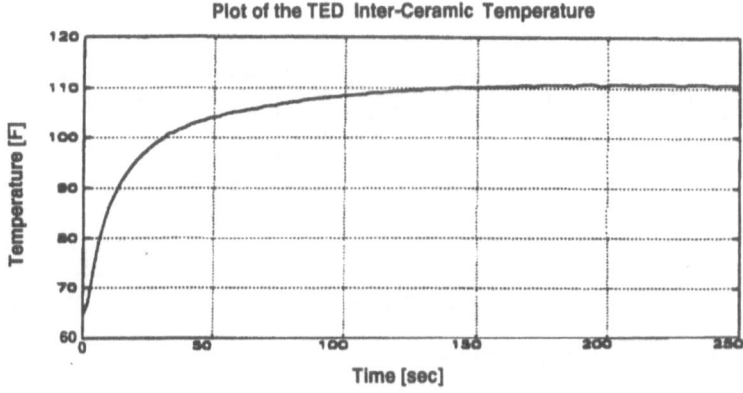

Figure 17. Inter-ceramic temperature of a TED.

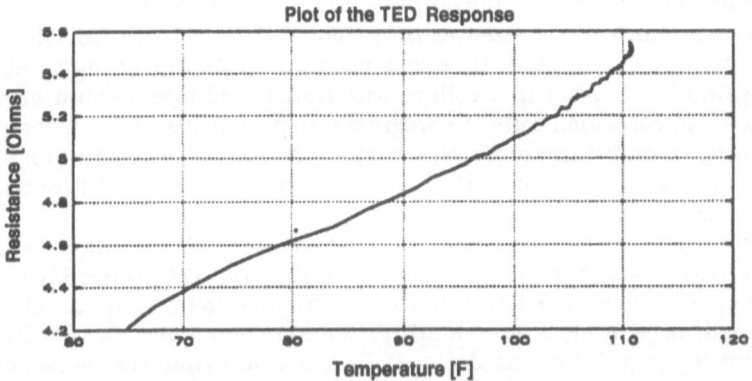

Figure 18. Dynamic resistance of a TED.

It can be seen that a TED is a dynamic device whose resistance depends on the inter-ceramic temperature. Consequently, lowering inter-ceramic temperature will cause a decrease in the TED's resistance and, therefore, an increase in TED current for the same step-input voltage. This increase in current means that TED efficiency could improve drastically, making the TED a valid alternative energy source for refrigeration.

4.3 Real-Time System Set-Up
In order to perform a real-time experiment, a TED was set up as shown in Figure 19 with control circuitry providing the voltage to the TED, while the measurement circuitry provided a means of measuring the temperature of the water in the reservoir. The control algorithm, which uses A/D and D/A converters to govern data entering and leaving the computer, is actually a software controller based on classical control theory or fuzzy logic theory.

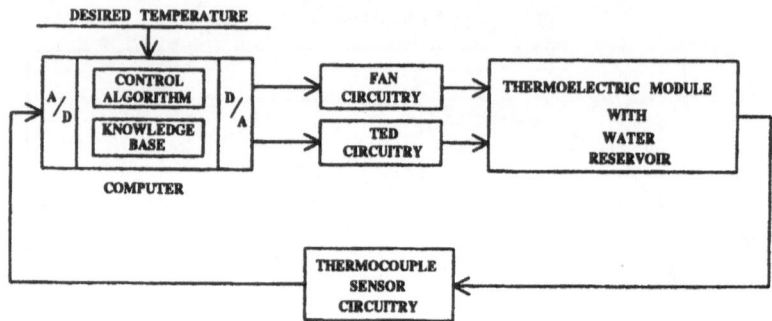

Figure 19. Block diagram of system set-up.

The measurement circuitry consists of a thermocouple sensor circuitry and an A/D converter[17] which is installed in the computer. Measurement circuitry uses a Type K thermocouple as a transducer transferring temperature into the voltage. A thermocouple was chosen because of its low cost, raggedness, and ease of implementation; however, the output of such a device is very low in magnitude and has a nonlinear response over the range of temperatures. To solve the low-magnitude problem, op-amp U1-c was set up in a non-inverting configuration with a gain of 300. After this voltage was transferred into a count via an A/D converter, a linearization software program was used to cancel the measurement-nonlinearity error. Furthermore, this program performs a software compensation for cold-junction compensation (CJC) of the thermocouple, which is necessary for accurate measurements.

The control circuitry consists of a fan circuitry, TED circuitry, and a D/A converter[12] which is installed in the computer. The D/A converter is used to transfer counts (determined by a control algorithm) into the range of voltages that can be easily used by electronic circuitry. Since fan nominal voltage is 12V at 0.5A of current, op-amp U1-a was designed in a non-inverting configuration with a gain of 1.2 [13] such that the input range of 0 to 10V is now transferred to a range of 0 to 12V. Op-amp U2-a is configured as a voltage follower, but is used as a motor driver because of its high current capabilities. Capacitor C_1 is used to filter out the noise produced by a fan (DC motor M_1). The purpose of the fan circuit is rapid heat removal from the heat sink, which is thermally connected to the hot plate of the TED. This action is necessary to improve TED efficiency.

The TED circuit was a linear amplifier for a TED. Op-amp U1-b is configured as a non-inverting amplifier with a gain of 2.7. This transfers the D/A output range of 0 to 5V into a range of 0 to 13.5V. However, Darlington pair Q_1 lowers this voltage by 1.5 volts (each V_{BE} is 0.75 volts), making the TED module range 0 to 12V. Therefore, 5V at the output of DAC #1 corresponds to 12V and thus provides the maximum cooling effect on the TED.

4.4 Proportional-Integral Controller Design

In Close and Frederick [11], a dynamic model of a thermal system was developed with transfer function

$$G_p(s) = \frac{\dfrac{1}{C}}{s + \dfrac{1}{RC}} \tag{1}$$

where R is thermal resistivity (conduction resistance between surroundings), and C is thermal capacitance (stored heat). Equation (1) is supported by Figure 20, which shows the system's open-loop response to a step input (classical first-order response).

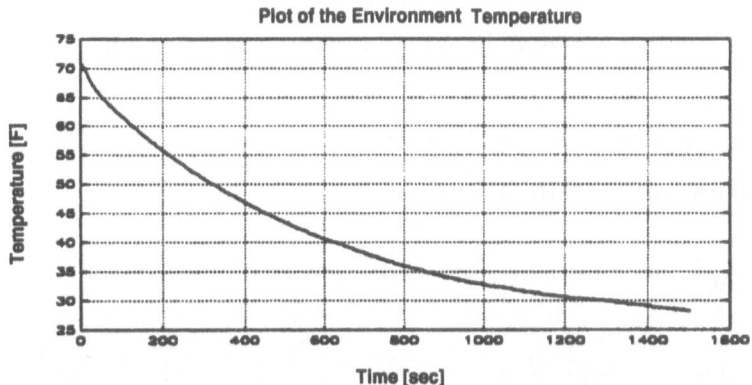

Figure 20. Open-loop response to the step input.

From Figure 21, rise time–i.e., the time the system takes to change 10% to 90% of initial value to steady-state value–is calculated to be 1000 seconds. Knowing that

$$t = RC \tag{2}$$

and using the fact that rise time for the first order system is governed by $T_R @ 2$, Equation (1) becomes

$$\cdot G_p(s) = \frac{\dfrac{1}{C}}{s + 0.002} \tag{3}$$

Let $K=1/C$, and apply steady-state theorem to (3) with $R(s) = 12/s$ and $ess=28$. Equation (3) becomes

$$. G_p(s) = \frac{0.00467}{s + 0.002} \tag{4}$$

However, using equation (4), the PI controller is governed by the following equation:

$$\frac{\overline{}}{\underline{}} \tag{5}$$

150

which implies that the controller will add to the system *Pole* at zero and *Zero* at the value of K_l/K_p. Therefore, the constraints in designating the values for K_l and K_p are:

1) $0 > K_l/K_p > 0.02$
2) $K_p = 1.9034$

The implementation algorithm for a PI controller was derived using Figure 21 as follows:

$$u(k)=K_p e(k)+K_l\{T_s e(k)+0.5T_s[e(k-1)-e(k)]\}$$

or

$$u(k)=K_p e(k)+0.5K_l T_s[e(k-1)+e(k)].\tag{3.6}$$

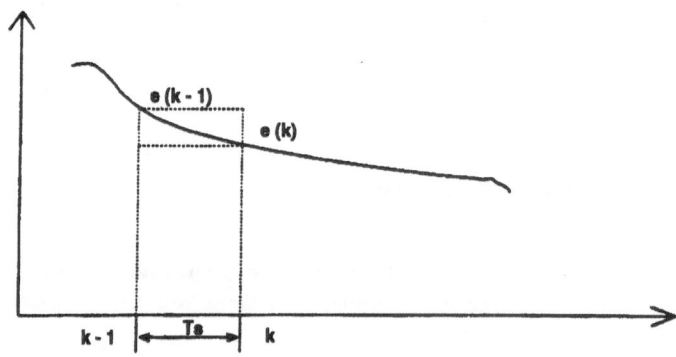

Figure 21. Error representation.

4.5 Fuzzy Logic Controller Design

Using the concepts of fuzzy membership functions and "expert" knowledge of TED behavior, it was decided to govern the FLC by magnitude and direction of error. Values of these variables determine the sign and the magnitude of the voltage change on the TED. Membership functions are shown in Figures 22 and 23.

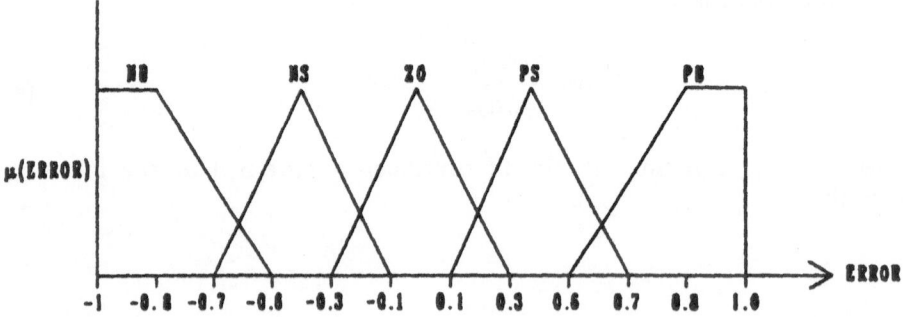

Figure 22. Error membership function.

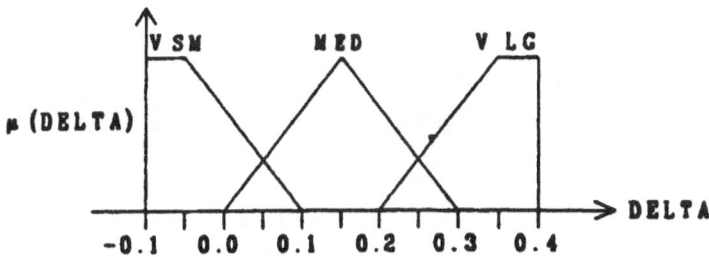

Figure 23. Delta membership function.

Fuzzy rules were devised by intuition and by previous experience with TEDs, as follows:

> Rule 1: *If ERROR is PB then DELTA is VLARGE.*
> Rule 2: *If ERROR is PS then DELTA is MED.*
> Rule 3: *If ERROR is ZO then DELTA is VSMALL.*
> Rule 4: *If ERROR is NS then DELTA is MED.*
> Rule 5: *If ERROR is NB then DELTA is VLARGE.*

Using a Fuzzy-C Compiler[14], membership functions and rules were then compiled into a subroutine called up by the main control algorithm. More detail information on this application can be obtained in reference 13.

4.6 Experiment and Results

Data for Figures 24 and 25 were achieved in real-time, and were graphed using Matlab[15] software. Error calculations were determined as a measured temperature versus desired temperature entered by the user.

Figure 25 shows that both the PI and the FLC show similar results for system error. Even though the FLC has a bigger undershoot, its steady-state error is smaller than that of the PI controller, which makes these compensators comparable in error performance. For a PI controller to achieve such a performance, its control input (shown in Figure 24) was fluctuating drastically from 0 to 10V, with an average of 5V. The FLC, on the other hand, had a much smoother control curve, with an average of about 4 volts. Since criteria for the TED controller design require an energy-efficient system, the FLC's energy savings of 20 percent make it an obvious choice.

In this experiment, performance of a fuzzy logic controller was compared to the performance of a conventional PI controller. A mathematical model of the system, though not available, was achieved through system identification techniques. This mathematical model is a necessary ingredient in the design of a conventional controller. On the other hand, when implemented on the lowest level, the FLC does not need a mathematical model but does need the availability of a human expert who can specify the rules underlying the system's behavior.

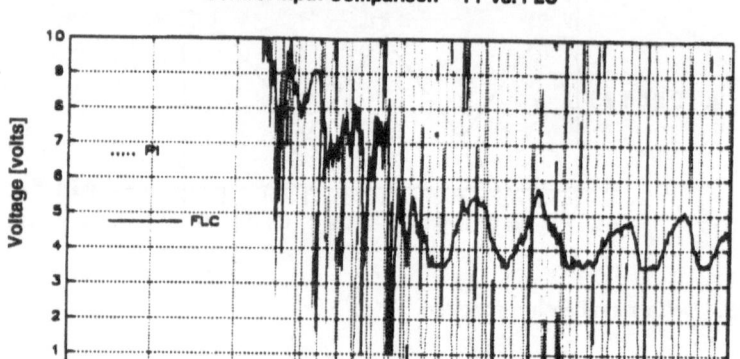

Figure 24. Control input - PI vs. FLC.

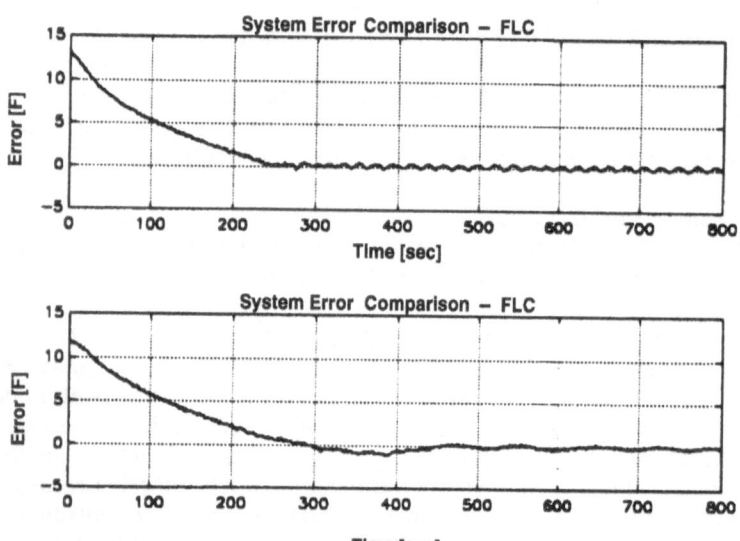

Figure 25. System error - PI vs. FLC.

5 Conclusion

This paper presents a few hardware applications of fuzzy logic and intelligent control at the University of New Mexico's CAD Laboratory for Intelligent and Robotic Systems. The following are some individual conclusions on the projects described in this paper.

The power generation system case study showed that fuzzy logic can be used in power systems control. The main result, however, is that this method provides a simple means of designing a controller for a nonlinear system without complicated calculations and expensive hardware. In some cases, the fuzzy controller was shown to provide marked improvements over non-fuzzy controllers in terms of overshoot suppression and speed of response.

Hardware implementation of the fuzzy control of the model train does offer an improved method of controlling acceleration. The desired amount of improvement was not achieved, but improvement over an "open throttle" was observed. It should be noted that in this hardware experiment, no feedback was used from the surface of the liquid on the trailing wagon behind the locomotive unit, and yet a smoother movement was demonstrated from rest to maximum velocity and back to rest. The small scale of this experiment affected what could be observed, and the problems discussed in the previous section dampened the potential results. Further refinements of this implementation of the fuzzy acceleration control for a model train are underway.

Fuzzy control of the thermoelectric refrigeration system showed that a marked improvement can be obtained in the settling time of the responses. The temperature of the chamber varied in an oscillatory mode using crisp control, while no significant sustained oscillation could be observed in the fuzzy logic case.

REFERENCES

1. Jamshidi, M., N. Vadiee and T. Ross (eds.) *Fuzzy Logic and Control: Hardware and Software Applications*. PH Series on Environmental and Intelligent Manufacturing Systems (M. Jamshidi, Series Editor), Vol. 2. Prentice Hall Publishing Company, Englewood Cliffs, NJ (1994).

2. Aminzadeh, F., and M. Jamshidi (eds.). *Soft Computing*, PH Series on Environmental and Intelligent Manufacturing Systems (M. Jamshidi, Series Editor), Vol. 2. Prentice Hall Publishing Company, Englewood Cliffs, NJ (1994).

3. Hiyama, T., and C. M. Lim. "Application of Fuzzy Logic Control Scheme for Stability Enhancement of a Power System." *IFAC Symposium on Power Systems and Power Plant Control* (Aug. 1989), Singapore.

4. Hsu, Y.Y. and C.H. Cheng. "Design of Fuzzy Power System Stabilisers for Mulitmachine Power Systems." IEE Proceedings, May 1990, Vol. 137, pp. 233-238.

5. Hassan, M.A.M., O.P. Malik, and G.S. Hope. "A Fuzzy Logic Based Stabilizer for a Synchronous Machine." *IEEE Trans. on Energy Conversion* (Sept. 1991), pp. 407-413.

6. David, A. K., and Rongda, Z., "An Expert System with Fuzzy Sets for Optimal Planning." *IEEE Trans. on Power Systems* (Feb. 1991), pp. 59-65.

7. Tomsovic, K. "A Fuzzy Linear Programming Approach to the Reactive Power/Voltage Control Problem." *IEEE Trans. on Power Systems* (Feb. 1992), pp. 287-293.

8. ITI FerroTec. *Thermoelectric Product Catalog and Technical Reference Manual.* Chelmsford, MA. 1992.

9. O'Geary, D. and T. O'Geary. "Development of a Reliable, Efficient Non-CFC Based A/C System." Proc. Environmental Conscious Manufacturing, Santa Fe, NM (1991), pp. 183-191.

10. Rogers, S. "Infinite Band Controller with a Fuzzy Tuner." ISA Trans., Vol. 31, No. 4 (1992), pp. 19-24.

11. Close, C. M. and D. K. Frederick. *Modeling and Analysis of Dynamic Systems.* Houghton Mifflin Company, New York (1978).

12. Phillips, C. L. and R. D. Harbor. *Feedback Control Systems.* Prentice-Hall, Englewood Cliffs, NJ (1988).

13. D. Barak. "Real-Time Fuzzy Logic-Based Control of Industrial Systems." MS Thesis, CAD Laboratory for Intelligent and Robotic Systems, Dept. EECE, University of New Mexico, Albuquerque, NM (May, 1993).

14. Togai InfraLogic, Inc. *Fuzzy-C-Development Systems User's Guide.* Bell Helicopter Textron, Inc., Fort Worth, TX (1992).

15. Mathworks Inc. *Matlab for Windows User's Guide.* Natick, MA., December 1991.

Modeling and Analysis of Switched-Mode Hybrid Systems Driven by Threshold Events

T. Niinomi and B.H. Krogh

Carnegie Mellon University
Dept. of Electrical and Computer Engineering
Pittsburgh, PA 15213-3890
e-mail:krogh@cmu.edu

Abstract. We consider a class of continuous-time hybrid systems in which the interface between the continuous-state dynamics and discrete-state dynamics is defined by threshold-crossing events. Discrete-state transitions are forced by events that occur when continuous signals cross specified thresholds. The discrete states, in turn, influence the continuous system dynamics through condition signals that are inputs to the continuous state equations. To apply discrete-state methods to analysis and synthesis of such systems, it is necessary to create a discrete-state model that accepts the event language generated by the hybrid system. This is not possible in general, however, because the threshold event language can be nonregular. We present an algorithm for computing a sequence of finite-state automata which accept regular languages that are increasingly refined outer approximations to the threshold event language. The definitions and algorithm are developed and illustrated for a simple two-integrator continuous system with arbitrary rates and 0-1 thresholds. The paper concludes with a discussion of several open problems for future research.

1 Introduction

Recently there has been considerable interest in the modeling and analysis of hybrid systems, that is, systems with both discrete and continuous state variables. One theme of research reported in the literature is to extend methods for verification of discrete state systems to hybrid systems (Alur et al. 1993). Another theme has been the qualitative analysis of the types of behaviors that can be exhibited when discrete and continuous dynamics interact (Ramadge 1990; Chase, Serrano, Ramadge 1993). There have also been some attempts to formulate strategies for controller synthesis (Nerode and Kohn, 1992; Antsaklis and Stiver and Lemmon, 1993). One general conclusion that emerges from the current research on hybrid systems is that even very simple systems can exhibit complex behaviors. Verification problems can be undecidable, simple feedback structures can exhibit chaotic behavior, and control synthesis will probably always require the introduction of approximations and heuristics to avoid intractable computations.

Because of the difficulties that arise when dealing with general general hybrid systems, it appears that progress must be made by studying particular classes of hybrid systems as a guide to directions for developing analysis and synthesis methods for more complex systems. In this paper we define a class of hybrid systems that represents a type of interaction between discrete and continuous behaviors that commonly arises in practice. In particular, we consider the modeling and analysis of systems in which a discrete controller switches the continuous dynamics among various "modes" according to threshold-crossing events generated by outputs from the continuous system. Applications include gain scheduling controllers, fault-detection and recovery, responses to emergency conditions, and rule-based methods for switching controlled behaviors in response to discrete sensor information.

We model these threshold-event-driven switched-mode hybrid systems using the formalism of *condition/event (C/E) systems* (Sreenivas and Krogh, 1991) with signal interfaces to continuous dynamics as proposed by Krogh (1993). For this class of hybrid systems, state transitions in the discrete dynamics (a finite state C/E system) are forced by events that occur when output signals from the continuous system cross specified thresholds. The output from the C/E system is a condition signal which selects the operating mode for the continuous system. The class of hybrid systems being considered is defined formally in the following section. In section 3 we define the *threshold event language* that characterizes the sequential behavior of the hybrid system. Section 4 introduces the notion of finite-state automata which accept languages that are outer approximations to the threshold event language for a given system. We present an algorithm for generating iteratively increasingly refined outer approximations. The results of this algorithm are illustrated for a simple two-integrator example in Section 5. The concluding section discusses directions for future research.

2 Switched-Mode Hybrid Systems with Threshold Events

Figure 1 illustrates the three components of the hybrid system model: the continuous dynamics, the C/E system dynamics, and the zero-crossing detector that generates the threshold events. The state space of this hybrid system is defined as $S = \Re^{n_x} \times Q$, where n_x is the number of real-valued continuous state variables and Q is a finite set of discrete states. A *state trajectory* $(x(\bullet), q(\bullet))$ is defined on $[0, \infty)$ as a solution following dynamic equations where the *continuous-state trajectory* $x : [0, \infty) \to \Re^{n_x}$ is continuous, the *discrete-state trajectory* $q : [0, \infty) \to Q$ is right continuous (a condition signal), and the *initial state* (x_0, q_0) is an element of a given set of possible initial states $S_0 \subseteq S$.

The C/E system dynamics in Fig. 1 are defined by the *discrete state transition equation*

$$q(t) = \delta_{C/E}(q(t^-), v(t)), \qquad q(0) = q_0 \tag{1}$$

and the *condition output equation*

$$u(t) = \phi(q(t)) \qquad (2)$$

where

$\delta_{C/E} : Q \times (V \cup \{0\}) \rightarrow Q$, *discrete state transition function*

$\phi : Q \rightarrow U$, *condition output function*

$V := \{0, 1\}^{n_v} - \{0\}$, finite set of *event vectors* (0 indicates the absence of an event)

U: finite set of *conditions*

$v : [0, \infty) \rightarrow V \cup \{0\}$, *threshold event signal* (defined below)

$u : [0, \infty) \rightarrow U$, mode-switching *condition signal*.

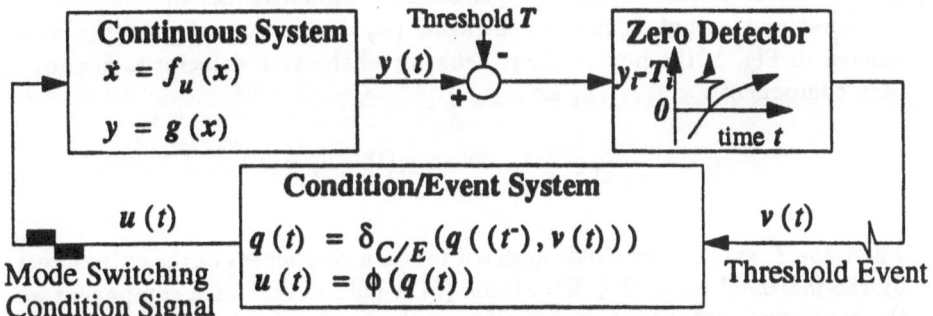

Fig. 1. Block diagram of switched-mode, threshold-event-driven hybrid system.

The discrete-state system defined above is a special case of the general C/E systems defined by Sreenivas and Krogh (1991) which include both condition signals and event signals as input and output signals. For the particular case considered here, the discrete-state transitions are deterministic. State transitions occur only at the event instants, that is, when $v(t)$ is nonzero. This implies $\delta_{C/E}(q, 0) = q, \forall q \in Q$. Properties of general C/E signals and systems are discussed in (Sreenivas and Krogh, 1991).

The continuous dynamics in Fig. 1 are defined by standard continuous state and output equations of the form

$$\dot{x}(t) = f_{u(t)}(x(t)), \qquad x(0) = x_0 \qquad (3)$$

and

$$y(t) = g(x(t)) \qquad (4)$$

where

$f_u : \Re^{n_s} \to \Re^{n_s}$, *continuous-state derivative function* for each $u \in U$

$g : \Re^{n_s} \to \Re^{n_v}$, *continuous output function*

$y : [0, \infty) \to \Re^{n_v}$, *continuous output signal*

The switched-mode behavior of the continuous dynamics is created by the dependence of the continuous-state derivative function on the condition signal $u(\bullet)$ which selects the particular state derivative function at each point in time. We assume the continuous state differential equations have unique, well-behaved solutions on any interval.

The *threshold event signal* $v(\bullet)$ is generated by the zero-detector block in Fig. 1. The *zero detector* generates an event in a component of its event output signal at any instant when the corresponding component of the incoming continuous signal is zero. Since the zero detection is performed componentwise, the dimensions of the input and output signals are the same ($n_y = n_v$). Formally, for the signals defined in Fig. 1, the input-output behavior of the zero detector is defined for each component $i = 1, \ldots, n_y$ as

$$v_i(t) = \begin{cases} 1, & \text{when } y_i(t) - T_i = 0 \\ 0, & \text{otherwise} \end{cases} \tag{5}$$

where the T_i are the *thresholds* specified for each component of the output vector by the *threshold vector* $T \in \Re^{n_v}$. From (5) it follows that each component g_i of the continuous output function g along with the threshold T_i defines a smooth manifold M_i in the continuous state space defined by

$$M_i = \{x \in \Re^{n_s} | g_i(x) - T_i = 0\}. \tag{6}$$

An event is generated every time the continuous state trajectory "hits" one or more of these manifolds. The resulting event causes a deterministic transition in the state of the discrete system, which in turn switches the mode of the continuous dynamics.

To have well defined state trajectories and C/E signals for the class of hybrid systems defined above, we make the following assumption:

A1. On any finite interval of time, $(t_1, t_2) \subset [0, \infty)$, $\{t \in (t_1, t_2) | v(t) \neq 0\}$ is finite.

Assumption A1 would be violated if the continuous state trajectory remained on one of the manifolds M_i for a finite interval of time, or the continuous state trajectory converged asymptotically into an intersection of two or more manifolds, as illustrated in Fig. 2.

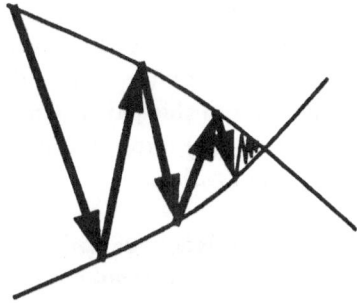

Fig. 2. Trajectory converging to intersection of manifolds, leading to a violation of assumption A1.

3 Threshold Event Languages

In applications, specifications and performance requirements for hybrid systems are often discrete in nature; that is, the qualitative properties of interest can be expressed in terms of discrete states, rather than continuous states. Furthermore, one is typically interested in the discrete behavior of the hybrid system for a whole class of state trajectories, rather than the system behavior for a single trajectory. In these cases, purely discrete models are most useful for evaluating system properties or synthesizing controllers, particularly when the controllers are discrete-state by nature, as is the case for programmable logic controllers.

In this paper we are interested in the discrete behaviors of hybrid systems as characterized by the sequences of threshold events that are generated for given sets of initial states. To study the sequences of threshold events that can be generated by any initial state $(x_0, q_0) \in S_0$, we map the timed sequences of threshold events characterized by the threshold event signal $v(\bullet)$ into an untimed sequence of *threshold event symbols* denoted by the finite set Θ which includes the *null event symbol*, ϵ. Given a labeling function $\Lambda : V \to \Theta - \{\epsilon\}$, we define the *trace* of a threshold event signal $v(\bullet)$, denoted by $tr(v(\bullet))$, to be the sequence of threshold event symbols,

$$tr(v(\bullet)) = \Lambda(v(t_1))\Lambda(v(t_2))\ldots. \tag{7}$$

where $0 < t_1 < t_2 < \ldots$ is the sequence of times for which $v(t) \neq 0$. To avoid having to deal with a mixture of finite-length and infinite-length sequences of threshold event symbols, we make the following two-part assumption:

A2. Given a hybrid system (1)-(5), and an initial state set S_0: (a) from any initial state in $(x_0, q_0) \in S_0$, the continuous state trajectory $x(\bullet)$ reaches some manifold M_i in finite time; and (b) for any $i = 1, \ldots, n_v$, if $x(t) \in M_i$ for some $t > 0$, then there exists some $i' = 1, \ldots, n_v$ and $t' > t$ such that $x(t') \in M_{i'}$.

Assumption A2 indicates that threshold events always occur, and that they will occur indefinitely. This means the threshold event symbol sequence associated with a threshold event signal will be an infinite-length sequence. We assume A2 throughout the remainder of this paper. Consequently, the trace operator tr maps threshold event signals $v(\bullet)$ into elements of Θ^ω, the set of infinite-length sequences of threshold event symbols.

We can now identify a formal language with a given hybrid system. In particular, we let $L_{S_0} \subseteq \Theta^\omega$ denote the *threshold event language* for the system defined as the set of all threshold event sequences corresponding to the possible threshold event signals generated by the hybrid system starting from some initial state in S_0. The sequential behaviors of the hybrid system can then be specified and verified in terms of the properties of language L_{S_0}.

Characterizing the language L_{S_0} is extremely difficult, if not impossible, for all but the most trivial examples. This is because the discrete behavior depends in general on the evolution of the continuous state trajectories, and computations of sets of these trajectories are fraught with numerical difficulties. Even when solving the continuous dynamic equations does not require numerical integration, as is the case with pure integrator dynamics, it becomes extremely difficult to represent accurately the evolution of the sets of state trajectories. Although it may be possible to use symbolic computation to circumvent the problems with numerical integration, the problem remains that the language L_{S_0} can be nonregular, making it impossible to characterize it with a finite state generator. An example of such case is given in section 5.

4 Approximating Threshold Event Languages

Given the above difficulties, one should evaluate whether a complete characterization of the language L_{S_0} is required. In many situations one may not need an exact representation of the set of system behaviors to determine whether that the system satisfies a particular specification. One can often settle for an imprecise representation that approximates the system dynamics provided it includes all critical behaviors. If a property holds for these critical behaviors, plus perhaps some other behaviors that are modeled but may never occur, that is sufficient to verify the property for the actual system. These practical considerations, combined with the extreme complexity of the behaviors of hybrid systems, lead to the concept of *approximating threshold event languages* defined in this section.

Given a threshold event language L_{S_0} and a language $L \subseteq \Theta^\omega$, we say L is an *outer approximation* of L_{S_0} if $L \supseteq L_{S_0}$. In this section we present an algorithm for obtaining a sequence of nondeterministic finite-state automata that accept languages (in the Buchi automata sense) that are increasingly refined outer approximations to a given threshold event language. Formally, a (Buchi) automaton B is a 5-tuple, $B = (\Sigma, \Theta, \delta, \sigma_0, \Sigma_F)$, where Σ is a finite set of states,

Θ is the finite set of threshold event symbols, $\delta : \Sigma \times \Theta \to 2^{\Sigma}$ is the partial state transition function (defined only for event symbols that are accepted at a given state), $\sigma_0 \in \Sigma$ is the initial state, and $\Sigma_F \subseteq \Sigma$ is the set of marker states. An infinite length sequence of symbols $w \in \Theta^{\omega}$ is *accepted* by the automaton B if:

(i) for $w = \theta_1 \theta_2 \ldots$ there exist $\sigma_1, \sigma_2, \ldots \in \Sigma$ such that $\sigma_k \in \delta(\sigma_{k-1}, \theta_k), \forall k = 1, 2, \ldots$; and

(ii) there exist $k_1 < k_2 < \ldots$ such that $\sigma_{k_j} \in \Sigma_F, \forall j = 1, 2, \ldots$.

Given an automaton B, the language of infinite-length sequences accepted by B is denoted by $L(B)$. Our objective is to create a sequence of automata, $B_1, B_2, \ldots$ such that for the given threshold event language L_{S_0} we have $L(B_1) \supseteq L(B_2) \supseteq \ldots \supseteq L_{S_0}$. In the case when there exists a finite-state automaton that accepts the language L_{S_0}, we would like to obtain an automaton B_n after a finite number n of iterations such that $L(B_n) = L_{S_0}$.

We introduce the following definitions and notation to describe the algorithm for computing the sequences of automata described above given a hybrid system as defined in Sec. 2. We define first, for each $i = 1, \ldots, n_v$, Q_i as the set of all possible discrete states reached when manifold M_i is reached, that is,

$$Q_i = \{q \in Q | \text{ there exists } q' \in Q, v \in V \text{ such that } v_i = 1 \text{ and } q = \delta(q', v)\}.$$

The sets of conditions which corresponds to states in each Q_i are defined as

$$U_i = \{u \in U | u = \phi(q) \text{ for some } q \in Q_i\}.$$

Any event caused by the continuous state trajectory at manifold M_i will lead to a transition to a discrete state in Q_i, which in turn will generate a mode defined by one of the elements of U_i.

Since we are interested in the state of the hybrid system only at the instants when the continuous-state trajectory encounters one of the manifolds, we define the set of all possible states at the manifolds as $\mathcal{M} = \{\bigcup_{i=1}^{n_v}(M_i \times Q_i)\}$. In the following we use the partition of $\mathcal{M}$ defined by the sets $\pi_{v,q}$, for $v \in V$, $q \in Q$, where

$$\pi_{v,q} = \{(x, q') \in \mathcal{M} | (x, q') \in \bigcap_{\{i | v_i = 1\}} (M_i \times Q_i), \quad q' = q\}$$

In words, $\pi_{v,q}$ is a distinct subset of a manifold, or the intersections of manifolds, with an associated discrete state, where v is the threshold event signal value generated when a continuous state trajectory hits the set of continuous states in $\pi_{v,q}$ and q is the discrete state entered when v occurs.

To represent the forward propagation of a set of hybrid states, we define the mapping $\mathfrak{S} : \mathcal{M} \cup S_0 \to \mathcal{M}$ where, in words, $\mathfrak{S}$ maps a hybrid state $(x, q) \in \mathcal{M}$ into the state in $\mathcal{M}$ corresponding to the first encounter with $\mathcal{M}$ along the state trajectory starting from (x, q). $\mathfrak{S}$ is extended to sets of states in the normal manner, and $\mathfrak{S}^{-1}$ denotes the inverse mapping, or pre-image, of $\mathfrak{S}$.

A set of hybrid states $R \subseteq M$ is said to be an *infinitely recurring set* if for any state trajectory starting at an initial condition $(x_0, q_0) \in S_0$ there exists an infinite sequence of time instants, $0 < t_1 < \ldots$, such that $(x(t_k), q(t_k)) \in R$, for all $k = 1, 2, \ldots$. In other words, any state trajectory starting in S_0 passes through R infinitely often. Since we assume there are an infinite number of threshold events along any state trajectory (A2), it follows that there will always be some infinitely recurring set as defined above. In particular, M is always an infinitely recurring set. In the sequel, we assume the initial state set S_0 is an infinitely recurring set.

Our procedure for constructing a finite state automaton that accepts an outer approximation of L_{S_0} is based on a two-stage analysis of the state trajectories that generate prefixes of strings in L_{S_0}. The final result is an automaton B_N that accepts all strings whose prefixes correspond exactly to the prefixes of strings in L_{S_0}. The correspondence is exact up to the threshold event that is generated by the Nth time the state trajectory passes through the infinitely recurring set S_0.

Stage one of the algorithm creates a tree in which each node corresponds to a threshold event along a feasible state trajectory. Each node is identified by a triple (l, m, n), where l is the number of events generated so far from the root node, m is an index on the nodes in the level l of the tree, and n is the number of times an events along the from the root node were generated by passing through S_0. The tree is created by computing subsets of the partition sets $\pi_{v,q}$ for each node that uniquely identify the subsets of continuous and discrete states that generate particular threshold event sequences from the root node. These subsets, denoted by $S^{l,m,n} \subseteq M$, are labels for the nodes. The tree is generated recursively as described in the following paragraphs.

The root node is associated with S_0 for which no triple is assigned. The nodes for the initial level of the tree ($l = 0$) are created for each nonempty set of the form $S^{0,m,0} = S_0 \cap \pi_{v,q}$; that is, the intersection of the initial set S_0 with each of the partition sets is computed and a node is created and an index m assigned for each nonempty intersection. Applying operator $\Im$ to each set $S^{0,m,0}$, the nodes at the next level of the tree are defined by the intersections $\Im(S^{0,m,0}) \cap \pi_{v,q}$. These new nodes are labeled $S^{1,m,0}$ if they are not subsets of S_0; and they are labeled $S^{1,m,1}$ if they are subsets of S_0, with m assigned as a distinct index for each new node. The construction of the tree continues in this manner for subsequent levels of the tree, and the iteration along any branch is terminated when $n = N$ at a node. This iterative procedure is illustrated in Fig.3. Eventually all branches are terminated since S_0 is assumed to be an infinitely recurring set.

The second stage of the algorithm starts at each of the terminal nodes in the tree from stage one and creates a single branch back to the root node. This decomposition of the intermediate nodes is created by applying the operator $\Im^{-1}$ iteratively from each terminal node to the initial node in the tree as illustrated in Fig. 4.

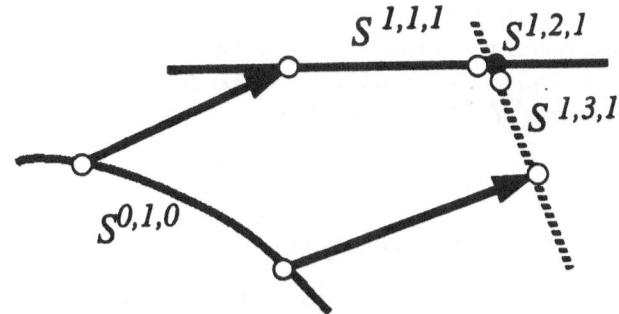

Fig. 3. Forward propagation of manifold-discrete state sets.

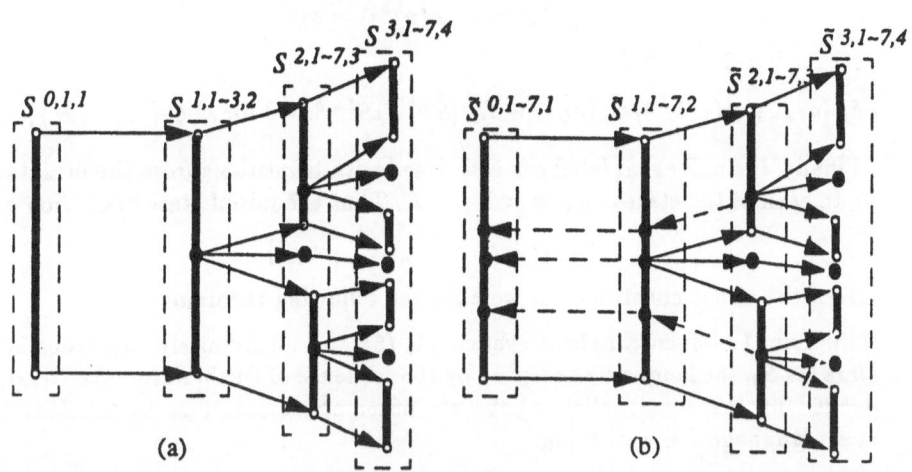

Fig. 4. Backward refinement of manifold discrete state sets: (a) forward propagation (b) backward refinement.

The refinements of the intermediate nodes are labeled by subsets of states $\tilde{S}^{l,k}$, where l is defined as before and k is an index assigned to the terminal node from which the refinement is initiated. Let K denote the number of terminal nodes in the tree from stage one and let L_k denote the length (maximum value of l) of the branch leading to terminal node k for $k = 1, \ldots, K$. The node labels in stage two are generated recursively for each k, for descending values of l, by the equations:

$$\tilde{S}^{L_k,k} = S^{L_k,m,N} \tag{8}$$

and, for $l = L_k, \ldots, 1$,

$$\tilde{S}^{l-1,k} = \mathfrak{S}^{-1}(\tilde{S}^{l,k}) \tag{9}$$

Associated with each $\tilde{S}^{l,k}$ for $k = 1, \ldots K, l = 1, \ldots, L_k$, there is a unique threshold event vector $v^{l,k}$ corresponding to the event vector generated when the continuous-state trajectory encounters the manifolds on which $\tilde{S}^{l,k}$ lies. Given $\tilde{S}^{l,k}$ the components of $v^{l,k}$ are defined by

$$v_i^{l,k} = \begin{cases} 1, & \text{if } P_{r_\mathcal{M}}(\tilde{S}^{l,k}) \subseteq \mathcal{M}_i \\ 0, & \text{otherwise} \end{cases} \tag{10}$$

where $P_{r_\mathcal{M}}$ denotes the projection onto the manifold set $\mathcal{M}$.

The Nth Buchi automation is defined as follows. The set of states is defined as

$$\Sigma_N = \{\sigma_{l,k} | l = 0, \ldots, L_k - 1; k = 1, \ldots, K\} \cup \{\sigma_0\}$$

The states $\sigma_{0,k}, k = 1 \ldots, K$, constitute the set of final marked states $\Sigma_{N,F}$, since they correspond to the partition of the infinitely recurring set S_0. The state transition function δ_N is defined for $k = 1, \ldots, K, l = 1, \ldots, L_k - 1$ as

$$\delta_N(\sigma_{l-1}, \Lambda(v^{l,k})) = \sigma_{l,k}$$

and for $k = 1, \ldots, K, l = L_k$ as

$$\delta_N(\sigma_{l,k-1}, \Lambda(v^{L_k,k})) = \{\sigma_{0,k'} \in \Sigma_N | \tilde{S}^{0,k'} \cap \tilde{S}^{L_k,k} \neq 0 \text{ for } k' \in \{1, \ldots, K\}\}. \tag{11}$$

Finally, the null event label ϵ is associated with transitions from the initial state σ_0 to each of the states $\sigma_{0,k}, k = 1, \ldots, K$. Thus the initial state transition set is

$$\delta_N(\sigma_0, \epsilon) = \{\sigma_{0,1}, \ldots, \sigma_{0,K}\}.$$

Given the above construction, we have the following theorem.

Theorem 1. Given a hybrid system (1)-(5) and an infinitely recurring initial state set S_0, the languages accepted by the sequence of Buchi automata $B_1, B_2, \ldots$ defined by the construction above are outer approximations to the threshold event language L_{S_0} satisfying:

(i)
$$Tr_N(L_{S_0}, S_0) = Tr_N(L(B_N), S_0),$$

where $Tr_N(L, S_0)$ denotes the set of finite length strings that are the prefixes to strings in a language $L \in \Theta^\omega$ associated with the given hybrid system truncated at the Nth occurrence of an event generated by passing through S_0, and

(ii) $L(B_1) \supseteq L(B_2) \supseteq \cdots \supseteq L_{S_0}$.

Proof. We first show that $\forall N, L(B_N) \supseteq L_S$. Given $(x_0, q_0) \in S_0$, let the sequence of states

$$(x_0, q_0)(x_1, q_1)(x_2, q_2)\ldots \tag{12}$$

be defined by the iteration $(x_{i+1}, q_{i+1}) = \Im(x_i, q_i)$. It follows from the definitions of the node labels $\tilde{S}^{l,k}$ that if $(x_0, q_0) \in \tilde{S}^{0,k}$ for $k_1 \in \{1, \ldots, K\}$ then for $l = 1, \ldots, L_{k_1}, (x_l, q_l) \in \tilde{S}^{l,k_1}$. Moreover, the threshold events for the trajectory are exactly v^{l,k_1} for $l = 1, \ldots, L_{k_1}$.

From the definition of state transition at (11), it follows that for some $k_2 \in \{1, \ldots K\}, (x_{L_{k_1}}, q_{L_{k_1}}) \in \tilde{S}^{0,k_2} \cap \tilde{S}^{L_{k_1},k_1}$. Therefore, $(x_{L_{k_1}}, q_{L_{K_1}}) \in \tilde{S}^{0,k_2}$. We have then $(x_{L_{k_1}+l}, q_{L_{k_1}+l}) \in \tilde{S}^{l,k_2}$ for $l = 1, \ldots, L_{k_2}$.

Continuing in this manner, it follows from the definition of δ_N that a state trajectory exists in B_N that accepts the events generated by the state sequence (12). Therefore,

$$L(B_N) \supseteq L_{S_0}$$

Statement (i) follows immediately from the construction above for the first L_k terms in the state sequence (12). Statement (ii) follows from the observation that the trees developed in stage one of the algorithm are identical up to the nodes generated for N. Therefore, the refinements leading to B_{N+1} are refinements of the nodes leading to B_N which implies (ii). $\square$

5 Two-integrator Example

We illustrate the calculation of approximating finite state machines using the following simple example of a hybrid system. Consider two uncoupled integrators which each integrate at a constant rate until the integrator state hits a threshold. The threshold events switch the sign of the integration rate, creating infinite cyclic behavior for each integrator. Let us distinguish the two integrators by subscript $i = 1, 2$. The differential equation for each integrator is $\dot{x}_i = f_{u_i}(x) = (-1)^{u_i} \rho_i$ where $u_i \in \{0, 1\}$ is the condition signal applied to the integrator i, and $\rho_i \in \Re^+$ is the rate of integration. Thus, when u_i is 0, x_i increases, and when u_i is 1, x_i decreases. Without loss of generality, we assume $\rho_1 = 1$. The output functions $g_{ij}(x)$ for $j = 1, 2$ are defined as $g_{i1}(x) = -x_i$, $g_{i2}(x) = x_i - 1$. The components of event vector defined by $v = \begin{bmatrix} v_{11} & v_{12} & v_{21} & v_{22} \end{bmatrix}^t$ are generated as $v_{ij} = 1 \leftrightarrow g_{ij}(x) = 0$.

The C/E system is defined by discrete state vector $q(t) = [\, q_1(t) \; q_2(t)\,]'$ and the state transition is defined as

$$q_i(t) = p_i(q_i(t^-), v_i(t)) = \begin{cases} 1, & \text{if } (q_i(t^-) = 0) \wedge (v_{i2} = 1) \\ 0, & \text{if } (q_i(t^-) = 1) \wedge (v_{i1} = 1) \\ q_i(t^-), & \text{otherwise} \end{cases}$$

The condition output of the C/E system which determines the modes of the integrators is defined as $u_i = q_i$. Figure 5 shows the state trajectory for a single integrator starting with zero initial state. When we consider two integrators, the continuous state moves in the unit square on $x_1 - x_2$ plane shown as Figure 6.

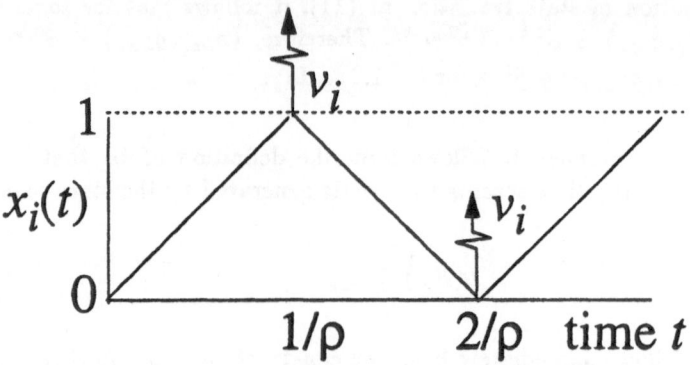

Fig. 5. Continuous state trajectory of an integrator starting from initial state $x_i(0) = 0$.

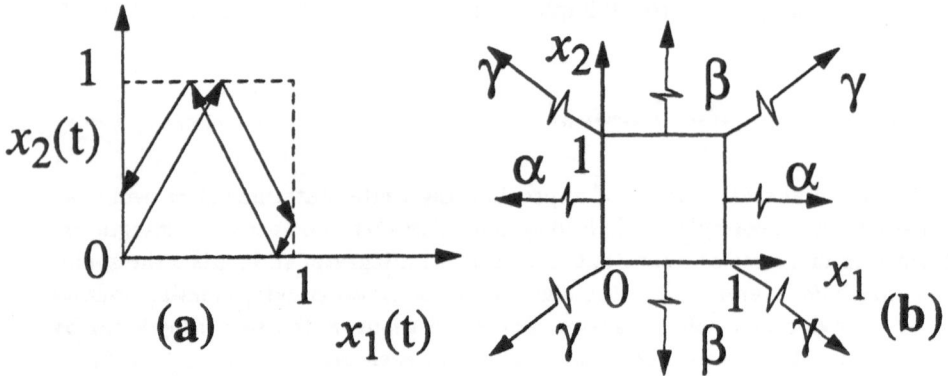

Fig. 6. Trajectory of double integrators and threshold manifolds for each events

Defining the set of threshold event symbols as $\Theta = \{\alpha, \beta, \gamma, \epsilon\}$, the event labeling function is defined by

$$\Lambda(v) = \begin{cases} \alpha & \text{if } (v_{11} + v_{12}) \neq 0 \text{ and } (v_{21} + v_{22}) = 0 \\ \beta & \text{if } (v_{11} + v_{12}) = 0 \text{ and } (v_{21} + v_{22}) \neq 0 \\ \gamma & \text{if } (v_{11} + v_{12})(v_{21} + v_{22}) \neq 0 \end{cases}$$

In words, α occurs when the first integrator state hits a limit and the second does not, β occurs when the second integrator state hits a limit and the first does not, and γ occurs whenever both integrator states hit limits simultaneously.

The initial state set S_0 is defined as the set of all possible states beginning on any of the threshold manifolds. Thus

$$S_0 = [0] \times [0, 1] \cup [1] \times [0, 1] \cup [0, 1] \times [0] \cup [0, 1] \times [1].$$

Proposition 2. *The two integrator example satisfies the assumptions A1 and A2.*

Proof. From the definition of the hybrid system, if v_{i1} is observed at time t when $x_i = 0$, the discrete state changes to 0 which means $\dot{x}_i = \rho_i$. Then $x(t+1/\rho_i) = 1$ and event v_{i2} is generated to turn the discrete state to 1 and $\dot{x}_i = -\rho_i$. By following similar steps, we conclude v_{i1} will occur at $t + 2/\rho_i$. Since $\rho_i > 0$ is constant for each integrator, the hybrid system satisfies assumption A1. The above analysis also shows that A2 also holds because starting at any state, the continuous trajectory will be on some manifold within $1/\rho_i$ time units. $\quad\square$

When the rate ρ_2 is rational the language becomes regular because the continuous state trajectory eventually comes back to the same place. This implies that all strings of L_{S_0} eventually become periodic. In case of an irrational rate, L_{S_0} can be nonregular, as shown in the proof of the following proposition.

Proposition 3. *For the two integrator example, if ρ_2 is irrational and $\rho_1 = 1$ then L_{S_0} is nonregular.*

To prove above proposition, we introduce following lemma.

Lemma 4. *Let η be irrational, $\alpha \in (0, 1)$ and R be any natural number, then for any $\epsilon > 0$, there exist natural numbers j, n,r such that $r > R$ and*

$$\alpha - \epsilon < j - n\frac{\lfloor r\eta \rfloor}{r} < \alpha + \epsilon \tag{13}$$

Proof of Lemma 4. Jacobi's theorem states that each orbit of the translation $\phi : x + \eta \pmod 1), \eta \in \Re$, is everywhere dense in $[0, 1)$ if and only if η is irrational (Arnold and Avez, 1968). Therefore, we can find j and n such that

$$\alpha - \frac{\epsilon}{2} < j - n\eta < \alpha + \frac{\epsilon}{2} \tag{14}$$

and for the above n, we can find r such that

$$0 < r\eta - \lfloor r\eta \rfloor < \frac{\epsilon}{2n}. \tag{15}$$

Moreover, r can be arbitrarily large, so choose $r > R$. Dividing (15) by r, subtracting η, multiplying the result by n and adding j gives

$$j - n\eta < j - n\frac{\lfloor r\eta \rfloor}{r} < j - n\eta + \frac{\epsilon}{2r}. \tag{16}$$

Applying (14) to the upper and lower bounds in (16) gives

$$\alpha - \frac{\epsilon}{2} < j - n\frac{\lfloor r\eta \rfloor}{r} < \alpha + \epsilon \tag{17}$$

which implies (13). $\quad\square$

Proof of Proposition 3. By the acceptance criterion for Buchi automata, if a language $L \in \Theta^\omega$ is regular, then there is necessarily a string $w \in L$ which is eventually periodic. That is, there are finite length strings $w', w_p \in \Theta^*$ such that $w = w'(w_p)^\omega$. We prove proposition 3 by assuming some trajectory for the two integrator example leads to such a string, and then demonstrate this leads to a contradiction when ρ_2 is irrational.

For $(x_0, q_0) \in S_0$, let $t_0 \in [0, \infty)$ be the first instant at which the continuous state trajectory is on the manifold $[0] \times [0, 1]$, that is $x_1(t_0) = 0$. Suppose that the string of event symbols w generated by the trajectory $(x(\bullet), q(\bullet))$ is periodic. If this is the case, we can choose finite length strings w', w_p such that $w = w'(w_p)^\omega$ where the last symbols in w' and w_p are generated by the continuous state trajectory hitting $[0] \times [0, 1]$. This implies that there exist some $t_1 > t_0$ and $T > 0$ such that (i) $x_1(t_1 + kT) = 0$ for all $k = 0, 1 \ldots$, and (ii) the symbols generated during any period T following t_1 are w_p. The last symbol in w' is generated by $x(t_1) = 0$. Note that $t_1 - t_0$ and T are integers since $\rho_1 = 1$.

Consider the number of times the state trajectory encounters the manifolds $[0, 1] \times [0]$ or $[0, 1] \times [1]$ (i.e., $x_2(t) \in \{0, 1\}$) starting from times t_1 and $t_2 = t_1 + T$ in a given duration Δ. These numbers are given by

$$N_1(\Delta) = \lfloor (t_1 + \Delta)\rho_2 \rfloor - \lfloor t_1 \rho_2 \rfloor$$

for $t_1 < t < t_1 + \Delta$, and

$$N_2(\Delta) = \lfloor (t_2 + \Delta)\rho_2 \rfloor - \lfloor t_2 \rho_2 \rfloor.$$

for $t_2 < t < t_2 + \Delta$.

Since the event symbol sequence is periodic following time t_1, if Δ is an integer, it must be the case $N_1(\Delta) = N_2(\Delta)$. We show, however, that there always exists some integer n such that

$$N_1(n) \neq N_2(n) \tag{18}$$

which is the desired contradiction since (x_0, q_0) was arbitrary.

Consider the values of $x_2(\bullet)$ at times t_1 and t_2 given by $x_2(t_1) = (t_1 - t_0)\rho_2 - \lfloor (t_1 - t_0)\rho_2 \rfloor$ and $x_2(t_2) = (t_2 - t_0)\rho_2 - \lfloor (t_2 - t_0)\rho_2 \rfloor$, respectively. We first note $x_2(t_1) \neq x_2(t_2)$ because $x_2(t_1) = x_2(t_2)$ would imply ρ_2 is rational, since $t_1 - t_0$ and $t_2 - t_0$ are integers. Suppose $x_2(t_1) < x_2(t_2)$ (the other case being equivalent for this proof). By lemma 4, we can choose j, n, r such that

$$x_2(t_1) < j - n\frac{\lfloor r\rho_2 \rfloor + 1}{r} < j - n\frac{\lfloor r\rho_2 \rfloor}{r} < x_2(t_2) \tag{19}$$

It follows for $N_1(n)$,

$$
\begin{aligned}
N_1(n) &= \lfloor (t_1 + n)\rho_2 \rfloor - \lfloor t_1 \rho_2 \rfloor \\
&= \lfloor n\rho_2 + x_2(t_1) \rfloor \\
&\leq \lfloor n\rho_2 + j - n\frac{\lfloor r\rho_2 \rfloor + 1}{r} \rfloor \\
&< j
\end{aligned}
$$

and for $N_2(n)$,

$$
\begin{aligned}
N_2(n) &= \lfloor (t_1 + n)\rho_2 \rfloor - \lfloor t_1\rho_2 \rfloor \\
&= \lfloor n\rho_2 + x_2(t_1) \rfloor \\
&\geq \lfloor n\rho_2 + j - n\frac{\lfloor r\rho_2 \rfloor}{r} \rfloor \\
&\geq j
\end{aligned}
$$

Therefore, $N_1(n) < j \leq N_2(n)$, which implies (18) to show the contradiction. □

We have implemented a computer program which generates finite state approximation of two integrator problem. Choosing $\rho_2 = 0.5$, the program identified finite state representation of the language generator as well as the set of partitions of manifolds as Fig. 7. When we choose $\rho_2 = 1.41421356$, the program obtained finite state approximation of generator as Fig. 8. In both cases the automata have been reduced to minimal state machines from the result given by the two-stage algorithm.

6 Discussion

This paper presents an algorithm for computing finite-state approximations to languages generated by threshold events in hybrid systems. The algorithm was implemented for the two-integrator example in Sec. 5. Numerical implementations of this algorithm will always give approximate results. Thus the concept of outer approximations needs to be incorporated into the computational implementation so that numerical truncation does not create incorrect representations of the reachable states.

There are several directions for further research. Conditions under which assumptions A1 and A2 are satisfied can be derived in terms of the equations defining the switching manifolds and the continuous-state dynamics. In applications the conditions for assumption A1 should characterize the requirements to avoid chattering in the switching rules. Conditions guaranteeing assumption A2 are basically criteria for state space reachability.

We are currently investigating potential applications for language approximations. It is of interest to determine what types of specifications can be verified using the approximate language representations. A more challenging problem is the use of such approximations in the synthesis of discrete controllers for hybrid systems. To use approximations for controller synthesis, the input-output structure of the original system need to be included in the definition of the finite state automata defining the approximate behaviors.

170

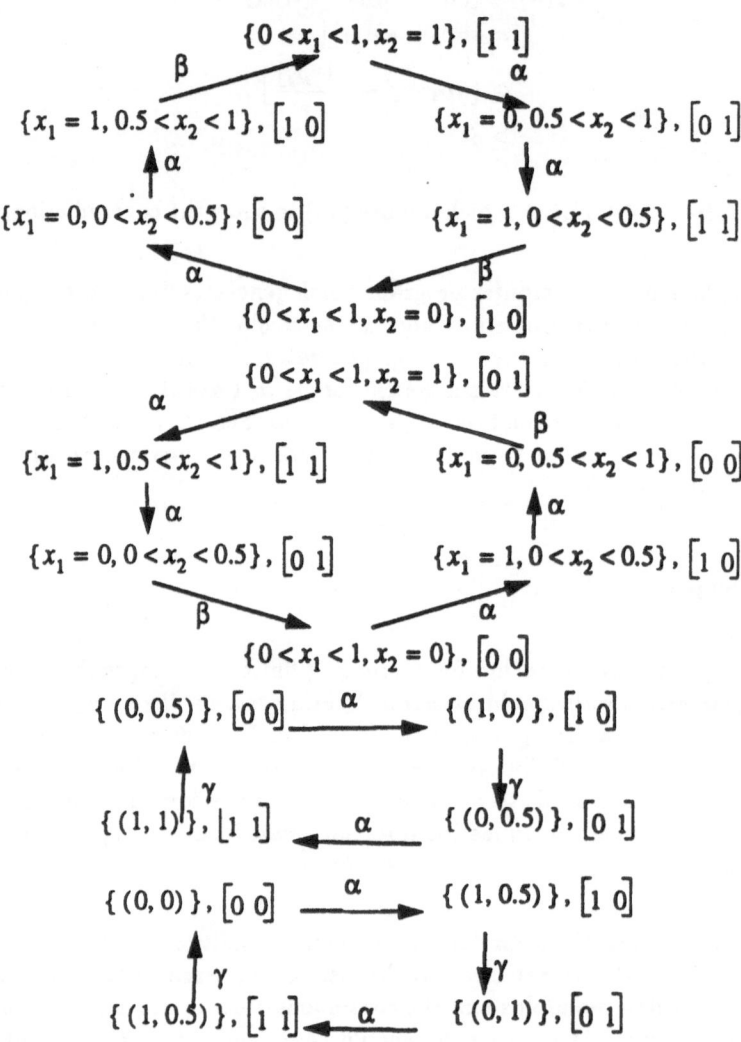

Fig. 7. Exact finite state generator for two integrator example with $\rho_2 = 0.5$

Table 1:

Hybrid State	Event	Discrete State	Intervals
1	γ	[0 0]	[0 0]
	γ	[0 1]	[0 1]
	γ	[1 0]	[1 0]
	γ	[1 1]	[1 1]
2	α	[0 0]	[0 1]-[0 0]
	α	[0 1]	[0 1]-[0 0]
	α	[1 0]	[1 1]-[1 0]
	α	[1 1]	[1 1]-[1 0]
3	β	[1 0]	[0.292893 1]
	β	[1 1]	[0.707107 1]
	β	[0 0]	[0.292893 0]
	β	[1 0]	[0.707107 0]
4	β	[1 1]	[0.292893 1]-[0 1]
	β	[1 1]	[1 1]-[0.707107 1]
	β	[0 0]	[0.292893 0]-[0 0]
	β	[1 0]	[1 0]-[0.707107 0]
5	β	[1 1]	[1 1]-[0.292893 1]
	β	[1 1]	[0.707107 1]-[0 1]
	β	[0 0]	[1 0]-[0.292893 0]
	β	[1 0]	[0.707107 0]-[0 0]

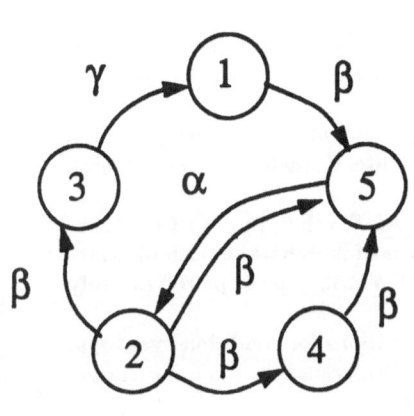

Fig. 8. Finite-state generator for outer approximation language for two-integrator example when $\rho_2 = 1.41421356$ for N=1.

References

Alur, R., T.A. Hensinger, and Ho,P.H.: Hybrid automata: an algorithmic approach to the specification, *Hybrid Systems*, pp.209-229, Springer Verlag, 1993.

Antsaklis, P.J., J.A. Stiver, and M. Lemmon, Hybrid system modeling and autonomous control systems, *Hybrid Systems*, Springer-Verlag; pp.366-392, 1993.

Arnold, V.I. and A. Aves, Ergodic Problems of Classical Mechanics, W.A. Benjamin, Inc., New York, p. 115-116, 1968 Chase, C., J. Serrano, and P.J. Ramadge, Periodicity and chaos from switched flow systems, *IEEE Trans. on Automatic Control*, pp. 70-83, Jan 1993.

Krogh, B.H., Condition/Event signal interfaces for block diagram modeling and analysis of hybrid systems, it Proceedings 1993 Symposium on Intelligent Control, pp. 180-185, Aug 1993.

Nerode, A., and W. Kohn, Models for hybrid systems: automata, topologies, stability. Technical Report 93-11, Mathematical Science Institute, Cornell University, March 1993.

Nerode, A., Linear automaton transformations, *Proc. AMS*, vol. 9, pp. 541-544, 1958

Ramadge,P., On the periodicity of symbolic observations of piecewise smooth discrete-time systems, *IEEE Trans. on Automatic Control*, Vol.35, No. 7, p807-713, July 1990.

Sreenivas, R.S. and B.H. Krogh, On discrete state realizations for condition event systems, *Proceedings Twenty-Eighth Annual Allerton Conference on Communication, Control and Computing*, Oct 1990, pp.465-74, 1991.

This article was processed using the LaTeX macro package with LLNCS style

Perfect Regulation of Linear Multivariable Systems
— a Low-and-High-Gain Design

Zongli Lin[1], Ali Saberi[2], Peddapullaiah Sannuti[3], Yacov Shamash[4]

[1] Department of Applied Mathematics, State University of New York at Stony Brook, Stony Brook, NY 11794-3600
[2] School of Electrical Engineering and Computer Science, Washington State University, Pullman, WA 99164-2752
[3] Department of Electrical and Computer Engineering, Rutgers University, P.O. Box 909, Piscataway, NJ 08855-0909
[4] College of Engineering and Applied Sciences, State University of New York at Stony Brook, Stony Brook, NY 11794-2200

Abstract. This paper is concerned with the design of 'Low-and-High-Gain' feedback controllers in order to achieve an objective called *perfect regulation* (p.r.). Necessary and sufficient conditions for the existence of state feedback controllers that achieve p.r. are derived for general linear multivariable time-invariant systems. More importantly, an explicit eigenstructure assignment procedure by state feedback that achieves p.r. is developed.

1   Introduction

Optimization of a cost criterion is a classical design tool. Design philosophy based on optimization has a rich and long history. For example, linear quadratic control problems (LQCP) belong to such a design philosophy. In an LQCP, the cost criterion is an infinite horizon integral whose integrand consists of a quadratic function in state and control vectors. The quadratic aspect of such a cost function is rooted in the well known least squares method for infimization of measurement error. A non-negative definite LQCP is the LQCP in which the quadratic integrand is non-negative definite. For a non-negative definite LQCP, the cost criterion can be reformulated or thought of as the L_2 norm of a controlled output vector. A non-negative definite LQCP is said to be *regular* when the coefficient matrix of the input in the controlled output vector is injective, otherwise it is said to be *singular*.

In the LQCP literature, a *perfect regulation* (p.r.) problem refers to the case when a state feedback controller is utilized and when the resulting infimum of the cost criterion (the L_2 norm of the controlled output) is zero. If solvable, in general, the p.r. problem requires a sequence or a family of state feedback control laws so that one can select a control law from the family such that the resulting L_2 norm of the controlled output is arbitrarily small. Obviously, the nature of state feedback gain required to achieve p.r., whenever it is possible, depends very much on the nature of the controlled output. If none of the controlled output variables are directly influenced by the control signal, and if there are no invariant zeros on the imaginary axis, then the state feedback is of *high-gain* type. This is so, since in this case the level of control signal does not explicitly contribute to the cost criterion, one can force the closed-loop system to have fast response so that the resulting cost criterion is arbitrarily small or zero. On the other hand, if the controlled output does not depend at all on the state, and it depends only on the input vector with its coefficient matrix being injective, then the controller is bound to be of *low-gain* type. In general, the controller for p.r. is a combination of low and high-gain type. Two issues arise in connection with a p.r. problem; the first is arriving at the existence conditions, i.e., the conditions under which a p.r. problem is solvable, and the second is a method of designing a family of controllers that achieve p.r.. The existence conditions are presently known for the case when the the controlled output does not have any direct feedthrough from the control input. As we shall prove in this paper, for general systems where the controlled output could have feedthrough from the control input, a p.r. problem is solvable if and only if the given system is right invertible and has all its invariant zeros in the closed left half complex plane. Regarding the methods of designing controllers that achieve p.r., there exists two methods; an Algebraic Riccati Equation (ARE) based method ([4] and [2]), and a direct eigenstructure assignment methodology [3].

Francis [2], using continuity arguments, approaches the design problem, via a 'regularization technique'. He provides an ARE based method of solving the p.r. problem when there is no feedthrough from the input to the controlled output. To be explicit, in this method, a sequence or a family of state feedback controllers is produced by solving an ARE parameterized in a parameter, say ε. For each specified $\varepsilon \neq 0$, one has to solve the ARE in order to obtain a member of the family of controllers. As ε tends to zero, the corresponding value of the cost criterion tends to zero. Thus, the method requires 'repetitive' solutions of AREs. Such an ARE based method has a major numerical problem in that the solution of the concerned ARE, especially for small ε, is in general numerically cumbersome and becomes 'stiff' owing to the low and high-gain nature of the state feedback gain. Besides 'stiffness', there are other issues such as the lack of freedom in assigning the resulting asymptotically infinite eigenstructure. These issues are discussed carefully elsewhere [11].

Kimura [3] approaches the design for p.r. from the perspective of direct closed-loop pole and eigenvector assignment. He does this for a number of reasons, e.g., (1) to alleviate the inherent computational difficulties of an ARE based method, (2) to gain freedom or flexibility in assigning the asymptotically infinite eigenstructure, and (3) to gain insights into the problem. However, one limitation of Kimura's method is that it excludes *a priori* the presence of invariant zeros of the given system on the imaginary axis. Kimura [3] does this by cleverly restricting himself to a class of parameterized state feedback gains having the property that as the gain tends to infinity the limits of all the resulting closed-loop root loci remain in the open left half complex plane. Kimura acknowledges the latent difficulties associated with the presence of invariant zeros on the imaginary axis, especially those having nontrivial Jordan blocks.

In this paper, we propose another direct eigenstructure assignment procedure to achieve p.r.whenever it is achievable. Unlike the existing literature, our design procedure is applicable to any general linear multivariable system. That is, it allows the the presence of invariant zeros on the imaginary axis. The design presented here can be viewed as an extension and generalization of our earlier 'asymptotic time-scale and eigenstructure assignment' (ATEA) design scheme [9]. Unlike the existing ATEA design scheme, where only high-gain feedback is involved, both the concepts of low and high-gains are utilized here. Also, the present ATEA method accommodates the presence of invariant zeros on the imaginary axis. Some attributes of our design scheme are enumerated below.

1. The design variable, i.e., state feedback gain, is parameterized directly in terms of a tuning parameter, say ε. The design equations for the gain can be solved without explicitly requiring values for ε. This implies that, unlike other methods such as ARE based designs where parameterization is implicit, no 'repetitive' solutions of our design equations are necessary as ε changes. In this sense, our design is a 'one-shot' design and thus ε truly acts as a tuning parameter. That is, when the controller is implemented by either hardware or software, the value of the tuning parameter can be adjusted either on line or off line so as to meet the required level of regulation.

2. Our design equations are developed using several subsystems of the given system. In this sense our design is decentralized. Such a decentralized method of design, reduces the computational complexity of designing a large scale system. Also, by adopting a standard method of design for each subsystem, the mechanics of performing the design are simplified. The computations required for each subsystem design do not involve arbitrarily small or large numbers. This implies that as the tuning parameter ε decreases, our design does not face any 'stiffness' problem which inherently cripples other design methods owing to the interaction of various slow and fast dynamic phenomena.

3. The asymptotic directions of asymptotically infinite eigenvalues of the closed-loop system can be assigned freely by the designer. On the other hand, such a freedom does not exist in an ARE based design (see for further details [11]).

A basic tool used for our design is a representation of the given system in the form of a special coordinate basis (SCB) which displays clearly both the finite and the infinite zero structure of the given system [8], [10]. Such an SCB has been utilized as a tool in a number of recent papers dealing with a variety of control related topics.

The paper is organized as follows. Sect. 2 introduces a formal definition of p.r. problem, while Sect. 3 recalls a special coordinate basis (SCB) which is instrumental for our development. Necessary and sufficient conditions for p.r. are stated in Sect. 4. Finally, Sect. 5 develops the 'Low-and-High-Gain' design procedure.

Throughout the paper, A' denotes the transpose of A, I denotes an identity matrix while I_k denotes an identity matrix of dimension $k \times k$. $\lambda(A)$ denotes the set of eigenvalues of A. The open left half and the open right half complex plane are respectively denoted by $\mathbf{C}^-$ and $\mathbf{C}^+$ while $\mathbf{C}^0$ represents the imaginary axis.

2 Problem Statement

Consider the following linear multivariable time invariant system

$$\Sigma : \begin{cases} \dot{x} = Ax + Bu, \quad x(0) = x_0 \\ \\ z = Cx + Du \end{cases} \tag{1}$$

where $x \in \mathbf{R}^n$ is the state, $u \in \mathbf{R}^m$ is the input, and $z \in \mathbf{R}^p$ is the controlled output. Let us also consider an associated cost criterion,

$$J(x_0, u) = \int_0^\infty \|z(t)\|^2 dt \ . \tag{2}$$

We make a standing assumption that the pair (A, B) is stabilizable. Moreover, without any loss of generality, we assume that the matrices $[C, \ D]$ and $\begin{bmatrix} B \\ D \end{bmatrix}$ are of full rank.

As mentioned in the introduction, our objective here is to develop a direct eigenstructure assignment design procedure to solve what can be called a perfect regulation (p.r.) problem the precise definition of which is given below.

Definition 1 (Perfect Regulation). Consider the system Σ as given by (1) along with the associated cost $J(x_0, u)$ as given by (2). Then the perfect regulation via state feedback control is defined as follows. Find, if possible, a family of parameterized linear state feedback laws $u = -F(\varepsilon)x$ having the following properties:
  (i) There exists an $\varepsilon^* > 0$ such that for all $\varepsilon \in (0, \varepsilon^*]$, the closed-loop system comprising of Σ and $u = -F(\varepsilon)x$ is internally stable;
  (ii) For each $x_0 \in \mathbf{R}^n$, one has

$$J(x_0, u) \rightarrow 0 \ \text{ as } \ \varepsilon \rightarrow 0 \ .$$

3 Preliminaries

In this section we shall first recall a special coordinate basis (SCB) of linear time-invariant systems [8, 10]. Such a coordinate basis has a distinct feature of explicitly displaying the finite and infinite zero structures of a given system as well as other system geometric properties. It plays an important part in developing an asymptotic time-scale and eigenstructure assignment procedure (ATEA) given in Sect. 5.

Consider the system Σ given by (1). It can be easily shown that using singular value decomposition one can always find an orthogonal transformation U and a nonsingular matrix V that render the direct feedthrough matrix D into the following form,

$$\bar{D} = UDV = \begin{bmatrix} I_{m_0} & 0 \\ 0 & 0 \end{bmatrix} \tag{3}$$

where m_0 is the rank of D. Thus the system in (1) can be rewritten as

$$\begin{cases} \dot{x} = A\,x + [B_0 \quad \hat{B}_1] \begin{bmatrix} u_0 \\ \hat{u}_1 \end{bmatrix} \\ \begin{bmatrix} z_0 \\ \hat{z}_1 \end{bmatrix} = \begin{bmatrix} C_0 \\ \hat{C}_1 \end{bmatrix} x + \begin{bmatrix} I_{m_0} & 0 \\ 0 & 0 \end{bmatrix} \begin{bmatrix} u_0 \\ \hat{u}_1 \end{bmatrix} \end{cases} \tag{4}$$

where the matrices B_0, $\hat{B}_1$, C_0 and $\hat{C}_1$ have appropriate dimensions. We have the following lemma.

Lemma 2 (SCB). *Consider the system Σ given in (1) and characterized by a matrix quadruple (A, B, C, D). Then, there exist*

1. *coordinate free non-negative integers n_a, n_b, n_c, n_d, $m_d \leq m - m_0$ and q_i. $i = 1, \cdots, m_d$, and*
2. *non-singular state, output and input transformations Γ_S, Γ_O and Γ_I which take the given Σ into a special coordinate basis that displays explicitly both the finite and infinite zero structures of Σ.*

The special coordinate basis which is referred to as the SCB, is described by the following set of equations:

$$x = \Gamma_S \bar{x}, \quad z = \Gamma_O \bar{z}, \quad u = \Gamma_I \bar{u}$$

$$\bar{x} = [\bar{x}_a', \ \bar{x}_b', \ \bar{x}_c', \ \bar{x}_d']', \quad \bar{x}_d = [\bar{x}_1', \ \bar{x}_2', \ \cdots, \bar{x}_{m_d}']'$$

$$\bar{z} = [\bar{z}_0', \ \bar{z}_d', \ \bar{z}_b']', \quad \bar{z}_d = [\bar{z}_1, \ \bar{z}_2, \ \cdots, \bar{z}_{m_d}]'$$

$$\bar{u} = [\bar{u}_0', \ \bar{u}_d', \ \bar{u}_c']', \quad \bar{u}_d = [\bar{u}_1, \ \bar{u}_2, \ \cdots, \bar{u}_{m_d}]'$$

and

$$\dot{\bar{x}}_a = A_{aa}\bar{x}_a + B_{0a}\bar{z}_0 + L_{ad}\bar{z}_d + L_{ab}\bar{z}_b \tag{5}$$

$$\dot{\bar{x}}_b = A_{bb}\bar{x}_b + B_{0b}\bar{z}_0 + L_{bd}\bar{z}_d, \quad \bar{z}_b = C_b\bar{x}_b \tag{6}$$

$$\dot{\bar{x}}_c = A_{cc}\bar{x}_c + B_{0c}\bar{z}_0 + L_{cb}\bar{z}_b + L_{cd}\bar{z}_d + B_c[E_{ca}\bar{x}_a + \bar{u}_c] \tag{7}$$

and for each $i = 1, \cdots, m_d$,

$$\dot{\bar{x}}_i = A_{q_i}\bar{x}_i + B_{0i}\bar{z}_0 + L_{id}\bar{z}_d + B_{q_i}\left[E_{ia}\bar{x}_a + E_{ib}\bar{x}_b + E_{ic}\bar{x}_c + \sum_{j=1}^{m_d} E_{ij}\bar{x}_j + \bar{u}_i\right] \tag{8}$$

$$\bar{z}_i = C_{q_i}\bar{x}_i, \quad \bar{z}_d = C_d\bar{x}_d \tag{9}$$

and

$$\bar{z}_0 = C_{0a}\bar{x}_a + C_{0b}\bar{x}_b + C_{0c}\bar{x}_c + \sum_{j=1}^{m_d} C_{0j}\bar{x}_j + \bar{u}_0 . \tag{10}$$

Here the states $\bar{x}_a$, $\bar{x}_b$, $\bar{x}_c$ and $\bar{x}_d$ are respectively of dimensions n_a, n_b, n_c and $n_d = \sum_{i=1}^{m_d} q_i$, while $\bar{x}_i$ is of dimension q_i for each $i = 1, \cdots, m_d$. The control vectors $\bar{u}_0$, $\bar{u}_d$ and $\bar{u}_c$ are respectively of dimensions m_0, m_d and $m_c = m - m_0 - m_d$, while the controlled output vectors $\bar{z}_0$, $\bar{z}_d$ and $\bar{z}_b$ are respectively of dimensions $p_0 = m_0$, $p_d = m_d$ and $p_b = p - p_0 - p_d$. The matrices A_{q_i}, B_{q_i} and C_{q_i} have the following form:

$$A_{q_i} = \begin{bmatrix} 0 & I_{q_i - 1} \\ 0 & 0 \end{bmatrix}, \quad B_{q_i} = \begin{bmatrix} 0 \\ 1 \end{bmatrix}, \quad C_{q_i} = [1, 0, \cdots, 0] . \tag{11}$$

(Obviously for the case when $q_i = 1$, we have $A_{q_i} = 0$, $B_{q_i} = 1$ and $C_{q_i} = 1$.) Furthermore, the pair (A_{cc}, B_c) is controllable and the pair (A_{bb}, C_b) is observable. Also, assuming that x_i are arranged such that $q_i \leq q_{i+1}$, the matrix L_{id} has the particular form,

$$L_{id} = [L_{i1}, L_{i2}, \cdots, L_{i\,i-1}, 0, 0, \cdots, 0] .$$

Also, the last row of each L_{id} is identically zero.

Proof : The proof of this lemma can be found in [8] and [10]. □

We can rewrite the SCB given by Lemma 2 in a more compact form.

$$\tilde{A} := \Gamma_S^{-1}(A - B_0 C_0)\Gamma_S = \begin{bmatrix} A_{aa} & L_{ab}C_b & 0 & L_{ad}C_d \\ 0 & A_{bb} & 0 & L_{bd}C_d \\ B_c E_{ca} & L_{cb}C_b & A_{cc} & L_{cd}C_d \\ B_d E_a & B_d E_b & B_d E_c & A_d \end{bmatrix}$$

$$\tilde{B} := \Gamma_S^{-1}[B_0 \quad \hat{B}_1]\Gamma_I = \begin{bmatrix} B_{0a} & 0 & 0 \\ B_{0b} & 0 & 0 \\ B_{0c} & 0 & B_c \\ B_{0d} & B_d & 0 \end{bmatrix}$$

$$\tilde{C} := \Gamma_O^{-1}\begin{bmatrix} C_0 \\ \hat{C}_1 \end{bmatrix}\Gamma_S = \begin{bmatrix} C_{0a} & C_{0b} & C_{0c} & C_{0d} \\ 0 & 0 & 0 & C_d \\ 0 & C_b & 0 & 0 \end{bmatrix}$$

and

$$\tilde{D} := \Gamma_O^{-1}D\Gamma_I = \begin{bmatrix} I_{m_0} & 0 & 0 \\ 0 & 0 & 0 \\ 0 & 0 & 0 \end{bmatrix} .$$

In what follows, we state some important properties of the SCB which are pertinent to our present work.

Property 3. *We note that (A_{bb}, C_b) and (A_{q_i}, C_{q_i}) form observable pairs. Unobservability could arise only in the variables x_a and x_c. In fact, the given system Σ is observable (detectable) if and only if (A_{obs}, C_{obs}) is an observable (detectable) pair, where*

$$A_{obs} = \begin{bmatrix} A_{aa} & 0 \\ B_c E_{ca} & A_{cc} \end{bmatrix}, \quad C_{obs} = \begin{bmatrix} C_{0a} & C_{0c} \\ E_a & E_c \end{bmatrix}$$

$$E_a = [E'_{1a} \quad E'_{2a} \quad \cdots \quad E'_{m_d a}]', \quad E_c = [E'_{1c} \quad E'_{2c} \quad \cdots \quad E'_{m_d c}]' \ .$$

Similarly, (A_{cc}, B_c) and (A_{q_i}, B_{q_i}) form controllable pairs. Uncontrollability could arise only in the variables x_a and x_b. In fact, Σ is controllable (stabilizable) if and only if (A_{con}, B_{con}) is a controllable (stabilizable) pair, where

$$A_{con} = \begin{bmatrix} A_{aa} & L_{ab}C_b \\ 0 & A_{bb} \end{bmatrix}, \quad B_{con} = \begin{bmatrix} B_{0a} & L_{ad} \\ B_{0b} & L_{bd} \end{bmatrix} \ .$$

Property 4. *The given system Σ is right-invertible if and only if x_b and hence z_b are nonexistent, left-invertible if and only if x_c and hence u_c are nonexistent, invertible if and only if both x_c and x_b are nonexistent.*

Property 5. *The eigenvalues of A_{aa} are the invariant zeros of Σ.*

Property 6. *The integer m_0 is the number of infinite zeros of order zero, and for $i = 1, 2, \cdots, m_d$, q_i corresponds to q_i number of infinite zeros of order q_i. In other words, the system Σ has m_0 number of infinite zeros of order zero and $\sum_{i=1}^{m_d} q_i$ number of infinite zeros of order greater than or equal to 1.*

4 Existence Conditions for p.r.

In this section, we develop the existence conditions for p.r. We have the following theorem.

Theorem 7. *Consider the system Σ as given in (1) along with the associated cost $J(x_0, u)$ as given in (2). Then the perfect regulation via state feedback $u = -F(\varepsilon)x$ is achievable if and only if the given system Σ is right invertible and has all its invariant zeros located in the closed left half s-plane $\mathbb{C}^- \cup \mathbb{C}^0$.*

Proof. The proof of this theorem for the special case of D being zero is given in [2], and it utilizes a perturbation and regularization technique. We offer here an alternative proof not based on any perturbation and regularization techniques. Also, it considers the general case where in the matrix D is not necessarily zero. To do so, we first define an auxiliary H_2 optimal control problem. Consider an auxiliary system,

$$\Sigma_a : \begin{cases} \dot{x} = Ax + Bu + Iw \\ z = Cx + Du \end{cases} \tag{12}$$

where, as before, $x \in \mathbf{R}^n$ is the state, $u \in \mathbf{R}^m$ is the input, $z \in \mathbf{R}^p$ is the controlled output, and $w \in \mathbf{R}^n$ is an unknown disturbance. The H_2 optimal control problem is defined as minimizing the H_2-norm of the transfer function from w to z over all possible internally stabilizing state feedback laws. Let γ^* denote the infimum of such an optimization. It is then straightforward to verify that the perfect regulation problem defined here is solvable if and only if γ^* for the above auxiliary H_2 optimal control problem is zero. However. from [13], we know that

$$\gamma^* = \sqrt{\text{trace } P}$$

where $P \geq 0$ is the largest solution of a linear matrix inequality,

$$\begin{bmatrix} A'P + PA + C'C & PB + C'D \\ B'P + D'C & D'D \end{bmatrix} \geq 0 .$$

As such, we conclude that $\gamma^* = 0$ if and only if $P = 0$. We also recall now from [12] that $P = 0$ if and only if the given system Σ is right invertible and has all its invariant zeros located in the closed left half s-plane $\mathbf{C}^- \cup \mathbf{C}^0$. $\qquad \square$

5 A Direct Eigenstructure Assignment Design for p.r.

In this section, we construct a state feedback gain that achieves p.r. whenever it is achievable. As discussed in introduction, our method is a direct eigenstructure assignment procedure which utilizes both low and high gains. We start with the development of a 'Low-Gain' design procedure which follows concepts similar to those in [5] and [6] but is somewhat different from them. It is of paramount importance to our construction of 'Low-and-High-gain' design that achieves p.r. .

5.1 A 'Low-Gain' Design Procedure

We need the following preliminary result before we describe the 'Low-Gain' design method.

Lemma 8. *Consider a linear single input system in the controllable canonical form*

$$\dot{x} = Ax + Bu, \quad x \in \mathbf{R}^n, \, u \in \mathbf{R}$$

where

$$A = \begin{bmatrix} 0 & 1 & 0 & \cdots & 0 \\ 0 & 0 & 1 & \cdots & 0 \\ \vdots & \vdots & \vdots & \ddots & \vdots \\ 0 & 0 & 0 & \cdots & 1 \\ -a_n & -a_{n-1} & -a_{n-2} & \cdots & -a_1 \end{bmatrix}, \quad B = \begin{bmatrix} 0 \\ 0 \\ \vdots \\ 0 \\ 1 \end{bmatrix} .$$

Here as usual x is a state vector while u is a control variable. Assume that all the eigenvalues of A are in the closed left half s-plane $\mathbf{C}^- \cup \mathbf{C}^0$. Let $F(\varepsilon) \in \mathbf{R}^{1 \times n}$

be a state feedback gain such that $\lambda(A - BF(\varepsilon)) = -\varepsilon + \lambda(A)$. Then, there exists an $\varepsilon^* > 0$ such that for all $\varepsilon \leq (0, \varepsilon^*]$,

$$\|F(\varepsilon)\| \leq \kappa\varepsilon \tag{13}$$

$$\|F(\varepsilon)e^{(A-BF(\varepsilon))t}\| \leq \alpha\varepsilon e^{-\varepsilon t/2} \tag{14}$$

$$\|e^{(A-BF(\varepsilon))t}\| \leq \frac{\beta}{\varepsilon^{r-1}}e^{-\varepsilon t/2} \tag{15}$$

for some $\kappa > 0$, $\alpha > 0$ and $\beta > 0$ independent of ε, where r is the largest algebraic multiplicity of the eigenvalues of A.

Proof. See [5]. □

'Low-Gain' design method:

We now give the 'Low-Gain' design procedure for the linear system

$$\dot{x} = Ax + Bu, \quad x \in \mathbf{R}^n, \ u \in \mathbf{R}^m \tag{16}$$

where we assume that (A, B) is stabilizable and all the eigenvalues of A are in the closed-left half s-plane $\mathbf{C}^- \cup \mathbf{C}^0$. The 'Low-Gain' design we are proposing is carried out in three steps.

Step 1 : Find the state transformation T ([1]) such that $(T^{-1}AT, T^{-1}B)$ is in the following form,

$$T^{-1}AT = \begin{bmatrix} A_1 & A_{12} & \cdots & A_{1q} & 0 \\ 0 & A_2 & \cdots & A_{2q} & 0 \\ \vdots & \vdots & \ddots & \vdots & \vdots \\ 0 & 0 & \cdots & A_q & 0 \\ 0 & 0 & 0 & 0 & A_0 \end{bmatrix}, \quad T^{-1}B = \begin{bmatrix} B_1 & 0 & \cdots & 0 & * \\ 0 & B_2 & \cdots & 0 & * \\ \vdots & \vdots & \ddots & \vdots & \vdots \\ 0 & 0 & \cdots & B_q & * \\ B_{01} & B_{02} & \cdots & B_{0q} & * \end{bmatrix}$$

(where q is an integer) and for $i = 1, 2, \cdots, q$,

$$A_i = \begin{bmatrix} 0 & 1 & 0 & \cdots & 0 \\ 0 & 0 & 1 & \cdots & 0 \\ \vdots & \vdots & \vdots & \ddots & \vdots \\ 0 & 0 & 0 & \cdots & 1 \\ -a^i_{n_i} & -a^i_{n_i-1} & -a^i_{n_i-2} & \cdots & -a^i_1 \end{bmatrix}, \quad B_i = \begin{bmatrix} 0 \\ 0 \\ \vdots \\ 0 \\ 1 \end{bmatrix}.$$

Furthermore, the transformation T is such that all the eigenvalues of A_i are on the jw-axis and all the eigenvalues of A_0 have strictly negative real parts. Here and elsewhere $*$'s represent submatrices of less interest.

Step 2 : For each (A_i, B_i), let $F_i(\varepsilon) \in \mathbf{R}^{1 \times n_i}$ be the state feedback gain such that

$$\lambda(A_i - B_i F_i(\varepsilon)) = -\varepsilon + \lambda(A_i) \in \mathbf{C}^-.$$

Such an $F_i(\varepsilon)$ exists and is unique since (A_i, B_i) is a single input controllable pair.

Step 3 : Let

$$u = -F(\varepsilon)x \qquad (17)$$

where the state feedback gain matrix $F(\varepsilon)$ is given as

$$F(\varepsilon) = \begin{bmatrix} F_1(\varepsilon^{2^{q-1}(r_2+1)(r_3+1)\cdots(r_q+1)}) & 0 \\ 0 & F_2(\varepsilon^{2^{q-2}(r_3+1)(r_4+1)\cdots(r_q+1)}) \\ \vdots & \vdots \\ 0 & 0 \\ 0 & 0 \\ 0 & 0 \end{bmatrix}$$

$$\begin{matrix} \cdots & 0 & 0 & 0 \\ \cdots & 0 & 0 & 0 \\ \ddots & \vdots & \vdots & \vdots \\ \cdots & F_{q-1}(\varepsilon^{2(r_q+1)}) & 0 & 0 \\ \cdots & 0 & F_q(\varepsilon) & 0 \\ \cdots & 0 & 0 & 0 \end{matrix} \Bigg] T^{-1} \quad (18)$$

and where r_i is the largest algebraic multiplicity of the eigenvalues of A_i.

The parameterized state feedback gain $F(\varepsilon)$ as given by (18) has the following prominent property.

Theorem 9. *Consider the linear system as given by (16). Suppose that (A, B) is stabilizable and all the eigenvalues of A are in the closed-left half s-plane. Then we have the following properties:*

1. The closed-loop system matrix $A - BF(\varepsilon)$ is Hurwitz-stable for all $\varepsilon > 0$.
2. There exists an $\varepsilon^* > 0$ such that for all $\varepsilon \in (0, \varepsilon^*]$,

$$\|F(\varepsilon)\| \le \kappa\varepsilon \qquad (19)$$

$$\|e^{(A-BF(\varepsilon))t}\| \le \frac{\delta_0}{\varepsilon^{2^{q-1}(r_1+1)(r_2+1)\cdots(r_q+1)}} e^{-\varepsilon^{2^{q-1}(r_1+1)(r_2+1)\cdots(r_q+1)}t/2} \qquad (20)$$

$$\|F(\varepsilon)\, e^{(A-BF(\varepsilon))t}\| \le \delta_q \varepsilon e^{-\varepsilon t/2} + \delta_{q-1}\varepsilon^{r_q+2} e^{-\varepsilon^{2(r_q+1)}t/2} + \cdots$$
$$+ \delta_1 \varepsilon^{2^{q-2}(r_2+1)(r_3+1)\cdots(r_q+1)+1} e^{-\varepsilon^{2^{q-1}(r_2+1)(r_3+1)\cdots(r_q+1)}t/2} \qquad (21)$$

where κ and δ_i's are some ε independent positive constant numbers.

Proof. Item 1 is apparent. We proceed directly to the proof of Item 2 of the theorem. Note that (19) follows readily from (13) of Lemma 8. To show (20) and (21), we introduce the following state transformation

$$x = T\tilde{x}, \quad \tilde{x} = [\tilde{x}'_1, \tilde{x}'_2, \cdots, \tilde{x}'_q, \tilde{x}'_0]', \quad \tilde{x}_i \in \mathbf{R}^{n_i}, \quad \tilde{x}_0 \in \mathbf{R}^{n_0}$$

and write the closed-loop system dynamics in the new state $\tilde{x}$ as follows,

$$\dot{\tilde{x}}_i = (A_i - B_i F_i(\varepsilon))\tilde{x}_i + \sum_{k=i+1}^{q} A_{ik}\tilde{x}_k, \quad i = 1, 2, \cdots, q,$$

$$\dot{\tilde{x}}_0 = A_0\tilde{x}_0 - \sum_{k=1}^{q} B_{0k}F_k(\varepsilon^{2^{q-k}(r_{k+1}+1)(r_{k+2}+1)\cdots(r_q+1)})\tilde{x}_k$$

By Lemma 8, there exists an $\varepsilon^*_q \in (0, 1]$ such that for all $\varepsilon \in (0, \varepsilon^*_q]$,

$$\|F_q(\varepsilon)e^{(A_q-B_qF_q(\varepsilon))t}\| \le \alpha_q\varepsilon e^{-\varepsilon t/2}$$

$$\|e^{(A_q-B_qF_q(\varepsilon))t}\| \le \frac{\beta_q}{\varepsilon^{r_q-1}}e^{-\varepsilon t/2}$$

for some $\alpha_q > 0$ and $\beta_q > 0$.

We then have

$$\|F_q(\varepsilon)\tilde{x}_q\| = \|F_q(\varepsilon)e^{(A_q-B_qF_q(\varepsilon))t}\tilde{x}_q(0)\|$$
$$\le \alpha_q\varepsilon e^{-\varepsilon t/2}\|\tilde{x}_q(0)\| \le \alpha'_q\varepsilon e^{-\varepsilon t/2}\|\tilde{x}(0)\| \quad (22)$$

for some positive constant α'_q independent of ε. Also we have, for all $\varepsilon \in (0, \varepsilon^*_q]$,

$$\|\tilde{x}_q\| = \|e^{(A_q-B_qF_q(\varepsilon))t}\tilde{x}_q(0)\|$$
$$\le \frac{\beta_q}{\varepsilon^{r_q-1}}e^{-\varepsilon t/2}\|\tilde{x}_q(0)\| \le \frac{\beta'_q}{\varepsilon^{r_q-1}}e^{-\varepsilon t/2}\|\tilde{x}(0)\| \quad (23)$$

for some positive constant β'_q independent of ε.

Viewing $\tilde{x}_q$ as an input to the dynamics of $\tilde{x}_{q-1}$ and again using Lemma 8, we have, for some $\varepsilon^*_{q-1} \in (0, \varepsilon^*_q]$, and for all $\varepsilon \in (0, \varepsilon^*_{q-1}]$,

$$\|F_{q-1}(\varepsilon^{2(r_q+1)})\tilde{x}_{q-1}\| \le \|F_{q-1}(\varepsilon^{2(r_q+1)})e^{(A_{q-1}-B_{q-1}F_{q-1}(\varepsilon^{2(r_q+1)}))t}\tilde{x}_{q-1}(0)\|$$

$$+\|F_{q-1}(\varepsilon^{2(r_q+1)})\int_0^t e^{(A_{q-1}-B_{q-1}F_{q-1}(\varepsilon^{2(r_q+1)}))(t-\tau)}A_{q-1q}\tilde{x}_q(\tau)d\tau\|$$

$$\le \alpha_{q-1}\varepsilon^{2(r_q+1)}e^{-\varepsilon^{2(r_q+1)}t/2}\|\tilde{x}(0)\|$$

$$+\frac{\alpha_{q-1}\beta'_q\|A_{q-1q}\|}{\varepsilon^{r_q-1}}\varepsilon^{2(r_q+1)}\int_0^t e^{-\varepsilon^{2(r_q+1)}(t-\tau)/2}e^{-\varepsilon\tau/2}d\tau\|\tilde{x}(0)\|$$

$$\le \alpha_{q-1}\varepsilon^{2(r_q+1)}e^{-\varepsilon^{2(r_q+1)}t/2}\|\tilde{x}(0)\|$$

$$+\alpha_{q-1}\beta'_q\|A_{q-1q}\|\varepsilon^{r_q+3}e^{-\varepsilon^{2(r_q+1)}t/2}\int_0^t e^{-(\varepsilon-\varepsilon^{2(r_q+1)})\tau/2}d\tau\|\tilde{x}(0)\|$$

$$\leq \alpha_{q-1}\varepsilon^{2(r_q+1)}e^{-\varepsilon^{2(r_q+1)}t/2}\|\tilde{x}(0)\|$$
$$+4\alpha_{q-1}\beta'_q\|A_{q-1q}\|\varepsilon^{r_q+2}e^{-\varepsilon^{2(r_q+1)}t/2}\|\tilde{x}(0)\|$$
$$\leq \alpha'_{q-1}\varepsilon^{r_q+2}e^{-\varepsilon^{2(r_q+1)}t/2}\|\tilde{x}(0)\| \tag{24}$$

for some positive constant α'_{q-1} independent of ε, and similarly.

$$\|\tilde{x}_{q-1}\| \leq \|e^{(A_{q-1}-B_{q-1}F_{q-1}(\varepsilon^{2(r_q+1)}))t}\tilde{x}_{q-1}(0)\|$$
$$+\|\int_0^t e^{(A_{q-1}-B_{q-1}F_{q-1}(\varepsilon^{2(r_q+1)}))(t-\tau)}A_{q-1q}\tilde{x}_q(\tau)d\tau\|$$
$$\leq \frac{\beta_{q-1}}{\varepsilon^{2(r_q+1)(r_{q-1}-1)}}e^{-\varepsilon^{2(r_q+1)}t/2}\|\tilde{x}_{q-1}(0)\|$$
$$+\frac{\beta_{q-1}\beta'_q\|A_{q-1q}\|}{\varepsilon^{2(r_q+1)(r_{q-1}-1)}\varepsilon^{r_q-1}}\int_0^t e^{-\varepsilon^{2(r_q+1)}(t-\tau)/2}e^{-\varepsilon\tau/2}d\tau\|\tilde{x}(0)\|$$
$$\leq \frac{\beta_{q-1}}{\varepsilon^{2(r_q+1)r_{q-1}-1}}e^{-\varepsilon^{2(r_q+1)}t/2}\|\tilde{x}(0)\|$$
$$+\frac{4\beta_{q-1}\beta'_q\|A_{q-1q}\|}{\varepsilon^{2(r_q+1)r_{q-1}-1}}e^{-\varepsilon^{2(r_q+1)}t/2}\|\tilde{x}(0)\|$$
$$\leq \frac{\beta'_{q-1}}{\varepsilon^{2r_{q-1}(r_q+1)-1}}e^{-\varepsilon^{2(r_q+1)}t/2}\|\tilde{x}(0)\| \tag{25}$$

for some positive constant β'_{q-1} independent of ε.

Continuing in the same manner, we can show that for each $i = q-2, q-3, \cdots 1$. there exists $\varepsilon_i^* \in (0, \varepsilon_{i+1}^*]$ such that for all $\varepsilon \in (0, \varepsilon_i^*]$,

$$\|F_i(\varepsilon^{2^{q-1}(r_{i+1}+1)\cdots(r_q+1)})\tilde{x}_i\|$$
$$\leq \alpha'_i\varepsilon^{2^{q-i-1}(r_{i+1}+1)\cdots(r_q+1)+1}e^{-\varepsilon^{2^{q-i}(r_{i+1}+1)(r_{i+2}+1)\cdots(r_q+1)}t/2}\|\tilde{x}(0)\| \tag{26}$$

and

$$\|\tilde{x}_i\| \leq \frac{\beta'_i}{\varepsilon^{2^{q-i}r_i(r_{i+1}+1)(r_{i+2}+1)\cdots(r_q+1)-1}}e^{-\varepsilon^{2^{q-i}(r_{i+1}+1)(r_{i+2}+1)\cdots(r_q+1)}t/2}\|\tilde{x}(0)\| \tag{27}$$

for some positive constants α'_i and β'_i independent of ε.

Finally, noting that all the eigenvalues of A_0 are in the open left half s-plane, and using (22), (24) and (26), it follows readily that there exists an $\varepsilon_0^* \in (0, \varepsilon_1^*]$. such that for all $\varepsilon \in (0, \varepsilon_0^*]$,

$$\|\tilde{x}_0\| \leq \beta'_0 e^{-\varepsilon^{2^{q-1}(r_2+1)(r_3+1)\cdots(r_q+1)}t/2}\|\tilde{x}(0)\| \tag{28}$$

for some positive constant β'_0 independent of ε.

Taking $\varepsilon^* = \varepsilon_0^*$, we have that, for all $\varepsilon \in (0, \varepsilon^*]$, (22)-(28) hold.

Now, noting that

$$\tilde{x}(t) = e^{T^{-1}(A-BF(\varepsilon))Tt}\tilde{x}(0)$$

and

$$F(\varepsilon)T\tilde{x}(t) = F(\varepsilon)Te^{T^{-1}(A-BF(\varepsilon))Tt}\tilde{x}(0)$$

and using (22)-(28), we have, for all $\varepsilon \in (0, \varepsilon^*]$,

$$
\begin{aligned}
\|\tilde{x}(t)\| &= \|e^{T^{-1}(A-BF(\varepsilon))Tt}\tilde{x}(0)\| \\
&\leq \|\tilde{x}_0(t)\| + \sum_{i=1}^{q}\|\tilde{x}_i(t)\| \\
&\leq \beta_0' e^{-\varepsilon^{2^{q-1}}\prod_{j=2}^{q}(r_j+1)}t/2\|\tilde{x}(0)\| \\
&\quad + \sum_{i=1}^{q}\frac{\beta_i'}{\varepsilon^{2^{q-i}r_i\prod_{j=i+1}^{q}(r_j+1)-1}}e^{-\varepsilon^{2^{q-i}}\prod_{j=i+1}^{q}(r_j+1)}t/2\|\tilde{x}(0)\| \\
&\leq \frac{\sum_{i=0}^{q}\beta_i'}{\varepsilon^{2^{q-1}}\prod_{j=1}^{q}(r_j+1)}e^{-\varepsilon^{2^{q-1}}\prod_{j=1}^{q}(r_j+1)}t/2\|\tilde{x}(0)\|
\end{aligned} \tag{29}
$$

and

$$
\begin{aligned}
\|F(\varepsilon)\,T\,\tilde{x}(t)\| &= \|F(\varepsilon)Te^{T^{-1}(A-BF(\varepsilon))Tt}\tilde{x}(0)\| \\
&\leq \sum_{i=1}^{q}\|F_i(\varepsilon^{2^{q-i}}\prod_{j=i+1}^{q}(r_j+1))\tilde{x}_i(t)\| \\
&\leq (\alpha_q'\varepsilon e^{-\varepsilon t/2} + \alpha_{q-1}'\varepsilon^{r_q+2}e^{-\varepsilon^2(r_q+1)t/2} + \cdots \\
&\quad + \alpha_1'\varepsilon^{2^{q-2}(r_2+1)(r_3+1)\cdots(r_q+1)+1}e^{-\varepsilon^{2^{q-1}}(r_2+1)(r_3+1)\cdots(r_q+1)t/2})\|\tilde{x}(0)\|\ .
\end{aligned} \tag{30}
$$

Since (29) and (30) hold for all $\tilde{x}(0)$, it follows readily that

$$
\|e^{T^{-1}(A-BF(\varepsilon))Tt}\| \leq \frac{\sum_{i=0}^{q}\beta_i'}{\varepsilon^{2^{q-1}}\prod_{j=1}^{q}(r_j+1)}e^{-\varepsilon^{2^{q-1}}\prod_{j=1}^{q}(r_j+1)}t/2 \tag{31}
$$

and

$$
\begin{aligned}
\|F(\varepsilon)\,&T\,e^{T^{-1}(A-BF(\varepsilon))Tt}\| \\
&\leq \alpha_q'\varepsilon e^{-\varepsilon t/2} + \alpha_{q-1}'\varepsilon^{r_q+2}e^{-\varepsilon^2(r_q+1)t/2} + \cdots \\
&\quad + \alpha_1'\varepsilon^{2^{q-2}(r_2+1)(r_3+1)\cdots(r_q+1)+1}e^{-\varepsilon^{2^{q-1}}(r_2+1)(r_3+1)\cdots(r_q+1)t/2}\ .
\end{aligned}
$$

Hence

$$
\begin{aligned}
\|e^{(A-BF(\varepsilon))t}\| &= \|Te^{T^{-1}(A-BF(\varepsilon))Tt}T^{-1}\| \\
&\leq \frac{\delta_0}{\varepsilon^{2^{q-1}}\prod_{j=1}^{q}(r_j+1)}e^{-\varepsilon^{2^{q-1}}\prod_{j=1}^{q}(r_j+1)}t/2
\end{aligned} \tag{32}
$$

and

$$
\begin{aligned}
\|F(\varepsilon)\,e^{(A-BF(\varepsilon))t}\| &= \|F(\varepsilon)Te^{T^{-1}(A-BF(\varepsilon))Tt}T^{-1}\| \\
&\leq \delta_q\varepsilon e^{-\varepsilon t/2} + \delta_{q-1}\varepsilon^{r_q+2}e^{-\varepsilon^2(r_q+1)t/2} + \cdots \\
&\quad + \delta_1\varepsilon^{2^{q-2}(r_2+1)(r_3+1)\cdots(r_q+1)+1}e^{-\varepsilon^{2^{q-1}}(r_2+1)(r_3+1)\cdots(r_q+1)t/2}
\end{aligned} \tag{33}
$$

where

$$\delta_0 = \|T\|\|T^{-1}\| \sum_{i=0}^{q} \beta'_i$$

and

$$\delta_i = \|T^{-1}\|\alpha'_i, \quad i = 1, 2, \cdots, q .$$

This shows (20) and (21). $\qquad\qquad\qquad\qquad\qquad\qquad\qquad\qquad\qquad\qquad$ $\square$

5.2 'Low-and-High-Gain' Design For p.r.

In this section, we assume that p.r. for the given system Σ is achievable, i.e, we assume that Σ is stabilizable, and moreover, it is right invertible and has all its invariant zeros located in the closed left half s-plane.

Our goal here of course is to design a family of state feedback laws, parameterized in ε, which achieves perfect regulation for the system Σ. This family of state feedback laws is constructed in the following steps.

'Low-and-High-Gain' design method:

Step 1 (Construction of the SCB of Σ): Perform a nonsingular state, input and controlled output transformation on the system Σ. That is, let

$$x = \Gamma_S \bar{x}, \ z = \Gamma_O \bar{z}, \ u = \Gamma_I \bar{u}$$

such that the system Σ can be written in the following SCB form,

$$\bar{x} = [\bar{x}'_a, \bar{x}'_c, \bar{x}'_d]', \ \bar{x}_d = [\bar{x}'_1, \bar{x}'_2, \cdots, \bar{x}'_{m_d}]', \ \bar{x}_i = [\bar{x}_{i1}, \bar{x}_{i2}, \cdots, \bar{x}_{iq_i}]'$$
$$\bar{z} = [\bar{z}'_0, \bar{z}'_d], \ \bar{z}_d = [\bar{z}_1, \bar{z}_2, \cdots, \bar{z}_{m_d}]', \ \bar{z}_i = \bar{x}_{i1}$$
$$\bar{u} = [\bar{u}'_0, \bar{u}'_d, \bar{u}'_c]', \ \bar{u}_d = [\bar{u}_1, \bar{u}_2, \cdots, \bar{u}_{m_d}]'$$

and

$$\dot{\bar{x}}_a = A_{aa}\bar{x}_a + B_{0a}\bar{z}_0 + L_{ad}\bar{z}_d \qquad\qquad (34)$$

$$\dot{\bar{x}}_c = A_{cc}\bar{x}_c + B_{0c}\bar{z}_0 + L_{cd}\bar{z}_d + B_c[E_{ca}\bar{x}_a + \bar{u}_c] \qquad\qquad (35)$$

$$\dot{\bar{x}}_i = A_{q_i}\bar{x}_i + B_{0i}\bar{z}_0 + L_i\bar{z}_d + B_{q_i}[E_{ia}\bar{x}_a + E_{ic}\bar{x}_c + \sum_{j=1}^{m_d} E_{ij}\bar{x}_j + \bar{u}_i] .$$

$$i = 1, 2, \cdots, m_d \qquad\qquad (36)$$

$$\bar{z}_0 = C_{0a}\bar{x}_a + C_{0c}\bar{x}_c + \sum_{j=1}^{m_d} C_{0j}\bar{x}_j + \bar{u}_0 . \qquad\qquad (37)$$

We note that the state $\bar{x}_b$ is not present as Σ is right invertible.

Step 2 (Construction of a parameterized low-gain matrix $F_a(\varepsilon)$): By the Property 3 of the SCB, the pair $(A_{aa}, [B_{0a}, L_{ad}])$ is stabilizable. Moreover, by Property 5 of the SCB, the eigenvalues of A_{aa} are the invariant zeros of the system Σ and, hence, are all located in the closed left half s-plane. Hence, following the 'Low-Gain' design method, one can design a feedback gain $F_a(\varepsilon)$ for the pair $(A_{aa}, [B_{0a}, L_{ad}])$ such that,

(a) The matrix $A_{aa} - [B_{0a}, L_{ad}]F_a(\varepsilon)$ is Hurwitz-stable for all $\varepsilon > 0$;

(b) There exists an $\varepsilon_a^* \in (0, 1]$ such that for all $\varepsilon \in (0, \varepsilon_a^*]$,

$$\|F_a(\varepsilon)\| \le k\varepsilon \tag{38}$$

$$\|e^{(A_{aa}-[B_{0a},L_{ad}]F_a(\varepsilon))t}\|$$
$$\le \frac{\delta_0}{\varepsilon^{2^{q-1}(r_1+1)(r_2+1)\cdots(r_q+1)}} e^{-\varepsilon^{2^{q-1}(r_1+1)(r_2+1)\cdots(r_q+1)}t/2} \tag{39}$$

$$\|F_a(\varepsilon)\, e^{(A_{aa}-[B_{0a},L_{ad}]F_a(\varepsilon))t}\|$$
$$\le \delta_q \varepsilon e^{-\varepsilon t/2} + \delta_{q-1}\varepsilon^{r_q+2} e^{-\varepsilon^{2(r_q+1)}t/2} + \cdots$$
$$+ \delta_1 \varepsilon^{2^{q-2}(r_2+1)(r_3+1)\cdots(r_q+1)+1} e^{-\varepsilon^{2^{q-1}(r_2+1)(r_3+1)\cdots(r_q+1)}t/2} \tag{40}$$

where q and r_i's are integers, and $k > 0$ and $\delta_i > 0$'s are constant numbers, all independent of ε.

For later use, denote

$$r_a = 2^{q-1}(r_1+1)(r_2+1)\cdots(r_q+1)$$

and partition the matrix $F_a(\varepsilon)$ as

$$F_a(\varepsilon) = [F_{a0}'(\varepsilon), F_{ad}'(\varepsilon)]' = [F_{a0}(\varepsilon)', F_{a1}'(\varepsilon), F_{a2}'(\varepsilon), \cdots, F_{am_d}'(\varepsilon)]'$$

where $F_{a0}(\varepsilon) \in \mathbf{R}^{m_0 \times n_a}$ and for each $i = 1, 2, \cdots, m_d$, $F_{ai}(\varepsilon) \in \mathbf{R}^{1 \times n_a}$.

Step 3 (Construction of a parameterized low-and-high gain matrix $F(\varepsilon)$): By the SCB lemma, the pair (A_{cc}, B_c) is controllable, hence one can choose a feedback gain matrix F_c such that $A_{cc} - B_c F_c$ is Hurwitz-stable and has a chosen set of eigenvalues. Also, choose F_i such that $A_{q_i} - B_{q_i} F_i$ is Hurwitz-stable. The existence of such gain a matrix F_i' is guaranteed by the special form of (A_{q_i}, B_{q_i}). For further use, let the first element of F_i be F_{i1}.

Next a composite static state feedback gain is formed for the system Σ. This state feedback gain takes the form of

$$F(\varepsilon) = \Gamma_I \begin{bmatrix} F_{u_0}(\varepsilon) \\ F_{u_d}(\varepsilon) \\ F_{u_c} \end{bmatrix} \Gamma_S^{-1} \tag{41}$$

where

$$F_{u_0} = [C_{0a} + F_{a0}(\varepsilon) \quad C_{0c} \quad C_{01} \quad C_{02} \quad \cdots C_{0m_d}]$$

$$F_{u_d} = \begin{bmatrix} F_{u_1}(\varepsilon) \\ F_{u_2}(\varepsilon) \\ \vdots \\ F_{u_{m_d}}(\varepsilon) \end{bmatrix}$$

$$F_{u_c} = [E_{ca} \quad F_c \quad 0 \quad 0 \quad \cdots \quad 0]$$

and for $i = 1$ to m_d,

$$F_{u_i} = \left[E_{ia} + \frac{F_{i1}}{\tilde{\varepsilon}^{q_i}} F_{ai} \quad E_{ic} \quad E_{i1} \quad E_{i2} \quad \cdots \quad E_{ii} + \frac{F_i}{\tilde{\varepsilon}^{q_i}} S_{q_i}(\tilde{\varepsilon}) \quad \cdots \quad E_{im_d} \right]$$

$$S_{q_i}(\tilde{\varepsilon}) = \text{Diag}\{1, \tilde{\varepsilon}, \tilde{\varepsilon}^2, \cdots, \tilde{\varepsilon}^{q_i-1}\}, \ \tilde{\varepsilon} = \varepsilon^{6r_a+1} \ .$$

This concludes the description of a 'Low-and-High-Gain' design method that constructs a parameterized gain $F(\varepsilon)$.

Next we choose a family of state feedback laws, parameterized in ε, as

$$u = -F(\varepsilon)x \tag{42}$$

where $F(\varepsilon)$ is as given by (41).

The following theorem establishes that the family of state feedback laws as given by (42) indeed achieves p.r. for the system Σ.

Theorem 10. *Consider the system Σ as given in (1) along with the associated cost $J(x_0, u)$ as given in (2). Also, assume that Σ is right invertible and has all its invariant zeros located in the closed left half s-plane. Then, the family of state feedback laws as given by (42) achieves p.r. for the system Σ.*

Proof. With the state feedback laws (42), the closed-loop system in the special coordinate basis can be written as,

$$\dot{\bar{x}}_a = A_{aa}\bar{x}_a + B_{0a}\bar{z}_0 + L_{ad}\bar{z}_d \tag{43}$$

$$\dot{\bar{x}}_c = (A_{cc} - B_c F_c)\bar{x}_c + B_{0c}\bar{z}_0 + L_{cd}\bar{z}_d \tag{44}$$

$$\dot{\bar{x}}_i = A_{q_i}\bar{x}_i + B_{0i}\bar{z}_0 + L_i\bar{z}_d + B_{q_i}[-\frac{F_{i1}}{\tilde{\varepsilon}^{q_i}}F_{ai}\bar{x}_a - \frac{F_i}{\tilde{\varepsilon}^{q_i}}S_{q_i}(\tilde{\varepsilon})],$$
$$i = 1, 2, \cdots, m_d \tag{45}$$

$$\bar{z}_0 = -F_{a0}(\varepsilon)\bar{x}_a. \tag{46}$$

We next consider the following scaling and redefinition of variables,

$$\tilde{x} = [\tilde{x}'_a, \ \tilde{x}'_c, \ \tilde{x}'_d]', \ \tilde{x}_a = \bar{x}_a, \ \tilde{x}_c = \bar{x}_c \tag{47}$$

$$\tilde{x}_d = [\tilde{x}'_1, \ \tilde{x}'_2, \ \cdots, \ \tilde{x}'_{m_d}]', \ \tilde{x}_i = [\tilde{x}_{i1}, \ \tilde{x}_{di2}, \cdots, \tilde{x}_{iq_i}]' \tag{48}$$

$$\tilde{x}_{i1} = \bar{x}_{i1} + F_{ai}\bar{x}_a, \ i = 1, 2, \cdots, m_d \tag{49}$$

$$\tilde{x}_{ij} = \tilde{\varepsilon}^{j-1}\bar{x}_{ij}, \ j = 2, 3, \cdots, q_i, \ i = 1, 2, \cdots, m_d \tag{50}$$

and rewrite the closed-loop system (43)-(46) as,

$$\dot{\tilde{x}}_a = (A_{aa} - [B_{0a}, L_{ad}]F_a(\varepsilon))\tilde{x}_a + L_{ad}\tilde{z}_d, \ \tilde{z}_d = \bar{z}_d - F_{ad}(\varepsilon)\tilde{x}_a \tag{51}$$

$$\dot{\tilde{x}}_c = (A_{cc} - B_c F_c)\tilde{x}_c + L_{cd}\tilde{z}_d - B_{0c}F_{a0}(\varepsilon)\tilde{x}_a - L_{cd}F_{ad}(\varepsilon)\tilde{x}_a \tag{52}$$

$$\dot{\tilde{x}}_d = \frac{1}{\varepsilon^{6r_a+1}}A_d^c\tilde{x}_d + D_{da}(\varepsilon)\tilde{x}_a + D_{dd}(\varepsilon)\tilde{x}_d \tag{53}$$

$$\bar{z}_0 = -F_{a0}(\varepsilon)\tilde{x}_a \tag{54}$$

where

$$A_d^c = \mathrm{Diag}\{A_{q_1} - B_{q_1}F_1, A_{q_2} - B_{q_2}F_2, \cdots, A_{q_{m_d}} - B_{q_{m_d}}F_{m_d}\}$$

and is Hurwitz-stable, and $D_{da}(\varepsilon)$ and $D_{dd}(\varepsilon)$ are some ε dependent matrices of appropriate dimensions satisfying

$$\|D_{da}(\varepsilon)\| \le d_{da}, \quad \|D_{dd}(\varepsilon)\| \le d_{dd}, \quad \forall \varepsilon \in (0, \varepsilon_a^*] \tag{55}$$

for some positive constants d_{da} and d_{dd} independent of ε.

To continue on our proof, let us first choose

$$V_a(\tilde{x}_a; \varepsilon) = \tilde{x}_a' P_a(\varepsilon)\tilde{x}_a \tag{56}$$

as a Lyapunov function candidate for the dynamics,

$$\dot{\tilde{x}}_a = (A_{aa} - [B_{0a}, L_{ad}]F_a(\varepsilon))\tilde{x}_a \tag{57}$$

where $P_a(\varepsilon)$ is the positive definite solution of the Lyapunov equation

$$(A_{aa} - [B_{0a}, L_{ad}]F_a(\varepsilon))'P_a(\varepsilon) + P_a(\varepsilon)(A_{aa} - [B_{0a}, L_{ad}]F_a(\varepsilon)) = -\varepsilon^{3r_a}I \ . \tag{58}$$

More specifically,

$$P_a(\varepsilon) = \int_0^\infty \varepsilon^{3r_a} e^{(A_{aa}-[B_{0a},L_{ad}]F_a(\varepsilon))'t} e^{((A_{aa}-[B_{0a},L_{ad}]F_a(\varepsilon))t} dt \tag{59}$$

from which and (39) we have that, for all $\varepsilon \in (0, \varepsilon_a^*]$,

$$\|P_a(\varepsilon)\| \le \delta_0^2 \varepsilon^{r_a} \int_0^\infty e^{-\varepsilon^{r_a}t}dt = \delta_0^2 \int_0^\infty e^{-\tau}d\tau = \delta_0^2 \ . \tag{60}$$

We next choose

$$V_c(\tilde{x}_c) = \tilde{x}_c' P_c \tilde{x}_c \tag{61}$$

as a Lyapunov function candidate for the dynamics

$$\dot{\tilde{x}}_c = (A_{cc} - B_c F_c(\varepsilon))\tilde{x}_c \tag{62}$$

where P_c is the positive definite solution of the Lyapunov equation

$$(A_{cc} - B_c F_c)'P_c + P_c(A_{cc} - B_c F_c) = -I \ . \tag{63}$$

We also choose

$$V_d(\tilde{x}_d) = \tilde{x}_d' P_d \tilde{x}_d \tag{64}$$

as a Lyapunov function candidate for the dynamics

$$\dot{\tilde{x}}_d = \frac{1}{\varepsilon^{6r_a+1}} A_d^c \tilde{x}_d \tag{65}$$

where P_d is the positive definite solutions of the Lyapunov equation

$$(A_d^c)'P_d + P_d A_d^c = -I \ . \tag{66}$$

Finally, for the dynamics of $\tilde{x}_a$, $\tilde{x}_c$ and $\tilde{x}_d$ as given by the system (51)-(53), we form a Lyapunov function candidate as

$$V(\tilde{x}_a, \tilde{x}_c, \tilde{x}_d; \varepsilon) = V_a(\tilde{x}_a; \varepsilon) + \varepsilon^{3r_a} V_c(\tilde{x}_c) + V_d(\tilde{x}_d) . \tag{67}$$

Using (38), (55), (58), (60), (63) and (66), the derivative of V along the trajectories of (51)-(53) can be evaluated as

$$\dot{V} \le -[\|\tilde{x}_a\| \ \|\tilde{x}_c\| \ \|\tilde{x}_d\|] \begin{bmatrix} \varepsilon^{3r_a} & -\alpha_{12}\varepsilon^{3r_a+1} & -\alpha_{13} \\ -\alpha_{12}\varepsilon^{3r_a+1} & \varepsilon^{3r_a} & -\alpha_{23}\varepsilon^{3r_a} \\ -\alpha_{13} & -\alpha_{23}\varepsilon^{3r_a} & \frac{1}{\varepsilon^{6r_a+1}} - \alpha_{33} \end{bmatrix} \begin{bmatrix} \|\tilde{x}_a\| \\ \|\tilde{x}_c\| \\ \|\tilde{x}_d\| \end{bmatrix} \tag{68}$$

for some positive numbers α_{ij}'s independent of ε. It is then straightforward to verify that there exists an $\varepsilon_b^* \in (0, \varepsilon_a^*]$ such that for all $\varepsilon \in (0, \varepsilon_b^*]$,

$$\dot{V} \le -\alpha_1 \varepsilon^{3r_a}(\|\tilde{x}_a\|^2 + \|\tilde{x}_c\|^2 + \|\tilde{x}_d\|^2) \tag{69}$$

for some positive constant α_1 independent of ε. This shows that the linear system (51)-(53) is stable. Moreover, because of (56), (60), (61) and (64),

$$\dot{V} \le -\alpha_2 \varepsilon^{3r_a} V, \quad \forall \varepsilon \in (0, \varepsilon_b^*] \tag{70}$$

for some ε independent constant α_2. This in turn shows that

$$V(\tilde{x}_a(t), \tilde{x}_c(t), \tilde{x}_d(t); \varepsilon) \le e^{-\alpha_2 \varepsilon^{3r_a} t} V(0), \quad \forall \varepsilon \in (0, \varepsilon_b^*] \tag{71}$$

where we note that

$$V(0) \le (\|P_a(\varepsilon)\| + \|P_c\| + \|P_s\|)\|\tilde{x}(0)\|^2, \quad \forall \varepsilon \in (0, \varepsilon_b^*] .$$

It now follows from (64) that

$$\|\tilde{x}_d(t)\| \le \alpha_3 e^{-\alpha_4 \varepsilon^{3r_a} t}\|\tilde{x}(0)\|, \quad \forall \varepsilon \in (0, \varepsilon_b^*] \tag{72}$$

where $\alpha_4 = \alpha_2/2$ and α_3 is some positive constant independent of ε. Viewing $\tilde{z}_d$ as an input to the dynamics of $\tilde{x}_a$, (51), recalling that $\tilde{z}_d = [\tilde{x}_{11}, \tilde{x}_{21}, \cdots, \tilde{x}_{p1}]'$ and using (39) and the majorization (72), it is straightforward to show that there exists an $\varepsilon_c^* \in (0, \varepsilon_b^*]$ such that

$$\|\tilde{x}_a(t)\| \le \frac{\alpha_5}{\varepsilon^{2r_a}} e^{-\alpha_4 \varepsilon^{3r_a} t}\|x(0)\|, \quad \forall \varepsilon \in (0, \varepsilon_c^*] \tag{73}$$

for some positive number α_5 independent of ε.

Next, viewing $D_{da}(\varepsilon)\tilde{x}_a + D_{dd}(\varepsilon)\tilde{x}_d$ as an input to the dynamics of $\tilde{x}_d$, (53), it is easy to show that there exists an $\varepsilon_d^* \in (0, \varepsilon_c^*]$ such that,

$$\|\tilde{x}_d(t)\| \le [\alpha_6 e^{-t/\varepsilon^{6r_a+1}} + \alpha_7 \varepsilon^{4r_a+1} e^{-\alpha_4 \varepsilon^{3r_a} t}]\|x(0)\|, \quad \forall \varepsilon \in (0, \varepsilon_d^*] \tag{74}$$

for some positive numbers α_6 and α_7 independent of ε. Using (40) and the majorization (74), it follows from (51) that there exists an $\varepsilon_e^* \in (0, \varepsilon_d^*]$ such that,

$$\|F_a(\varepsilon)\tilde{x}_a(t)\| \leq \alpha_8 [\varepsilon^{3r_a} e^{-\alpha_4 \varepsilon^{3r_a} t} + \varepsilon\varepsilon e^{-\varepsilon t/2} + \varepsilon^{r_p+2} e^{-\varepsilon^{2(r_p+1)} t/2} + \cdots$$
$$+ \varepsilon^{2^{p-2}(r_2+1)\cdots(r_p+1)+1} e^{-\varepsilon^{2^{p-1}(r_2+1)\cdots(r_p+1)} t/2}]\|x(0)\|, \ \forall \varepsilon \in (0, \varepsilon_e^*] \ (75)$$

where α_8 is some positive number independent of ε.

Recalling that $\bar{z}_d = \tilde{z}_d - F_{ad}(\varepsilon)\tilde{x}_a$ and $\bar{z}_0 = -F_{a0}(\varepsilon)\tilde{x}_a$, it follows readily from (74) and (75) that

$$\|\bar{z}(t)\| \leq \alpha_9 [e^{-t/\varepsilon^{6r_a+1}} + \varepsilon^{3r_a} e^{-\alpha_4 \varepsilon^{3r_a} t} + \varepsilon\varepsilon e^{-\varepsilon t/2} + \varepsilon^{r_p+2} e^{-\varepsilon^{2(r_p+1)} t/2} + \cdots$$
$$+ \varepsilon^{2^{p-2}(r_2+1)\cdots(r_p+1)+1} e^{-\varepsilon^{2^{p-1}(r_2+1)\cdots(r_p+1)} t/2}]\|x(0)\|, \ \forall \varepsilon \in (0, \varepsilon_e^*] \ (76)$$

for some positive constant α_9 independent of ε, and hence

$$J(x_0, u) = \int_0^\infty \|z(t)\|^2 dt$$
$$= \int_0^\infty \|\Gamma_O \bar{z}(t)\|^2 dt \leq \|\Gamma_O\| \int_0^\infty \|\bar{z}(t)\|^2 dt \to 0, \text{ as } \varepsilon \to 0. \quad (77)$$

□

6 Conclusions

After deriving the necessary and sufficient conditions to achieve a design objective called *perfect regulation* (p.r.), we developed a direct eigenstructure assignment design procedure to achieve p.r.. Unlike other methods reported in the literature, the design procedure developed here accommodates the presence of invariant zeros on the imaginary axis. Thus it treats general linear multivariable systems without any priori conditions or exceptions on them. That is, as long as the design objective of p.r. is achievable for the given system, our design procedure achieves it. The heart of our design procedure is an appropriate parameterization of state feedback gain F with a tuning parameter ε. The state feedback gain $F(\varepsilon)$ consists of as its components both low as well as high gains. Thus the design method is termed as 'Low-and-High-Gain' design. The procedure used to construct the components of $F(\varepsilon)$ does not need explicit values of the parameter ε. In this sense, ε truly acts as a tuning parameter, and can be adjusted either off-line or on-line to achieve performance as close as required to the ideal design objective. Also, the procedure used to construct $F(\varepsilon)$ utilizes some decentralized subsystems of the given system, and in doing so avoids inherent numerical difficulties associated with other methods such as those based on Riccati equations. Utilizing the m-file for constructing the SCB [7], the 'Low-and-High-Gain' design procedure developed here is readily implemented in Matlab.

References

1. C.T. Chen, *Linear System Theory and Design*, Holt, Rinehart and Winston, New York, 1984.
2. B.A. Francis, "The optimal linear-quadratic time-invariant regulator with cheap control," *IEEE Trans. Auto. Contr.*, Vol. 24, pp. 616-621, 1979.
3. H. Kimura, "A new approach to the perfect regulation and the bounded peaking in linear multivariable control systems," *IEEE Transactions on Automatic Control*, Vol. AC-26, pp. 253-270, 1981.
4. H. Kwakernaak and R. Sivan, *Linear optimal control systems*, John Wiley, New York, 1972.
5. Z. Lin, *Global and Semi-Global Control Problems for Linear Systems Subject to Input Saturation and Minimum-Phase Input-Output Linearizable Systems*, Ph.D Dissertation, Washington State University, May 1994.
6. Z. Lin and A. Saberi, "Semi-global exponential stabilization of linear systems subject to 'input saturation' via linear feedbacks," *Systems & Control Letter*, Vol. 21, pp. 225-239, 1993.
7. Z. Lin, A. Saberi and B.M. Chen, *Linear Systems Toolbox*, (Commercially available through *A.J. Controls Inc.* Seattle, Washington.) Washington State University Report No. EE/CS 0097, 1991.
8. P. Sannuti and A. Saberi, "A special coordinate basis of multivariable linear systems – finite and infinite zero structure, squaring down and decoupling," *International J. Control*, Vol. 45, pp. 1655-1704, 1987.
9. A. Saberi and P. Sannuti, "Time-scale structure assignment in linear multivariable systems using high-gain feedback," *International Journal of Control*, Vol. 49, No. 6, pp. 2191-2213, 1989.
10. A. Saberi and P. Sannuti, "Squaring down of non-strictly proper systems," *International Journal of Control*, Vol. 51, No. 3, pp. 621-629, 1990.
11. A. Saberi, B.M. Chen and P. Sannuti, "Theory of LTR for non-minimum phase systems, Recoverable Target loops, Recovery in a subspace – Part II: Design," *International Journal of Control*, Vol. 53, No.5, pp. 1117-1160, May, 1991.
12. A. Saberi, P. Sannuti and B.M. Chen, H_2 *Optimal Control*, to be published by Prentice Hall, Simon & Schuster International Group.
13. A.A. Stoorvogel, A. Saberi and B.M. Chen, "Full and reduced order observer based controller design for H_2-optimization," *International Journal of Control*, Vol. 58, No. 4, pp. 803-834, 1993.

This article was processed using the LaTeX macro package with LLNCS style

Manifold Based Feedback Design[*]

K. David Young

Lawrence Livermore National Laboratory, University of California, Livermore, California 94550 U.S.A.

Abstract. In this paper, we present an unified view of a variety of feedback control design approaches under the framework of manifold based feedback design. We examine the role of manifolds in system reduction and feedback design which exploits time-scale separations. Design specified manifolds for a number of different control design objectives are introduced, using high gain and variable structure feedback as the mean of enforcing desirable closed loop behavior. Using robotics applications, the utility of manifold based feedback design is demonstrated.

1 Motivation

The notion of manifold originates from differential geometry and has been widely adopted in studying qualitative behavior of dynamic systems. In control theory, we have also utilized manifolds to facilitate the development of a variety of control design techniques. Nevertheless, in many cases, the use of manifold is implicit, and in some cases, may even be unintentional. In this paper, we hope to unify a number of seemingly unconnected pieces of research on feedback control design methods under the general framework of manifold based feedback control design.

2 Mechanical System on Manifold

In order to provide a physical connection to the notion of manifold, we introduce the following mechanical system with holonomic constraints:

$$M(q)\ddot{q} + H(\dot{q}, q) = u + K(q)^T \lambda, \qquad q \in \mathbb{R}^n \tag{1}$$

$$\phi(q) = 0, \qquad \phi \in \mathbb{R}^r \tag{2}$$

$$K(q) \doteq \frac{\partial \phi}{\partial q} \tag{3}$$

where $q, \dot{q}$ are the generalized displacement and velocity vectors respectively, and $M(\cdot)$ is the moment of inertia matrix, $H(\cdot, \cdot)$ denotes the effects of cross axis coupling and other mechanical loads, u is a vector of generalized forces, and

---

[*] This paper was prepared during the author's professional leave with the Institute of Industrial Science, University of Tokyo, Tokyo 106, JAPAN which is locally sponsored by the Foundation for the Promotion of Industrial Science.

λ is a vector of Lagrange multipliers which characterizes the reaction force due to the constraints.

If the mechanical system models a robotic manipulator and the r dimensional manifold is representable in a Cartesian coordinate system, then (2) may describe a physical surface with which the end effector is in contact, and $K(q)^T\lambda$ is the reaction torque due to the reaction force acting at the end effector. We can further define task space coordinates in additional to the generalized displacements to provide more physical meaning to the constraint manifold. Let p be a task space vector whose dimension is no greater than that of q, then forward kinematics of the robotic manipulator define a nonlinear transformation from q to p and the associated Jacobian matrix:

$$p = p(q), \tag{4}$$

$$J \doteq \frac{\partial p}{\partial q} \tag{5}$$

The constraint manifold can then be expressed as

$$\phi(p) = 0, \qquad \phi \in \mathbb{R}^r \tag{6}$$

For example, a visualizable constraint in the Cartesian coordinate frame is a circle with radius r in the X-Y plane . It is described by the scalar equation,

$$\phi(p) = x^2 + y^2 - r^2 = 0, \tag{7}$$

and p is a vector in the Cartesian coordinate frame whose origin is at the base of the robotic manipulator:

$$p \doteq [x, \, y, \, z]^T. \tag{8}$$

For the robotic manipulator with the circle constraint, the end effector's trajectory on this circle is motion on the manifold, whereas if the end effector is not in contact with the circle, the motion is outside the manifold. For a robotic task whose goal is to move the end effector from a point inside the circle to follow the circumference of the circle, the resulting trajectory is characterized by motion which converges to the manifold, i.e., $\phi < 0$, and after the manifold is reached, motion continues on the manifold $\phi = 0$. Thus with respect to constraint manifold, the system's motion is decomposable into two phases: a reaching phase, and a constrained motion phase.

During the reaching phase, the reaction force is zero, however, once the system reaches the manifold, a reaction force is developed. Suppose we let the generalized force u be zero, and assume that the system motion is constrained on the manifold, then the Lagrange multiplier λ can be computed from the condition

$$\ddot{\phi} = KM^{-1}(-H + K^T\lambda) + \dot{K}\dot{q} = 0 \tag{9}$$

Note that in order for the reaction force to be uniquely defined, the matrix $KM^{-1}K^T$ must be nonsingular. Now consider $\phi(q) = 0$ as a virtual manifold, i.e., there is no physical contact, thus no reaction force is developed. For the robotic manipulator with the circle constraint, this would be the case if the end effector follows an imaginary circle in the air instead of a physical circle constraint. Nevertheless it is possible to derive the corresponding generalized force $u(t)$ using the equivalent control method for variable structure systems [1], and define the sliding mode manifold as

$$\sigma = \dot{\phi}(q) = 0 \tag{10}$$

The equivalent control $u_{eq}(t)$ is computed from $\dot{\sigma} = 0$:

$$\dot{\sigma} = \ddot{\phi} = KM^{-1}(-H + u_{eq}) + \dot{K}\dot{q} = 0 \tag{11}$$

which has the same form as (9) with u_{eq} equating to the reaction torque $K^T \lambda$.

So far we have only discussed about the necessary condition and the associated forces, either internal reaction forces or equivalent external forces, for motion on a manifold to exist. In order for motion to reach the manifold, and continue on it, control action is generally required to guarantee the stability of motion with respect to the manifold $\phi(q) = 0$. The associated design problem is often formulated as a stability synthesis problem with $\phi(q) = 0$ as an equilibrium point using, for example, Lyapunov method. Moreover, additional efforts are required to ensure that the motion on the manifold converges to the system's equilibrium point ($q = 0, \dot{q} = 0$), i.e., motion on the manifold is stable. In order to provide a better understanding of the issues relating to the control design for the reaching and the constrained motion phase, we introduce displacement dependent transformations to decompose the system's dynamics into an unconstrained and constrained subsystems. The motivation for using these transformation comes from the $2n$th dimensional model for the constrained motion which is obtained by substituting the solution for the Lagrange multiplier in (9) into (1) with $u = 0$:

$$M\ddot{q} + [I - MK^T(KM^{-1}K^T)^{-1}KM^{-1}]H = -K^T(KM^{-1}K^T)^{-1}\dot{K}\dot{q} \tag{12}$$

where the q dependence of M and K, and the $q, \dot{q}$ dependence of H are abbreviated. We further define

$$C(q) \doteq KM^{-1}, \quad B(q) \doteq K^T \tag{13}$$

and note that $CB = KM^{-1}K^T$ is invertible. Furthermore,

$$P(q) = B(CB)^{-1}C \tag{14}$$

is a projection matrix, i.e., $PP = P$ for any q. This projection matrix has a similar form to the one for reducing the order of a linear time-invariant singularly perturbed system,

$$\mu\dot{x} = (\mu A + BC)x, x \in \mathbb{R}^n, \quad B \in \mathbb{R}^{n \times m}, C \in \mathbb{R}^{m \times n}, \quad CB \text{ is invertible} \tag{15}$$

when the parasitic parameter μ tends to zero. The following transformations are introduced in [2]:

$$x = Nz + B(CB)^{-1}s \tag{16}$$

$$s = Cx, \quad z = Mx \tag{17}$$

$$MB = 0, \ MN = I_{n-m}, \ CN = 0, \quad M \in \mathbb{R}^{n-m \times n}, N \in \mathbb{R}^{n \times n-m} \tag{18}$$

in which, z is the component of x in the null space of C, and s is the component in the range space of B. For this system, the manifold is $s = Cx = 0$. We adopt similar transformations [1] to reveal the dynamics of the reaching and the constrained motion phase as follows:

$$M\dot{q} = \mathcal{N}(q)\dot{q}_N + \mathcal{B}(q)\dot{q}_c \tag{19}$$

$$\dot{q}_N = \mathcal{M}(q)M(q)\dot{q} \tag{20}$$

$$\dot{q}_c = \mathcal{C}(q)M(q)\dot{q} = K(q)\dot{q} = \dot{\phi} \tag{21}$$

$$\mathcal{M}(q)\mathcal{B}(q) = 0, \quad \mathcal{C}(q)\mathcal{N}(q) = 0, \quad \mathcal{M}(q)\mathcal{N}(q) = I_{n-r}, \quad \forall q \tag{22}$$

The "normalized" generalized velocity $M\dot{q}$ is decomposed into a component $\dot{q}_N$ in the null space of KM^{-1}, and a component $\dot{\phi}$ in the range space of K^T. Since $\dot{\phi} = 0$ is implied by $\phi = 0$, i.e., motion is on the constraint manifold, $\dot{q}_N$ can be interpreted as a generalized velocity of the constrained motion phase. We define a corresponding generalized displacement ψ to complete the transformed model:

$$\psi \doteq q_N \tag{23}$$

The generalized force u and the coupling force $H(\cdot, \cdot)$ can also be decomposed similarly:

$$u = \mathcal{N}u_N + \mathcal{B}u_c \tag{24}$$

$$H = \mathcal{N}H_N + \mathcal{B}H_c \tag{25}$$

Applying the above transformation to (1) yields the following unconstrained and constrained subsystems:

$$\ddot{\psi} = (\dot{\mathcal{M}} + \mathcal{M}\dot{M}M^{-1})(\mathcal{N}\dot{\psi} + \mathcal{B}\dot{\phi}) + u_c - H_N \tag{26}$$

$$\ddot{\phi} = \dot{K}M^{-1}(\mathcal{N}\dot{\psi} + \mathcal{B}\dot{\phi}) - C\mathcal{B}H_c + C\mathcal{B}u_N + C\mathcal{B}\lambda \tag{27}$$

The above model is applicable for both the reaching and the constrained motion phase provided that we add the condition that

$$\lambda = 0, \quad \text{if } \phi \neq 0 \tag{28}$$

For the reaching phase, (27) is expressed directly in terms of the "manifold" variable ϕ. On the other hand, by setting $\dot{\phi} = \phi = 0$, i.e., motion is on the manifold, the system order is reduced from $2n$ to $2(n-r)$, and the governing dynamic model for the constrained motion is (26) with $\phi = 0$. Note that, in accordance with Mechanics, the reaction force which is expressed in terms of λ does not affect the motion on $\phi = 0$.

In connection with the design of feedback control for the reaching and the constrained motion phase, two independent control design problems are identified: The reaching phase dynamics are controlled with the force component u_N. The design goal is to regulate $\phi(t)$ to zero, i.e., the manifold is a stable equilibrium point. For the constrained motion phase, the design goal is to track some desired trajectory $\psi_d(t)$ on the manifold with the force component u_c. We shall evoke the manifold based feedback control design concept to solve these two problems. While the desired objective is to be on the constrained manifold, the manifold itself is not sufficient for specifying the desired behavior of the reaching phase. Instead, we introduce a manifold in the $(\phi, \dot{\phi})$ space as follows:

$$\sigma_N = \dot{\phi} + C_N(\phi - \phi^*) = 0 \qquad (29)$$

where C_N is a diagonal matrix with positive diagonal elements, and ϕ^* is a constant vector whose components are positive. We can assume without loss of generality that $\phi(t) < 0$ defines the half space such that motion is unconstrained, thus the initial condition $\phi(t_o) < 0$ is assumed. Clearly, if the motion is steered onto this manifold, $\phi(t)$ satisfies

$$\dot{\phi} = -C_N(\phi - \phi^*) \qquad (30)$$

and since all the components $\phi_i(t), i = 1, \ldots, r$ are exponentially converging to positive steady state values, there exists t^* such that $\phi(t^*) = 0$. The dynamic behavior in the other half space $\phi \geq 0$ depends on the "stiffness" of the constraint manifold. If it is infinitely stiff, i.e., with finite force, the deformation of the surface which is defined as the deviation from $\phi = 0$ such that $\phi_i(t) > 0$ is infinitely small, then motion on the manifold $\phi = 0$ satisfies the condition:

$$\lim_{\phi_i \to 0+} \dot{\phi}_i \phi_i < 0, \quad i = 1, \ldots, r \qquad (31)$$

On the other hand, the constraint manifold may be compliant, i.e., motion in the domain $\phi > 0$ is finite and its behavior is determined by the force – displacement dynamics of the surface deformation. In any case, in the domain $\phi \leq 0$, convergence towards $\phi = 0$ is guaranteed by (31).

For the constrained motion phase design, we specify a manifold with respect to the tracking error $e_\psi \doteq \psi(t) - \psi_d(t)$:

$$\sigma_C = \dot{e}_\psi - C_c e_\psi = 0 \qquad (32)$$

with C_c being a Hurwitz matrix. The feedback control for the force component u_c should then be designed such that $\psi(t)$, which characterized motion in the constraint manifold, is on this design manifold. Clearly, tracking error is guaranteed to converge to zero, as can been seen from the following equation

$$\dot{e}_\psi = C_c e_\psi \qquad (33)$$

which governs the motion on $\sigma_C = 0$.

3 System Reduction and Feedback Control

We now return to the role of manifold in feedback control design, and in particular, the use of reduced order system models for control design. The development of feedback control design which exploits the existence of time scale separation in the plant dynamics is based on the theory of singular perturbation. The plant model of full order contains a parasitic parameter μ:

$$\dot{x}_1 = A_{11}x_1 + A_{12}x_2 + B_1u \qquad (34)$$

$$\mu\dot{x}_2 = A_{21}x_1 + A_{22}x_2 + B_2u \qquad (35)$$

The reduced order model is formally obtained by letting $\mu = 0$ in (35), and the resulting algebraic equation is solved to eliminate the state x_2. While there exists a number of feedback control designs using μ as a perturbation parameter, the underlying principle is to specify a manifold in the (x_1, x_2) space:

$$s(x_1, x_2, u(x_1, x_2)) = A_{21}x_1 + A_{22}x_2 + B_2u(x_1, x_2) = 0 \qquad (36)$$

such that the reaching and the constrained motion phase dynamic behavior meets the system performance requirements. We shall examine two different feedback control designs using singular perturbation theory [3]. The first one is the so-called reduced order, or zeroth order design which only uses the system's reduced order model:

$$\dot{\bar{x}}_1 = A_o\bar{x}_1 + B_o\bar{u} \qquad (37)$$

$$A_o = A_{11} - A_{12}A_{22}^{-1}A_{21}, \quad B_o = B_1 - A_{12}A_{22}^{-1}B_2 \qquad (38)$$

The design derives a feedback control

$$\bar{u} = F_o\bar{x}_1 \qquad (39)$$

for the reduced order plant (37). The reduced order controller is implemented as

$$u = F_o x_1 \qquad (40)$$

A direct substitution of (40) into (34) and (35) yields a homogeneous singularly perturbed system:

$$\dot{x}_1 = (A_{11} + B_1F_o)x_1 + A_{12}x_2 \qquad (41)$$

$$\mu\dot{x}_2 = (A_{21} + B_2F_o)x_1 + A_{22}x_2 \qquad (42)$$

The desired manifold is

$$s_r = (A_{21} + B_2F_o)x_1 + A_{22}x_2 = 0 \qquad (43)$$

We shall transform the above system into a form where the reaching and the constrained motion phase can be identified. Let

$$\sigma_r \doteq A_{22}^{-1}s_r = A_{22}^{-1}(A_{21} + B_2F_o)x_1 + x_2 \qquad (44)$$

Then (41) and (42) can be written as

$$\dot{x}_1 = (A_o + B_oF_o)x_1 + A_{12}\sigma_r \qquad (45)$$

$$\mu\dot{\sigma}_r = \mu A_{r1}x_1 + (A_{22} + \mu A_{r2})\sigma_r \qquad (46)$$

where

$$A_{r1} = A_{22}^{-1}(A_{21} + B_2 F_o)[(A_{11} + B_1 F_o) - A_{12} A_{22}^{-1}(A_{21} + B_2 F_o)] \qquad (47)$$
$$A_{r2} = A_{22}^{-1}(A_{21} + B_2 F_o) A_{12} \qquad (48)$$

We note that the manifold $\sigma_r = 0$ and the given manifold $s_r = 0$ coincide. With the usual assumption that the fast dynamics of the open loop plant are asymptotically stable, i.e., $\text{Re}\lambda(A_{22}) < 0$, the reaching phase dynamics are characterized by fast motion converging to an $O(\mu)$ neighborhood of the equilibrium point $s_r = 0$ in the sense that for any finite initial condition $s_r(t_o) = O(1)$,

$$\|s_r(t)\| \le O(\mu), \quad \text{for } t \ge t_o + O(\mu) \qquad (49)$$

With $\sigma_r = 0$ in (45), the dynamics of x_1 are identical to the desired reduced order closed loop system (37) and (39). This motion also coincides with the constrained motion on the manifold $s_r = 0$. However, it is important to point out that for finite μ, the manifold $s_r = 0$ is reached only asymptotically with $t \to \infty$. Likewise,

$$\|x_1(t) - \bar{x}_1(t)\| \le O(\mu), \quad \text{for } t \ge t_o + O(\mu) \qquad (50)$$
$$\lim_{t \to \infty} x_1(t) = \bar{x}_1(t) \qquad (51)$$

Thus, in the reduced order design, the system is assumed to be initially on the manifold $s_r = 0$. The design of the feedback control (39) is basically for regulating the behavior of the system on the constraint manifold. The reaching phase dynamics are inherently asymptotically stable, and the reduced order feedback control preserves the fast time scale convergence to the constraint manifold by providing only slow time scale feedback actions.

The second feedback control design is the so-called two-time-scale composite control design. Two control design problems are solved independently: The first one is identical to the reduced order design given by (37) and (39). The second design derives a feedback control

$$u_f = F_f x_f \qquad (52)$$

for the fast subsystem

$$\dot{x}_f = A_{22} x_f + B_2 u_f \qquad (53)$$

The composite controller is implemented as

$$u = F_1 x_1 + F_f x_2 \qquad (54)$$
$$F_1 \doteq F_o + F_f A_{22}^{-1}(A_{21} + B_2 F_o) \qquad (55)$$

The resulting closed loop singularly perturbed system is

$$\dot{x}_1 = (A_{11} + B_1 F_1) x_1 + (A_{12} + B_1 F_f) x_2 \qquad (56)$$
$$\mu \dot{x}_2 = (A_{21} + B_2 F_1) x_1 + (A_{22} + B_2 F_f) x_2 \qquad (57)$$

and the desired manifold is

$$s_c = (A_{21} + B_2 F_1)x_1 + (A_{22} + B_2 F_f)x_2 = 0 \qquad (58)$$

Similarly to the transformation introduced in the reduced order design, we define

$$\sigma_c \doteq (A_{22} + B_2 F_f)^{-1} s_c \qquad (59)$$

Note that directly from the definition of s_c in (58):

$$\sigma_c = (A_{22} + B_2 F_f)^{-1}(A_{21} + B_2 F_1)x_1 + x_2 \qquad (60)$$

The subtlety of the composite control design lies in the following equality for σ_c:

$$\sigma_c = (A_{22})^{-1}(A_{21} + B_2 F_o)x_1 + x_2 \qquad (61)$$

Thus, the manifolds $\sigma_r = 0$ and $\sigma_c = 0$ both coincide with the given manifolds $s_r = 0$ and $s_c = 0$. With this expression, (56) and (57) can be written as

$$\dot{x}_1 = (A_o + B_o F_o)x_1 + (A_{12} + B_1 F_f)\sigma_c \qquad (62)$$
$$\mu\dot{\sigma}_c = \mu A_{c1}x_1 + (A_{22} + B_2 F_f + \mu A_{c2})\sigma_c \qquad (63)$$

where

$$A_{c1} = A_{22}^{-1}(A_{21} + B_2 F_o)[(A_{11} + B_1 F_1) - (A_{12} + B_2 F_f) \cdot$$
$$\cdot \, A_{22}^{-1}(A_{21} + B_2 F_o)] \qquad (64)$$

$$A_{c2} = A_{22}^{-1}(A_{21} + B_2 F_o)(A_{12} + B_2 F_f) \qquad (65)$$

We note that by comparing (45) and (46) with (62) and (63), the reduced order and composite control design yield similar dynamics for the reaching and the constrained motion phase. The fast transient behavior of the reaching phase to an $O(\mu)$ neighborhood of $s_c = 0$ is designed via the fast subsystem control design. The constrained motion on manifold $s_c = 0$ is governed by the desired reduced order dynamics as before.

The constraint manifold in systems with two-time-scale separation is basically determined from the open loop system characteristics. Its functional dependence on the control input u is invariant to the choice of feedback control designs which exploit the time scale separation. As we have shown with the two different feedback designs, the resulting feedback functionals modify the manifold in the state space (x_1, x_2) in order to affect the dynamic behavior in the reaching phase, and more importantly to regulate the motion on the manifold which is the dominant motion of the system.

DC Motor Drive Control Example - 1 The notion of a manifold in the context of a two-time-scale system is illustrated with the design of speed control for DC motor drives. The full order plant model is :

$$J\dot{\omega} = -B\omega + K_T i \qquad (66)$$

$$L_a \dot{i} = -R_a i - K_b \omega + v \qquad (67)$$

where L_a is the armature circuit inductance, R_a is the circuit resistance, i is the armature current, v is the applied voltage to the armature circuit terminal, K_T is the torque constant and K_b denotes the back e.m.f. voltage constant. The mechanical system parameters are J – the moment of inertia of the motor and load, and B is the equivalent damping coefficient. For DC motors with small inductances, we let the inductance be the parasitic parameter, i.e., $\mu \doteq L_a$. The given manifold $s = 0$ is

$$- R_a i - K_b \omega + v = 0 \qquad (68)$$

The constrained motion on the manifold is governed by the reduced order model:

$$\dot{\bar{\omega}} = a\bar{\omega} + b\bar{v} \qquad (69)$$

$$a = \left(-\frac{B}{J} - \frac{K_T K_b}{R_a} \right) \qquad (70)$$

$$b = \frac{K_T}{R_a} \qquad (71)$$

for any control input $\bar{v}$. This model is the commonly used first order mechanical system model for DC motor speed control design. Suppose the goal of the reduced order design is to provide asymptotic stability to the constrained motion, then the reduced feedback control has the form

$$\bar{v} = -k_\omega \bar{\omega} \qquad (72)$$

where the eigenvalue of the desired constrained motion is

$$-\frac{B}{J} - \frac{K_T(K_b + k_\omega)}{R_a} \qquad (73)$$

The reduced order controller implements speed feedback to improve the transient response of the mechanical subsystem

$$v = -k_\omega \omega \qquad (74)$$

The manifold $s = 0$ is a line in the (ω, i) plane:

$$i = -\frac{(K_b + k_\omega)}{R_a}\omega \qquad (75)$$

Note that the manifold $\sigma_r = 0$ is also represented by the same line.

For the composite control design, both speed and current feedback are implemented:

$$v = -\left(k_\omega + k_i \frac{(K_b + k_\omega)}{R_a} \right) \omega - k_i i \qquad (76)$$

The manifold $\sigma_c = 0$ is represented by the same line in the (ω, i) plane. Figure 1 shows the responses of the closed loop system using speed feedback only as specified in the reduced order design, and with feedback of both speed and current which are required in the composite control design. The constraint manifold – a line in the (ω, i) plane is also shown. Note that with the additional current feedback, the phase trajectory is forced closer to the constraint manifold, whereas with the reduced order design, the convergence to the manifold is dictated by the electric time constant of the open circuit armature.

4 Design Specific Manifold

In time scale separation feedback designs, the constraint manifold is determined a priori by the open loop plant dynamics. Generally, design specific manifolds can be introduced as part of the design process for many different purposes. We shall first introduce the output nulling manifold design problem which is related to Variable Structure Servomechanism (VS Servo) design [4].

4.1 VS Servo

The VS servo formulation follows closely to that of multivariable servomechanism. The plant and regulated output y are given by

$$\dot{x} = Ax + Bu + Ew \tag{77}$$
$$y = Cx + Fw \tag{78}$$

where u is the control input, and w is an external exogenous input whose dynamic model is

$$\dot{w} = Zw \tag{79}$$

and in general, $\text{Re}\lambda(Z) > 0$. Existence of a control input such that the design objective $y = 0$ is achievable requires that matrices X and U exist such that

$$AX + BU + E = XZ \tag{80}$$
$$CX + F = 0 \tag{81}$$

The open loop dynamics are more convenient expressed in terms of deviation variables

$$\Delta x \doteq x - Xw \tag{82}$$
$$\Delta u \doteq u - Uw \tag{83}$$

which satisfy

$$\dot{\Delta x} = A\Delta x + B\Delta u \tag{84}$$
$$\Delta y = C\Delta x \tag{85}$$

A zero deviation output $\Delta y = 0$ implies that the regulated output $y = 0$. Thus it is desirable to specify the output nulling manifold $\Delta y = 0$ as a constraint manifold, and design the feedback control such that the reaching phase terminates on this manifold, and the constrained motion is asymptotically stable. This is the design objective of the VS Servo. Under the condition that CB is nonsingular, the motion on the manifold is determined by the transmission zeros of the triple $\{C,A,B\}$, i.e., if RHP zeros exist, constrained motion becomes unstable. In this case, we must abandon the idea of using the zero output manifold as a constraint manifold. This however also implies that output can only be nulled if the state $\Delta x = 0$.

For systems with non-minimum phase zeros, we may adopt a singular control formulation to minimize a quadratic performance index on Δy:

$$J = \frac{1}{2} \int_0^\infty [\Delta y^T \Delta y] dt \tag{86}$$

The optimal singular manifold, i.e., the manifold on which the above index is minimized, coincides with the output nulling manifold if the system is minimal phase. For systems with RHP zeros, the optimal manifold is expressible in the form:

$$\sigma_o = \Delta y + H\Delta x = 0 \tag{87}$$

For a more explicit look at the underlying tradeoff between nulling the output and stabilizing the constrained motion on the optimal singular manifold, we introduce a state transformation such that in the transformed coordinates, (84) is in the so-called regular form:

$$\dot{x}_1 = A_{11}x_1 + A_{12}x_2, \quad x_1 \in \mathbb{R}^{n-m} \tag{88}$$

$$\dot{x}_2 = A_{21}x_1 + A_{22}x_2 + B_2u, \quad x_2 \in \mathbb{R}^m, u \in \mathbb{R}^m \tag{89}$$

$$y = C_1x_1 + C_2x_2, \quad y \in \mathbb{R}^m \tag{90}$$

where the matrix B has been row reduced such that B_2 is invertible. Note that we have dropped the Δ notation in these new variables. Our earlier condition on CB becomes C_2B_2 is invertible, which implies C_2 is also invertible. The optimal singular manifold in (87) can then be expressed as

$$\sigma_o = y + C_2^{-T}A_{12}^T P_s x_1 = 0 \tag{91}$$

where P_s is obtained as a limiting solution of a matrix Riccati differential equation due to the non-negativeness of the state weighting matrix in the equivalent LQ problem [4]. The constrained motion dynamics on $\sigma_o = 0$ can be derived from solving for x_2 in (91) and substituting into (88):

$$\dot{x}_1 = [A_{11} - A_{12}C_2^{-1}(C_1 + C_2^{-T}A_{12}^T P_s)]x_1 \tag{92}$$

With $P_s = 0$, the manifold $\sigma_o = 0$ coincides with the output nulling manifold, and the transmission zeros of $\{C, A, B\}$ are the eigenvalues of $A_{11} - A_{12}C_2^{-1}C_1$. For systems with minimum phase zeros, the constrained motion dynamics on

$y = 0$ are asymptotically stable. However, with RHP zeros, motion on the output nulling manifold is unstable. Thus, an alternate manifold $\sigma_o = 0$ is introduced such that the deviations from the output nulling manifold are minimized in the sense of the quadratic index. This deviation is measured in terms of the state variable of the constrained motion dynamics x_1, and it relates also to the shift of RHP zeros towards the LHP such that motion on $\sigma_o = 0$ is asymptotically stable.

Coupled Spring–Mass System Example The output nulling manifold idea and the associated issues with systems with RHP zeros can be illustrated with a simple coupled spring–mass system whose model is:

$$\ddot{q}_1 + k_1 q_1 - k_2(q_2 - q_1) = 0 \tag{93}$$
$$\ddot{q}_2 + k_2(q_2 - q_1) = u \tag{94}$$
$$y = \dot{q}_2 + c q_2, \quad c > 0 \tag{95}$$

where the mass has been normalized. The output nulling manifold defines the desirable transient dynamics for q_2 - the displacement of the second mass. The constrained motion dynamics are governed by

$$\ddot{q}_1 = -(k_1 + k_2)q_1 + k_2 q_2 \tag{96}$$
$$\dot{q}_2 = -c q_2 \tag{97}$$

Clearly, its eigenvalues include a pair of $j\omega$ axis poles which are the zeros of the system with output q_2. The output nulling manifold $y = 0$ contains the submanifold $q_2 = 0$. The optimal singular manifold in this case has the general form

$$\sigma_o = y + b_o \dot{q}_1 + k_o q_1 + k_* q_2 = 0 \tag{98}$$

As can be seen from

$$\ddot{q}_1 = -(k_1 + k_2)q_1 + k_2 q_2 \tag{99}$$
$$\dot{q}_2 = -(c + k_*)q_2 - b_o \dot{q}_1 - k_o q_1, \tag{100}$$

the dynamics on this manifold are stabilizable with proper choices of positive b_o, and k_* which provide the necessary damping to the first mass.

4.2 Feedback Control Design for Reaching Phase

In contrast to systems with two-time-scale, no a priori reaching phase dynamics exist in the plant. With the constraint manifold chosen in the process of the control design, the reaching phase dynamics must be synthesized with feedback control action. A family of feedback control design methods has been developed to handle this problem. Basically, these methods fall into two classes: high gain feedback control, and discontinuous feedback control with sliding mode.

High Gain Feedback For the system given in regular form (88) and (89), let the desired manifold be

$$\sigma = K_1 x_1 + x_2 = 0, \ \sigma \in \mathbb{R}^m \tag{101}$$

With high gain feedback control given by

$$u = gK_2\sigma \tag{102}$$

where g is a large scalar gain parameter, and K_2 an invertible matrix, the closed loop system is a singularly perturbed system with the reciprocal of g as the parasitic parameter, i.e., $\mu \doteq g^{-1}$,

$$\dot{x}_1 = A_{11}x_1 + A_{12}x_2, \quad x_1 \in \mathbb{R}^{n-m} \tag{103}$$
$$\mu\dot{x}_2 = (\mu A_{21} + B_2K_2K_1)x_1 + (A_{22} + B_2K_2)x_2, \ x_2 \in \mathbb{R}^m \tag{104}$$

The constrained motion dynamics can be derived by solving for x_2 in (101), and substituting into (103):

$$\dot{\bar{x}}_1 = (A_{11} - A_{12}K_1)\bar{x}_1 \tag{105}$$

The slow motion of the singular perturbed system is $O(\mu)$ close to the constrained motion $\bar{x}_1$:

$$\|x_1(t) - \bar{x}_1(t)\| = O(\mu), \text{ for } t \geq t_o + O(\mu) \tag{106}$$

For a controllable system, the eigenvalues of the constrained motion can be placed arbitrarily by solving a reduced order pole placement problem for the pair (A_{11}, A_{12}), i.e., the state x_2 is treated as the substituted control variable. Alternatively, an LQ formulation for the same reduced order system yields optimal constrained motion [2]. The fast motion, if designed to be asymptotically stable, is the reaching phase. The dominant eigenvalues in the x_2 dynamics are the eigenvalues of B_2K_2. Any desirable eigenvalues for the reaching phase can be obtained by a proper choice of K_2. Similarly to the two-time-scale design, the manifold $\sigma = 0$ is reached only asymptotically with $t \to \infty$, however, an $O(\mu)$ neighborhood of the manifold is reached for any finite initial condition $\sigma(t_o)$ in finite time:

$$\|\sigma(t)\| \leq O(\mu), \quad \text{for } t \geq t_o + O(\mu) \tag{107}$$

Discontinuous Feedback Control with Sliding Mode In order to reach the constraint manifold in finite time, we introduce discontinuous feedback control such that sliding mode occurs on the constraint manifold [5]. This design approach also appeals to nonlinear systems with control appears linearly:

$$\dot{x} = f(x) + G(x)u, \ x \in \mathbb{R}^n, \ u \in \mathbb{R}^m \tag{108}$$

With respect to the manifold

$$\sigma(x) = 0, \quad \sigma \in \mathbb{R}^m, \ \sigma^T = [\sigma_1, \dots, \sigma_m] \tag{109}$$

the component of the control input is discontinuous with respect to the component of σ:

$$u_i = \begin{cases} u_i^+(x), & \text{if } \sigma_i(x) > 0 \\ u_i^-(x), & \text{if } \sigma_i(x) < 0 \end{cases}, \quad u^T = [u_1, \ldots, u_m] \tag{110}$$

The feedback system, (108) and (110), is a variable structure system whose trajectories may include sliding mode – constrained motion on the manifold $\sigma = 0$ if the feedback functionals $u_i^+(\cdot)$ and $u_i^-(\cdot)$ are appropriately designed such that $\sigma = 0$ is an equilibrium point in the subspace σ. A critical question is the existence and uniqueness of sliding mode on the constraint manifold. Its answer lies in the existence and uniqueness of the equivalent control which is defined as the solution of u in the algebraic equation

$$\dot{\sigma} = S(x)[f(x) + G(x)u] = 0 \quad, \quad S(x) \doteq \frac{\partial \sigma(x)}{\partial x} \tag{111}$$

Evidently, this solution is given by

$$u_{eq} = -[[S(x)G(x)]^{-1}S(x)]f(x) \tag{112}$$

if the matrix $S(x)G(x)$ is invertible for all x. The constrained motion dynamics are also defined by the equivalent control. The so-called Equation of Sliding Mode is obtained from substituting (112) into (108):

$$\dot{x} = [I - G(x)[S(x)G(x)]^{-1}S(x)]f(x) \tag{113}$$

Recall that in Sect. 2 on Mechanical System on Manifold, we have encountered similar form of dynamics where a projection matrix defined by the manifold is responsible for system order reduction. In the present case, the projection matrix is

$$P(x) \doteq G(x)[S(x)G(x)]^{-1}S(x) \tag{114}$$

In order to establish a direct relationship between high gain feedback and discontinuous feedback with sliding mode, we consider the linear system in regular form (88) and (89), and let the manifold be given by (101). In this case, the equation of sliding mode (113) becomes

$$\dot{x}_1 = A_{11}x_1 + A_{12}x_2 \tag{115}$$
$$\dot{x}_2 = -K_1\dot{x}_1 \quad \Rightarrow \quad \dot{\sigma} = 0 \tag{116}$$

The decomposition which introduces the subspace components is given by

$$\begin{bmatrix} x_1 \\ x_2 \end{bmatrix} = \begin{bmatrix} I_{n-m} \\ -K_1 \end{bmatrix} x_1 + \begin{bmatrix} 0 \\ B_2 \end{bmatrix} \sigma \tag{117}$$

where x_1 is the component in the constraint manifold whose dynamics are governed by

$$\dot{x}_1 = (A_{11} - A_{12}K_1)x_1 \tag{118}$$

with $\sigma = 0$ guaranteed by the existence of sliding mode. With initial condition $\sigma(t_o)$ of $O(1)$, the desired manifold is reached in finite time $t^* > 0$ due to finite control input magnitudes. Thus the reaching phase is terminated in finite time, i.e., for $t \geq t_o + t^*$,

$$\sigma(t) = 0, \tag{119}$$

and $x_1(t)$ satisfies (118).

Design methods which guarantee the convergence to the manifold $\sigma = 0$ using discontinuous feedback control (110) are given in [6]. One approach is to use a Lyapunov function in σ,

$$V = \frac{1}{2}\sigma^T\sigma \tag{120}$$

and derive sufficient conditions for u_i's such that along the trajectories of (108)

$$\dot{V} < 0 \tag{121}$$

4.3 Inner-Outer Loop Manifold Design

This type of manifold based feedback design is most commonly found in the so-called inner-outer loop feedback configuration. The idea is intuitively appealing – we decoupled the feedback design into two sequential designs. First, an outer loop is designed using a system state variable as a fictitious control input. The resulting control action for the outer loop is then treated as a reference tracking input for the inner loop control design.

DC Motor Drive Control Example - 2 We utilize the previously introduced motor drive example to illustrate this manifold design idea. The armature induc-tance L_a is no longer assumed to be small – this can be due to either a reduction of the moment of inertia J, e.g., in miniature motor drives, or to an increase in the desired magnetic flux for large power rating drives. The outer loop involves the mechanical subsystem:

$$\dot{\omega} = -(B/J)\omega + (K_T/J)u_1 \tag{122}$$
$$u_1 = -k_\omega\omega \tag{123}$$

with the armature current u_1 as the outer loop control input, and speed feedback for speed regulation is designed. The inner loop is designed with u_1 as a reference tracking signal for the current i. This is where the manifold

$$\sigma = -i + u_1 = 0 \tag{124}$$

is specified. Clearly, perfect tracking is achieved if the armature current is con-strained to be on this manifold. The inner loop feedback control design is re-duced to a high gain current loop which is commonly found in current drive amplifiers for motion control applications. This high gain feedback loop is often implemented with armature voltage saturation,

$$v = v_{max}\text{sat}(\sigma) \tag{125}$$

$$\text{sat}(\sigma) = \begin{cases} g\sigma, & \text{if } |\sigma| \leq v_{max}/g \\ v_{max}\text{sgn}(\sigma), & \text{otherwise} \end{cases} \tag{126}$$

for the armature circuit dynamics

$$L_a\dot{i} = -R_a i - K_b \omega + v \qquad (127)$$

Alternatively, it can be implemented with a switching control voltage

$$v = v_{max}\text{sgn}(\sigma) \qquad (128)$$

This is a simplified form of the discontinuous feedback control with sliding mode where both $u^+(\cdot)$ and $u^-(\cdot)$ are specified as constant saturated voltages with opposite signs.

This design procedure can be generalized to the system in regular form (88), (89). The desired manifold for the inner loop design is dependent on u_1, the control input for the outer loop:

$$\sigma(x_1, x_2, u_1) = F_1 x_1 + F_2 x_2 + u_1 = 0 \qquad (129)$$

The inner loop feedback control $u(\cdot)$ is designed to guarantee reaching of this manifold using either high gain or discontinuous feedback. Let

$$x_2 = -F_2^{-1}(F_1 x_1 + u_1) \qquad (130)$$

be the fictitious control input for the outer loop dynamics

$$\dot{x}_1 = A_{11} x_1 + A_{12} x_2 \qquad (131)$$

The resulting outer loop feedback design is with respect to the system

$$\dot{x}_1 = (A_{11} - A_{12} F_2^{-1} F_1) x_1 + A_{12} F_2^{-1} u_1 \qquad (132)$$

Recall that since the pair (A_{11}, A_{12}) is controllable, the desired manifold can also be used to place the "open loop" eigenvalues for the outer loop design. However, transient dynamics can also be shaped with the new input u_1. Thus there seems to be apparent redundancy in this generalization, although the construction of this manifold follows the "natural" one found in Sect. 3 on System Reduction and Feedback Control. We shall return to this point in the following section.

4.4   Disturbance Rejection Manifold Design

Manifold based feedback designs can provide excellent disturbance rejection and insensitivity to parameter variations. Consider the following linear system with external disturbance and uncertain parameters:

$$\dot{x} = (\bar{A} + \Delta A)x + (\bar{B} + \Delta B)u + Ef(t), \quad x \in \mathbb{R}^n, \ u \in \mathbb{R}^m, \ f \in \mathbb{R}^r \quad (133)$$

The nominal system matrices $\bar{A}$ and $\bar{B}$ are subjected to uncertainties denoted by ΔA and ΔB. The external disturbance is an unknown, but bounded function of time, and the uncertain parameter matrices ΔA and ΔB are also bounded, i.e.,

$$\|f(t)\| \le \bar{f} < \infty, \quad \|\Delta A\| \le \bar{a} < \infty, \quad \|\Delta B\| \le \bar{b} < \infty \qquad (134)$$

If the structures of the disturbance matrix E and the uncertain matrices ΔA and ΔB satisfy the so-called Drazenovic condition [7]:

$$\mathcal{R}(E) \subseteq \mathcal{R}(\bar{B}) \Leftrightarrow \text{rank}(\bar{B} \vdots E) = \text{rank}(\bar{B}) \qquad (135)$$

$$\mathcal{R}(\Delta A) \subseteq \mathcal{R}(\bar{B}) \Leftrightarrow \text{rank}(\bar{B} \vdots \Delta A) = \text{rank}(\bar{B}) \qquad (136)$$

$$\mathcal{R}(\Delta B) \subseteq \mathcal{R}(\bar{B}) \Leftrightarrow \text{rank}(\bar{B} \vdots \Delta B) = \text{rank}(\bar{B}) \qquad (137)$$

then there exists disturbance rejection manifolds such that constrained motion on the manifold is totally insensitive to disturbance and uncertain parameters. Using the state transformation which brings (133) into the regular form,

$$\dot{x}_1 = (\bar{A}_{11} + \Delta A_{11})x_1 + (\bar{A}_{12} + \Delta A_{12})x_2 + \Delta B_1 u + E_1 f \qquad (138)$$
$$\dot{x}_2 = (\bar{A}_{21} + \Delta A_{21})x_1 + (\bar{A}_{22} + \Delta A_{22})x_2 + (\bar{B}_2 + \Delta B_2)u + E_2 f \qquad (139)$$

the impact of the above geometric conditions becomes transparent. The transformed nominal system matrices are $\bar{A}_{ij}$'s and $\bar{B}_2$. For regular form, $\bar{B}_1 = 0$ and $\bar{B}_2$ is nonsingular. Conditions (135), (136), (137) imply $E_1 = 0$, $\Delta A_{11} = \Delta A_{12} = 0$, and $\Delta B_1 = 0$ respectively. Thus, for systems satisfying the Drazenovic condition, we have in regular form

$$\dot{x}_1 = \bar{A}_{11}x_1 + \bar{A}_{12}x_2 \qquad (140)$$
$$\dot{x}_2 = (\bar{A}_{21} + \Delta A_{21})x_1 + (\bar{A}_{22} + \Delta A_{22})x_2 + (\bar{B}_2 + \Delta B_2)u + E_2 f \qquad (141)$$

The disturbance rejection manifolds are parameterized by

$$\sigma = F_1 x_1 + F_2 x_2 + u_1 = 0, \quad \sigma \in \mathbb{R}^m \qquad (142)$$

with any nonsingular F_2 matrix, and u_1 is a control input for the constrained motion dynamics. This type of manifolds has been introduced in the last section for inner-outer loop design. Disturbance rejection is in fact accomplished in the outer loop. The constrained motion dynamics are identical to the outer loop dynamics (132) with nominal system matrices:

$$\dot{x}_1 = (\bar{A}_{11} - \bar{A}_{12}F_2^{-1}F_1)x_1 + \bar{A}_{12}F_2^{-1}u_1 \qquad (143)$$

where invariance to disturbance and parametric uncertainties is evident. However, as it is indicated in (141), the feedback design for $u(\cdot)$, whose goal is to guarantee reaching of the desired manifold, will be influenced by the uncertain parameters and the external disturbance. The norm bounds assumed in (134) are utilized in the design of the discontinuous feedback functionals such that sliding mode exists on the manifold.

We shall examine the feasibility of extending the disturbance rejection manifold idea to systems where the Drazenovic condition is violated in (135) and (136), however, condition on ΔB - (137) remains to be satisfied. In this case, the constrained motion dynamics on $\sigma = 0$ will depend on the uncertain parameters and disturbance:

$$\dot{x}_1 = (\bar{A}_{11} + \Delta A_{11} - (\bar{A}_{12} + \Delta A_{12})F_2^{-1}F_1)x_1 + (\bar{A}_{12} + \Delta A_{12})F_2^{-1}u_1 + E_1 f \qquad (144)$$

This system has the general form of a system with external disturbance and parameter uncertainties. The dimension of the state $n' \doteq n - m$ is reduced, but the control input u_1 remains to be m-dimensional, and disturbance is still r-dimensional. A disturbance rejection manifold for the above system can be similarly defined as

$$\sigma' = F_1' x_1' + F_2' x_2' + u_1' = 0 \tag{145}$$

with F_2' nonsingular, and x_1' and x_2' are the new coordinates of x_1 after the system is transformed into regular form. Applying the Dranzenvic condition to the system (144), the resulting constrained motion dynamics on $\sigma' = 0$ are again invariant to disturbance and parameter uncertainties:

$$\dot{x}_1' = (\bar{A}_{11}' - \bar{A}_{12}' F_2'^{-1} F_1') x_1' + \bar{A}_{12}' F_2'^{-1} u_1', \quad x_1' \in \mathbb{R}^{n'-m} \tag{146}$$

We note that $\sigma' = 0$ is a submanifold of $\sigma = 0$, i.e., $\sigma' = 0 \Rightarrow \sigma = 0$. This procedure of constructing embedded disturbance rejection manifolds can, in principle, be applied until the last submanifold is the origin of the entire state space $x = 0$. If Drazenovic condition is not satisfied for any of the submanifolds, disturbance rejection can still be provided primarily in the multiple, successive reaching phases onto the embedded submanifolds. The use of discontinuous feedback control to reach the first manifold $\sigma = 0$ is feasible, however, for the subsequent submanifolds, discontinuous feedback control cannot be used since sliding mode cannot exist for a manifold which is not continuous. This point can be illustrated by considering a design with $\sigma = 0$ and its submanifold $\sigma' = 0$. In order to provide reaching onto $\sigma' = 0$, let $u_1(\cdot)$ be discontinuous on $\sigma' = 0$. However, from (142), $\dot{\sigma}$ is unbounded for any discontinuous function $u_1(\cdot)$. Thus, with discontinuous feedback control, sliding mode cannot exist on the manifold $\sigma = 0$. The alternative is to use strictly high gain feedback to provide fast convergence to a "boundary layer" neighborhood of the submanifold. Since the manifolds are embedded, multiple high gains must be designed such that the gain for the preceding submanifold is larger than that for the subsequent submanifold. In practice, since control inputs are nevertheless bounded, it would be difficult to physically construct a feedback control structure with more than a few embedded manifolds.

4.5   Disturbance Reconstruction Manifold Design

Manifold based feedback design can also be applied to the design of disturbance estimators in the presence of system parametric uncertainties [8]. We consider systems with external disturbance and uncertain parameters which are given in regular form, (138) and (139), and design a disturbance estimator

$$\dot{\hat{x}}_2 = \bar{A}_{21} x_1 + \bar{A}_{22} x_2 + \bar{B}_2 u + v \tag{147}$$

using a discontinuous feedback control $v(\cdot)$ to reach a desired manifold,

$$\sigma_e = x_2 - \hat{x}_2 = 0 \tag{148}$$

In contrast to the desired manifolds we have previously chosen for different design goals, we are not interested in the constrained motion dynamics. Instead, the primary goal of this design is to obtain the equivalent control of $v(\cdot)$ for the constrained motion. The equivalent control v_{eq} is computed from

$$\dot{\sigma}_e = \Delta A_{21} x_1 + \Delta A_{22} x_2 + \Delta B_2 u + E_2 f - v_{eq} = 0 \tag{149}$$

Clearly, this signal contains information on the the system's parameteric uncertainties and the external disturbance which is useful for feedback compensation. For systems which satisfy the Drazenovic condition (135),(136), and (137), feedback control u can be designed with both linear feedback and feedback compensation using this signal to provide disturbance rejection and robustness to parameter uncertainties in the closed loop system. Let

$$u = K_1 x_1 + K_2 x_2 - \bar{B}_2 v_{eq} \tag{150}$$

then the closed loop system becomes

$$\dot{x}_1 = \bar{A}_{11} x_1 + \bar{A}_{12} x_2 \tag{151}$$
$$\dot{x}_2 = (\bar{A}_{21} + \bar{B}_2 K_1) x_1 + (\bar{A}_{22} + \bar{B}_2 K_2) x_2 \tag{152}$$

We note that this system is on the constraint manifold $\sigma_e = 0$. During the reaching phase, the system behavior is driven by the external disturbance. Similar to the disturbance rejection manifold design, reaching of the desired manifold is nevertheless guaranteed by the design of $v(\cdot)$ in the presence of disturbance and parameter uncertainties.

Simple Pendulum Tracking Example The design procedure of the disturbance estimator and the feedback compensation control is illustrated with a tracking problem for a simple pendulum whose dynamic model is given as

$$\dot{x}_1 = x_2 \tag{153}$$
$$\dot{x}_2 = -\alpha \sin x_1 + bu + d(t) \tag{154}$$

The states x_1, x_2 are the angular position and velocity of the pendulum respectively, u is the control torque, and $d(t)$ may represent external torque due to an uncertain loading force applied on the pendulum. The parameter α is a function of the pendulum's mass and the distance of the center of mass from the axis of rotation, b is the inverse of moment of inertia. From mechanics, both α and b are positive and bounded. We assume that the nominal values $\bar{\alpha}$ and $\bar{b}$ are known.

A first order disturbance estimator is constructed as follows:

$$\dot{\hat{x}}_2 = v + \bar{b} u \tag{155}$$
$$v = h_R \text{sgn}(\sigma_e), \quad \sigma_e = x_2 - \hat{x}_2 \tag{156}$$

The gain parameter h_R is chosen such that sliding mode occurs on the manifold $\sigma_e = 0$. The linear feedback component is designed to guarantee a zero steady state tracking error for a ramp position reference. Thus, a Type 2 servomechanism is designed. The resulting feedback control with disturbance compensation is given by

$$u = -k_o x_o - k_1 x_1 - k_2 x_2 - z \tag{157}$$

$$\tau^3 z^{(3)} + 3\tau^2 \ddot{z} + 3\tau \dot{z} + z = v \tag{158}$$

$$\dot{x}_o = x_1 - x_{1r} \tag{159}$$

where x_o is a servo compensator state, and $x_{1r}(t)$ is the position reference input. The role of the third order Butterworth filter (158) is to remove any high frequency residual component in v which does not exist in v_{eq} theoretically, according to the method of equivalent control.

Simulation results for different operating environments of the pendulum show the adaptive nature of the disturbance estimator. The position reference is the origin. The first set of results shown in Figs. 2 and 3 assumes that the external disturbance is zero, but there is a 10% of error in b, i.e., $b = 1.1\bar{b}$. The reaching phase is indicated by the convergence of the uncertainty estimate which follows the actual uncertainty when the desired manifold is reached. The phase plane trajectory shows that the closed loop system behaves very closely to the designed Type 2 seromechanism. The second set of results shown in Figs. 4,5, and 6 assumes, in addition to the error in b as before, a sinusoidal disturbance exists. There is no visible difference between the phase plane trajectories for the two set of operating conditions. As expected, the feedback control torque has a sinusoidal component to compensate for the unexpected disturbance. The disturbance estimate is a composition of parameter uncertainties and external disturbance.

5 Robotic Manipulator Applications

One of the engineering applications in which manifold based feedback design offers distinctive advantages is robotic manipulator control. We shall first introduce this design approach to joint space path control which was first reported in [9][2]. We then present an application to task space path control which is based on recent work on Cartesian path control[10] where manifold based design is pursued in the spirit of servomechanism. Finally, we return to the constrained motion control design for mechanical systems on manifold which is first introduced in Sect. 2.

5.1 Joint Space Path Control

Robotic manipulators are complex, multi-degree-of-freedom mechanical systems. For high precision robotic operations, it is advantageous to eliminate the mechanical transmission between the motor and its mechanical load, thus direct

---

[2] This paper was an extended version of a report submitted for a graduate level control laboratory course project taught by Professor Petar Kokotovic at the University of Illinois, Urbana-Champaign in 1976

drive motors are preferred. For a robotic manipulator with direct drive actuated revolute joints, its dynamic behavior is modeled by the following set of nonlinear differential equations:

$$M(\theta)\ddot{\theta} + F(\dot{\theta}, \theta)\dot{\theta} + G(\theta)g = u \qquad (160)$$

where $\theta, \dot{\theta}, \ddot{\theta}$ are the joint angular displacement, velocity, and acceleration vectors respectively, g is the gravitational constant, $M(\cdot)$ is the moment of inertia matrix, $F(\cdot, \cdot)$ indicates the velocity dependent cross axis coupling, $G(\cdot)g$ is the gravity induced loading torque vector, and u is a vector of motor torques which actuate the joints. We consider herein only robotic manipulators which have six independently actuated joints. The six joints are kinematically mapped to the three rotations (pitch, roll, and yaw), and the three Cartesian coordinates of the end effector uniquely. Thus, both θ and u are 6×1 vectors. We first assume that the desired path of the robotic manipulator is defined in the joint space, whereas in the next section, we shall deal with designs in which desired path is defined in the task space – as it is commonly done in practice.

Let the desired path be given by $\theta_d(t)$ which is assumed to be twice differentiable, i.e., there is no abrupt change in the desired path acceleration. For the purpose of control design, we rewrite the above equation in terms of the path tracking error

$$e(t) \doteq \theta_d(t) - \theta(t), \qquad (161)$$
$$\ddot{e} = \ddot{\theta}_d + M(\theta)^{-1}[F(\dot{\theta}, \theta)\dot{\theta} + G(\theta)g - u] \qquad (162)$$

If the model of the robotic manipulator is accurate, then it is possible to synthesize a torque control using the modeling information and the desired path in the form

$$u_c = F(\dot{\theta}, \theta)\dot{\theta} + G(\theta)g + M(\theta)[\ddot{\theta}_d + K_v(\dot{\theta}_d - \dot{\theta}) + K_p(\theta_d - \theta)] \qquad (163)$$

The resulting closed loop system dynamics

$$\ddot{e} + K_v\dot{e} + K_pe = 0 \qquad (164)$$

can be then shaped with the control gain matrices K_v, K_p such that desirable path tracking error transient performance is attained. This control algorithm is the so-called computed torque method, a popular control design technique for robotic manipulators. However, any modeling inaccuracies, and any deviation of the actual weight of the object that the manipulator is transporting from the assumed weight will affect the dynamic behavior of the path tracking error.

For this reason, a control algorithm which is based on the Disturbance Rejection Manifold Design introduced earlier is preferred for high precision path tracking because of its strong disturbance rejection capability and robustness to modeling uncertainties. We first convert the path tracking dynamics into the regular form:

$$\dot{x}_1 = x_2, \quad x_1 \doteq \theta_d - \theta \qquad (165)$$
$$\dot{x}_2 = M(\theta)^{-1}[G(\theta)g + F(\theta, \dot{\theta})\dot{\theta}] + M(\theta)^{-1}(\ddot{\theta}_d - u), \quad x_2 \doteq \dot{\theta}_d - \dot{\theta} \qquad (166)$$

For robotic manipulators whose mechanical degree of freedom is equal to the actuated degree of freedom, the moment of inertia matrix $M(\theta)$ is nonsingular for any θ. Thus, the Drazenovic condition is satisfied although the manipulator dynamics are nonlinear, i.e.,

$$\text{rank}(M(\theta) \vdots G(\theta)g + F(\theta,\dot\theta)\dot\theta) = \text{rank}(M(\theta)), \forall\theta,\dot\theta \tag{167}$$

For joint path control, let the desired manifold be specified by

$$\sigma = x_2 - Cx_1 = 0, \tag{168}$$

where the matrix C is chosen such that the constrained motion dynamics governed by

$$\dot x_1 = Cx_1 \tag{169}$$

have desirable transient behavior. Furthermore, the matrix C can be chosen to be a diagonal matrix, in which case, the path tracking error for the individual joint is independent of the other joints. The equivalent control corresponding to the constraint manifold is

$$u_{eq} = M(\theta)(C\dot e - \ddot\theta_d) + F(\dot\theta,\theta)\dot\theta + G(\theta)g \tag{170}$$

Clearly, the required compensation is realized to cancel the effects of the velocity and gravity dependent couplings, however, precise model of the manipulator is not needed.

For the reaching phase control design, in addition to the Lyapunov method described in the last section, it is also possible to decouple the design into independent single submanifold reaching control designs. Corresponding to the ith component of the vector σ, $\sigma_i = 0$ is a one dimensional submanifold of $\sigma = 0$. A sufficient condition for the system's trajectory to reach and to continue on this submanifold is [5]

$$\sigma_i\dot\sigma_i < 0 \tag{171}$$

Using a nominal moment of inertia $\bar M(\theta)$ which is assumed to be known, a new control vector u' is introduced to achieve a nominally decoupled reaching phase control design,

$$u' = \bar M(\theta)^{-1}u \tag{172}$$

The time derivative of σ_i becomes

$$\dot\sigma_i = f_{i1}(\theta) + f_{i2}(\theta,\dot\theta)^T\dot\theta + f_{i3}(\theta)^T\ddot\theta_d - u_i' + \sum_{i\neq j} h_{ij}(\theta)u_j' \tag{173}$$

The magnitudes of the nonlinear functions f_{ij}'s can be upperbounded by

$$|f_{i1}(\theta) + f_{i2}(\theta,\dot\theta)^T\dot\theta + f_{i3}(\theta)^T\ddot\theta_d| \leq \alpha_i^T|\theta| + \beta_i^T|\dot\theta| + \gamma_i^T|\ddot\theta_d| + \delta_i \tag{174}$$

where $\alpha_i, \beta_i, \gamma_i$ are vectors with positive components and δ_i is a positive scalar, and the absolute value of a vector is defined as a vector of the absolute values of the components. Thus, a possible design for $u_i'(\cdot)$ such that condition (171) is satisfied can be expressed in terms of feedback gain vectors k_{pi}, k_{vi}, k_{ai} and a relay component κ_i in

$$u_i' = [k_{pi}^T|\theta| + k_{vi}^T|\dot{\theta}| + k_{ai}^T|\ddot{\theta}_d| + \kappa_i]\mathrm{sgn}(\sigma_i) \tag{175}$$

If the deviation of $\bar{M}(\cdot)$ from the actual moment of inertia is small, i.e.,

$$|h_{ij}(\theta)| \leq \epsilon_{ij} \tag{176}$$

where ϵ_{ij} are sufficiently small positive scalars, then the design parameters in (175) can be derived independently for each submanifold by assuming that the $h_{ij}(\cdot)$'s are zero. This approach was used in the feedback design for a two link manipulator experiment where sliding mode control was implemented and tested [11]. In [9], the so-called Hierarchy of Controls Method was applied to derive similar feedback control laws as (175) without making use of the nominal moment of inertia matrix. In this design, the submanifolds were designed such that, following a chosen sequence which ended with constrained motion on the manifold $\sigma = 0$, the system's motion is constrained on one submanifold at a time.

5.2 Task Space Path Control

Robotics operations which are typically specified in terms of the end effector's kinematic degrees of freedom present a dilemma to the control system designer since the manipulator dynamics are naturally defined with respect to the actuated degrees of freedom. The conventional design approach in robotics is to decouple the problem of path planning and joint servo control. The desired trajectory is chosen as a set of points in the manipulator's task space whose vector components are the three rotations and the three Cartesian coordinates of the end effector. The joint space points corresponding to the task space points are computed from the inverse kinematics of the robotic manipulator. Reference joint command input profiles are generated from two consecutive joint displacement reference points using either a trapezoidal or an S-curve profile which are commonly used in motor control. The joint servo control is designed such that with respect to the joint command profile, the tracking error is minimized. For robotic applications where precise end point dynamic tracking is critical, the aforementioned practice becomes extremely cumbersome due to the excessive overhead required to specify the desired task space path with high spatial density.

Using the manifold based feedback design approach, the task space path planning and joint servo control are integrated into a single control design algorithm, thus greatly simplifying the process. Furthermore, the resulting controller has the same strong disturbance rejection capability and robustness to modeling uncertainties as in the joint space path control design of the previous section. This design procedure has three steps:

Task Space Path Planning Given the desired end points of the task space vector, we construct an S-curve profile directly for each of its components. Let the task space vector be given by

$$p = [x^E, y^E, z^E, \psi_x, \psi_y, \psi_z]^T \tag{177}$$

where (x^E, y^E, z^E) is the Cartesian frame coordinate of the end effector, and ψ_x, ψ_y, ψ_z are the three rotation angles with respect to the axes of a Cartesian frame which is attached to the end effector. The desired end point vectors in this space is then denoted p^1 and p^2. For each task space component, a reference model is generated. Let x_i^r be the reference state for the ith task space component whose dynamics are governed by

$$\dot{x}_i^r = \begin{bmatrix} 0 & 1 & 0 \\ 0 & 0 & 1 \\ 0 & 0 & -\tau_i^{-1} \end{bmatrix} x_i^r + \begin{bmatrix} 0 \\ 0 \\ \tau_i^{-1} \end{bmatrix} w_i(t), \quad x_i^r \doteq \begin{bmatrix} \rho_i^r \\ v_i^r \\ a_i^r \end{bmatrix} \tag{178}$$

$$w_i(t) = \begin{cases} w_i^+, & \text{if } t \le \frac{T_i}{2} \\ w_i^-, & \text{if } t > \frac{T_i}{2} \\ 0, & \text{if } t \ge T_i \end{cases} \tag{179}$$

where

$$w_i^+ = \frac{2(p_i^2 - p_i^1)}{T_i}, \quad w_i^- = -w_i^+ \tag{180}$$

and p_i^1, p_i^2 are the desired end points. A time period of T_i is specified for the transition from the initial point to the final point. The desired position and velocity trajectory for the ith task space coordinate are $\rho_i^r(t)$ and $v_i^r(t)$ respectively. With the initial conditions, $\rho_i^r(t_1) = p_i^1$ and $v_i^r(t_1) = a_i^r(t_1) = 0$, this reference model generates an S-type profile for $\rho_i^r(t)$ which satisfies the desired end point conditions

$$\rho_i^r(t_1) = p_i^1, \quad \rho_i^r(t_2) = p_i^2, \quad t_2 \ge t_1 + T_i \tag{181}$$

and guarantees that the desired acceleration is continuous. The time constant τ_i should be chosen such that $\tau_i \ll T_i$. The S-curve profile degenerates to a trapezoidal profile if $\tau_i = 0$, i.e., without any smoothing of the reference input pulse w_i.

Task Space Servo From classical servomechanism design, the steady state position error depends on the nature of the reference tracking input, e.g., step or ramp input, and the number of poles of the open loop transfer function at the origin, i.e., the system type. The goal of the Task Space Servo is to construct a similar relationship between the task space position steady state error and the task space reference states such that the structure of the feedback controller for the desirable steady state behavior is determined. Let e^p be a vector of task space tracking error,

$$e^p \doteq p^r - p, \quad e^p \doteq [e_1^p, e_2^p \ldots, e_6^p]^T \tag{182}$$

$$(p^r) \doteq [\rho_1^r, \rho_2^r \ldots, \rho_6^r]^T \tag{183}$$

where p^r is the task space reference position vector. The desired task space servo dynamics for the ith component are chosen as

$$\dot{\xi}_i = \begin{bmatrix} 0 & 1 \\ 0 & 0 \end{bmatrix} \xi_i + \begin{bmatrix} 0 \\ 1 \end{bmatrix} e_i^p \tag{184}$$

$$\eta_i = h_i \xi_i + d_i e_i^p \tag{185}$$

$$\dot{e}_i^p = \eta_i + \dot{\rho}_i^r \tag{186}$$

which would provide zero steady state tracking error for the task space reference dynamics specified earlier. Similar servo dynamics apply to each of the task space components, and the linear feedback gains h_i's and d_i's are design parameters which are to be optimized independently for each component to yield the desirable transient tracking error response.

Desired Manifold Design The above task space servo dynamics can be considered as the desired dynamics of the robotic manipulator. We evoke the manifold based feedback design to enforce constrained motion on a manifold such that the desired task space servo is realized. The desired manifold is actually specified by the identity in (186) which can be rewritten in vector form as

$$\sigma = \dot{p} + \eta = 0, \quad \eta \doteq [\eta_1, \ldots, \eta_6]^T \tag{187}$$

The robotic manipulator's kinematics and inverse kinematics are defined by the following nonlinear transformations

$$p = p(\theta), \quad \theta = \theta(p) \tag{188}$$

and the associated Jacobian matrix is

$$J(\theta) \equiv \frac{\partial p}{\partial \theta} \tag{189}$$

is nonsingular for almost all θ, with the exception of the singularity points. The constrained motion dynamics on $\sigma = 0$ are the collection of uncoupled task space servo dynamics which are rewritten in vector form as follows:

$$\dot{\xi} = F\xi + Ge^p, \quad \xi = [\xi_1, \ldots, \xi_6]^T \tag{190}$$

$$\dot{p} = H\xi + De^p, \quad e^p = p^r - p \tag{191}$$

where F, G, H, D are diagonal matrices. In order for motion to exist on $\sigma = 0$, it is necessary to check the uniqueness of the corresponding equivalent control u_{eq} which is computed from

$$\dot{\sigma} = -J(\theta)M(\theta)^{-1}(F(\dot{\theta}, \theta) + G(\theta)g + u_{eq}) + $$
$$+ (K(\theta)\dot{\theta} - DJ(\theta))\dot{\theta} + HF\xi - HGp^r + D\dot{p}^r = 0 \tag{192}$$

$$K(\theta) \equiv \frac{\partial J}{\partial \theta} \tag{193}$$

With the exception of the singularity points, $M(\theta)J(\theta)^{-1}$ is invertible, thus u_{eq} exists. For task space control design, it is general practice to avoid the singularity points of the manipulator by path planning. Therefore, we shall assume that there exists feedback control such that constrained motion on the desired manifold can be realized.

Cartesian Path Tracking Example We shall use a two revolute joint manipulator with Cartesian path planning in a two dimensional plane $X - Y$ to demonstrate the servo design. The nomenclature of the manipulator's parameters are inferable from Fig. 7. The model's nonlinearities are given below using the same notations as in (160):

$$M(\theta) = \begin{bmatrix} m_{11} & m_{12} \\ m_{12} & m_2 r_2^2 \end{bmatrix} \tag{194}$$

$$m_{11} = (m_1 + m_2)r_1^2 + m_2 r_2^2 + 2m_2 r_1 r_2 c\theta_2 \tag{195}$$

$$m_{12} = m_2 r_2^2 + m_2 r_1 r_2 c\theta_2 \tag{196}$$

$$F(\theta, \dot{\theta})\dot{\theta} = \begin{bmatrix} -m_2 r_1 r_2 s\theta_2(\dot{\theta}_1^2 + 2\dot{\theta}_1\dot{\theta}_2) \\ m_2 r_1 r_s \theta_2 \dot{\theta}_2^2 \end{bmatrix} \tag{197}$$

$$G(\theta) = \begin{bmatrix} (m_1 + m_2)r_1 c\theta_1 + m_2 r_2 c\theta_{12} \\ m_2 r_2 c\theta_{12} \end{bmatrix} \tag{198}$$

with $c\theta_i \equiv \cos(\theta_i), i = 1, 2, c\theta_{12} \equiv \cos(\theta_1 + \theta_2)$. The kinematics are defined by

$$x^E = r_1 c\theta_1 + r_2 c\theta_{12} \tag{199}$$

$$y^E = r_1 s\theta_1 + r_2 s\theta_{12} \tag{200}$$

and the Jacobian matrix $J(\theta)$ is

$$J(\theta) \doteq \begin{bmatrix} j_{11} & j_{12} \\ j_{21} & j_{22} \end{bmatrix} = \begin{bmatrix} -(r_1 s\theta_1 + r_2 s\theta_{12}) & -r_2 s\theta_{12} \\ (r_1 c\theta_1 + r_2 c\theta_{12}) & r_2 c\theta_{12} \end{bmatrix} \tag{201}$$

which is singular for $\theta_2 = 0$, corresponding to a configuration where the first and second link are aligned. The task of the Cartesian path sliding mode servo design begins with the definition of the initial and final point of the Cartesian path, $(p_1^1, p_2^1), (p_1^2, p_2^2)$ respectively. If neither of these points nor the path connecting them satisfy the condition

$$x^E(t)^2 + y^E(t)^2 = (r_1 + r_2)^2 \tag{202}$$

then the singularity points are avoided. The physical parameters of this manipulator are: $m_1 = 1\text{Kg}$, $m_2 = 0.5\text{Kg}$, $r_1 = 0.5\text{meters}$, and $r_2 = 0.25\text{meters}$. Using the Cartesian reference model (178), we generate the reference Cartesian path between the point $(0.01, -0.7)$ and $(0.09, -0.6)$ using identical time constant $\tau_1 = \tau_2 = 0.1$ and transition time $T_1 = T_2 = 4\text{sec}$. for both the X and the Y axis. The desired Cartesian path is a line connecting between the initial and final points. As shown in Figs. 8 and 9, the reference time profiles are of the S-curve family. Fig. 10 shows the desired straight line path.

The next step is to design the feedback gains for the desired task space servo (185) to provide good transient error response. Beside the desired bandwidth and damping factor, effort to minimize integral windup is needed, typical for this type of servo. Closed loop eigenvalues are placed at $-5 \pm j8.66$ and -5. The two switching surfaces for variable structure control are constructed using the Jacobian matrix:

$$\sigma_1 = j_{11}(\theta)\dot{\theta}_1 + j_{12}(\theta)\dot{\theta}_2 + h_x\xi_x + d_1(\rho_1^r - x^E(\theta_1, \theta_2)) = 0 \qquad (203)$$

$$\sigma_2 = j_{21}(\theta)\dot{\theta}_1 + j_{22}(\theta)\dot{\theta}_2 + h_y\xi_y + d_2(\rho_2^r - y^E(\theta_1, \theta_2)) = 0 \qquad (204)$$

We assume herein that the joint angular displacements are measured, however, the end point's Cartesian coordinates are not sensed, but indirectly calculated from the forward kinematics.

Simulation results of the Cartesian path tracking servo show that the motion on the desired manifold provide the expected linear servo response despite of the nonlinear system dynamics for the task space. The actual vs. reference Cartesian path are plotted in Fig. 11. Figure 12 shows the Cartesian path tracking error responses which match with the desired servo responses. The overshoot results from the response of a nonzero initial Cartesian coordinate error to the servo system. Zero steady state error is achieved as designed.

5.3 Constrained Motion Control

In Sect. 2, we introduce the notion of manifold using a robotic manipulator which operates in a configuration space with physical constraint surfaces. The design of feedback control for the reaching phase and the constrained motion phase is shown to be decoupled. A desired manifold $\sigma_N = 0$ is introduced to obtain the desirable reaching phase dynamic behavior. While it is convenient to assume that the physical constraint surface is rigid, thereby it can be treated as holonomic constraints from the analytical mechanics perspective, physical constraint surfaces should be more realistically modeled with "local" dynamics about some nominal reference state. The notion of a deformable surface is useful for describing the dynamic phenomenon of motion when the holonomic constraints are violated in the forbidden region. When the robotic manipulator is in the unconstrained region of the robot's configuration space, the physical surface is undeformed and its dynamic state ϕ is at its equilibrium point $\phi = \phi_c \doteq 0$. As specified in Sect. 2, the half space $\phi < 0$ shall be the unconstrained domain. Note that the holonomic constraints are expressed as $\phi(\theta) = 0$ where the joint space angles θ are chosen as the generalized displacements q. If the holonomic constraints are satisfied, the dynamic state ϕ is in equilibrium. By adopting the surface deformation notion, we can deal with the motion in the half space $\phi > 0$ while using the equilibrium ϕ_c as the nominal reference. The "impedance" of the surface deformation is defined as an operator on the "deformation"

$$\Delta\phi \doteq \phi - \phi_c \qquad (205)$$

A linear operator which models the inertial, damping and stiffness effects of the surface can be expressed using a state space model with states $\Delta\phi$ and $\dot{\Delta\phi}$:

$$M_e \ddot{\Delta\phi} + B_e \dot{\Delta\phi} + K_e \Delta\phi = \lambda, \quad \text{for } \Delta\phi \geq 0 \tag{206}$$

where M_e is positive definite, and B_e, K_e are symmetric and positive definite matrices, and λ is the reaction force.

Using the two degrees of freedom planar robotic manipulator introduced in the Cartesian path tracking example, we shall illustrate the design of a feedback control system for driving the end effector from the unconstrained region onto a deformable physical constraint surface.

Circle Constraint Tracking Example The circle constraint is expressed in terms of the robotic manipulator variables introduced in Fig. 7 as

$$\phi = (x^E(\theta_1, \theta_2))^2 + (y^E(\theta_1, \theta_2))^2 - r^2 \tag{207}$$
$$= (r_1^2 + r_2^2 + 2r_1 r_2 c\theta_2) - r^2 = 0, \quad \phi \in \mathbb{R} \tag{208}$$

where $r < r_1 + r_2$. Thus the half space $\phi < 0$ denotes the unconstrained region of the configuration space. A more realistic model of the dynamic interactions between the end effector and the environment is chosen such that the impedance of the constraint surface is defined by

$$m_e \ddot{\phi} + k_e \phi = -\lambda, \quad \text{if } \phi \geq 0 \tag{209}$$

The dynamics of the robotic manipulator for any ϕ are governed by

$$M(\theta)\ddot{\theta} + F(\theta, \dot{\theta})\dot{\theta} + G(\theta)g = \begin{bmatrix} u_1 \\ u_2 \end{bmatrix} + \begin{bmatrix} 0 \\ 2(x^E(\theta)j_{12}(\theta) + y^E(\theta)j_{22}(\theta)) \end{bmatrix} \lambda^* \tag{210}$$

where u_1, u_2 are the joint control torques for joint 1 and 2 respectively, and

$$\lambda^* = \begin{cases} \lambda, & \text{if } \phi \geq 0 \\ 0, & \text{if } \phi < 0 \end{cases} \tag{211}$$

and λ denotes the reaction force of the constraint surface. From (208) and (209),

$$\lambda = 2m_e r_1 r_2 (s\theta_2 \ddot{\theta}_2 + c\theta_2 (\dot{\theta}_2)^2) + 2k_e r_1 r_2 (s\theta_2)\dot{\theta}_2 \tag{212}$$

Hence, the mechanical system which consists of the robotic manipulator and the constraint surface impedance has an additional inertia term and velocity dependent torques when the circle is deformed, i.e., $\phi \geq 0$. The desired manifold for the reaching phase of the circle constraint is chosen as

$$\sigma_N = \dot{\phi} + c_N(\phi - \phi^*) = 0, \quad c_N > 0, \phi^* > 0 \tag{213}$$

In accordance with (24), the decomposition of the control input is given by

$$u_1 = u_c \tag{214}$$
$$u_2 = 2(x^E j_{12} + y^E j_{22})u_N + \frac{m_{12}}{m_{11}} u_c \tag{215}$$

If we let $u_c = 0$, then the motion which satisfies the circle constraint is the free motion of the robotic manipulator, i.e., the end effector can be on any point of the circle. From (208), the circle constraint $\phi = 0$ is satisfied with

$$\theta_2 = \theta_2^* \doteq \cos^{-1}\left(\frac{r^2 - (r_1^2 + r_2^2)}{2r_1r_2}\right) \tag{216}$$

Thus the constrained motion dynamics attributable to the robotic manipulator are obtained from (210) by setting $\theta_2 = \theta_2^*$:

$$m_{11}^* \ddot{q}_1 - m_2 r_1 r_2 \theta_2^* \dot{\theta}_1 + ((m_1 + m_2) r_1 c\theta_1 + m_2 r_2 \cos(\theta_1 + \theta_2^*))g = u_1 \tag{217}$$
$$m_{11}^* \doteq (m_1 + m_2)r_1^2 + m_2 r_2^2 + 2m_2 r_1 r_2 c\theta_2^* \tag{218}$$

This equation prescribes the free motion of the robotic manipulator while the end effector is interacting with the circle constraint surface. The "local" dynamics of the deformable surface are dependent on the control of the reaching phase.

We now focus on the design of u_N which is the component for the control of the unconstrained motion. Let

$$u_N = \begin{cases} u_N(\theta, \dot{\theta})\mathrm{sgn}(\sigma_N)\,, & \text{if } \phi < 0 \\ 0\,, & \text{if } \phi \geq 0 \end{cases} \tag{219}$$

For this design, we are not directly controlling the dynamic response of the deformed surface, i.e., our design goal is satisfied if

$$\dot{\phi}\phi < 0\,, \quad \text{for } \phi < 0 \tag{220}$$

The functional $u_N(\cdot)$ is designed such that the above condition is satisfied. Thus, the desired manifold $\sigma_N(t) = 0, t \geq t_1$ is reached in finite time t_1. The equilibrium $\phi(t) = \phi_c = 0, t \geq t_2 > t_1$ is reached also in finite time t_2 since ϕ^* is positive and c_N is positive. Thus the physical constraint surface $\phi = 0$ is a submanifold of the desired manifold of the feedback design $\sigma_N = 0$. From

$$\dot{\sigma}_N = -m_e^{-1}(k_e\phi + \lambda) + c_N^2(\phi - \phi^*) = 0\,, \quad \text{for } \phi \geq 0 \tag{221}$$

the dynamics governing the surface deformation can be found,

$$m_e\ddot{\phi} + m_e c_N^2 \phi = -m_e c_N^2 \phi^*\,, \phi(t) \geq 0 \tag{222}$$

and the associated initial conditions are

$$\phi(t_c) = 0\,, \quad \dot{\phi}(t_c) = c_N\phi^* \tag{223}$$

Note that the reaction force which corresponds to these initial conditions is

$$\lambda\big|_{\sigma_N=0, \phi=0} = -m_e c_N^2 \phi^* \tag{224}$$

which depends on c_N and ϕ^*, both of which are design parameters. In order to complete the dynamic characterization of ϕ and $\dot{\phi}$, we write down the motion on the desired manifold $\sigma_N = 0$,

$$\dot{\phi} + c_N(\phi - \phi^*) = 0\,, \quad \phi(t) < 0 \tag{225}$$

Given the initial condition in the unconstrained region $\phi(t_o) \doteq \phi_o < 0$, the time-to-reach t_c, i.e., $\phi(t_c) = 0$, is computed from the solution to the above equation,

$$t_c = t_o - c_N^{-1} \ln \left(\frac{-\phi^*}{\phi_o - \phi^*} \right) \tag{226}$$

Thus, it is desirable to pick large positive c_N and small ϕ^* to obtain fast reaching of the circle constraint, while minimizing the initial reaction force. This tradeoff is clearly displayed in (224) and (226).

We simulate the free motion of the robotic manipulator in a configuration space where a deformable circle is the constraint surface. The radius r of the circle is 0.7 meters, and the equivalent mass and stiffness are $m_e = 0.01$ and $k_e = 1000$. Figures 13–15 show that this circle is "soft" in the sense that it is easily deformable : the manipulator is fully extended to 0.75 meters at one time instant despite of the constraint. Ideally, if the circle is infinitely stiff, i.e., $\phi = 0$, then joint 2 angular displacement is held at $\theta_2^* = 0.7813$rad.. From Fig. 14, θ_2 exceeds this value when the circle is deformed. For the feedback control design, we choose $c_N = 10$ and $\phi^* = 0.1$ to yield a time-to-reach $t_c = 0.1019$sec., and an initial reaction force of 0.1Newtons. The dynamic behavior of the manipulator under constrained motion control is shown in Figs. 16-18. Small deformations of the circle result from the controlled unconstrained motion. In the constrained region, the circle deforms according to the forced motion of a harmonic oscillator. The free motion of joint 1 is shown in Fig. 17.

6 Conclusions

We have shown in this paper that the notion of manifold has been successfully adopted in feedback control in a variety of control system design formulations, and for a multitude of goal specific design approaches. By consolidating all these different formulations and approaches, an unified framework for the design and analysis of feedback control systems with motion on manifolds is established. It is our hope that this framework will stimulate future generation control engineers to cast their design in terms of manifolds of desired motions, and to utilize the experience accumulated in the works that we have collected herein.

References

1. Young, K. D., "Sliding mode for constrained robot motion control," Chapter 8, *Variable Structure Control for Robotics and Aerospace Applications*, pp. 157-172, Elsevier Science Publishers, 1993.
2. Young, K-K. D. , P. V. Kokotovic, and V. I. Utkin, " A singular perturbation analysis of high gain feedback systems," *IEEE Trans. Autom. Contr.*, Vol. AC-22, pp. 931-938, 1977.
3. Kokotovic, P. V., H. K. Khalil, and J. O'Reilly, *Singular Perturbation Methods in Control: Analysis and Design*, Academic Press, 1986.

4. Young, K-K. D. and H. G. Kwatny, "Variable Structure Servomechanism Design and its Application to Overspeed Protection Control," *Automatica*, Vol.18, No. 4, pp. 385-400, 1982.
5. Utkin, V.I. *Sliding Mode and Their Applications*, Moscow: Mir 1978.
6. Utkin V. I.*Sliding Modes in Control Optimization*, Springer-Verlag 1992.
7. Drazenovic, B., "The invariance conditions in variable structure systems," *Automatica*, Vol.5, No. 3, pp. 287-295, 1969.
8. Young, K. D., and S. V. Drakunov,"Discontinuous frequency shaping compensation for uncertain dynamic systems," *Proceedings 12th IFAC World Congress*, Sydney, Australia, pp. 39-42, 1993.
9. Young, K-K. D.,"Controller Design for Manipulator Using Theory of Variable Structure Systems," *IEEE Transaction on Systems, Man, and Cybernetics*, Vol. SMC-8, February 1978, pp. 101-109, also appeared in *Robot Motion Planning and Control*, M. Brady, J. M. Hollerbach, T. L. Johnson, T. Lozano-Perez and M. T. Mason (editors), pp. 201-219, MIT Press, 1982, and *Tutorial on Robotics*, C. S. G. Lee, R. C. Gonzalez, and K. S. Fu (editors), pp. 251-259, IEEE Computer Society Press.
10. Young, K. D., "Cartesian path sliding mode servo design for robotics applications", *Proceedings First Asian Control Conference*, July 27-30, 1994, Tokyo, Japan, pp. 389-392, 1994.
11. Hashimoto H., K. Maruyama, and F. Harashima, "A microprocessor based robot manipulator control with sliding mode," *IEEE Transactions on Industrial Electronics*, Vol.IE-34, No.1, pp. 11-18, 1987.

This article was processed using the LaTeX macro package with LLNCS style

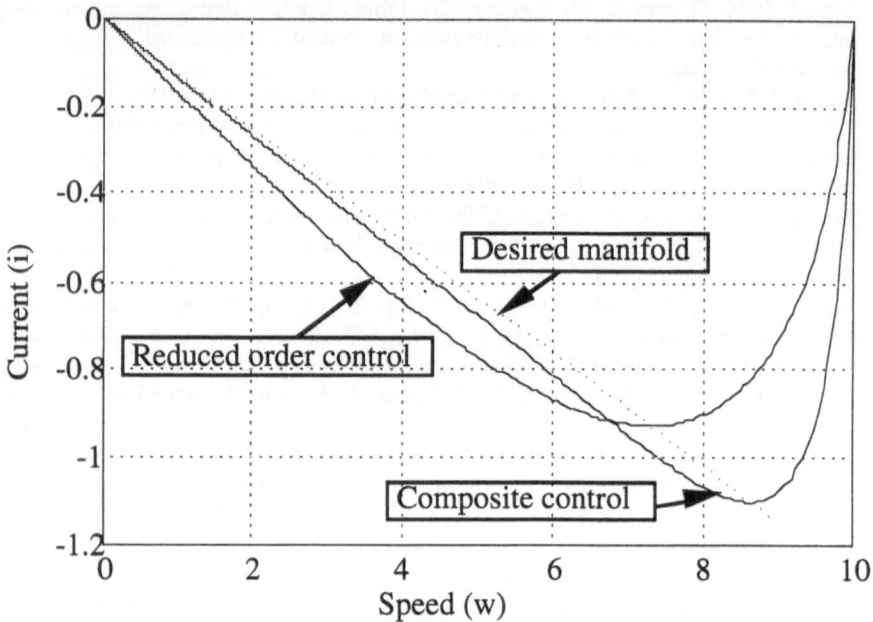

Fig. 1. The constraint manifold and the phase trajectories for the reduced order and the composite control design

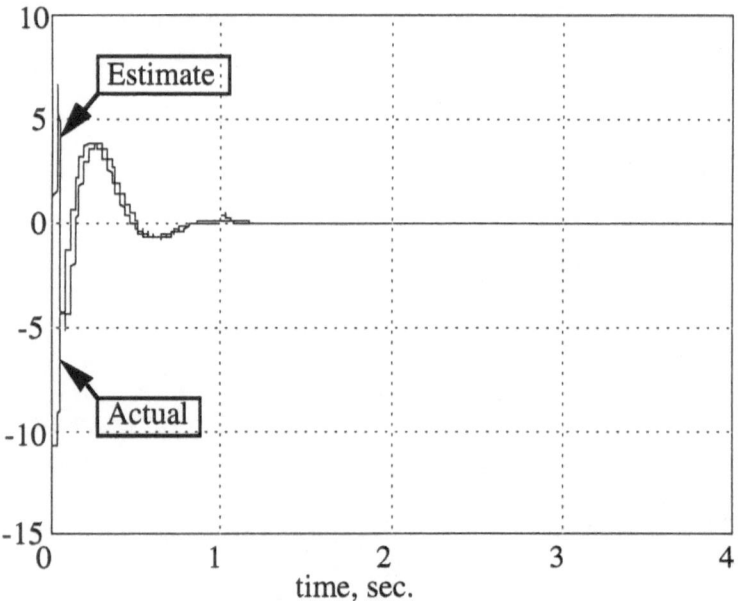

Fig. 2. Uncertainty estimate for $b = 1.1\bar{b}$

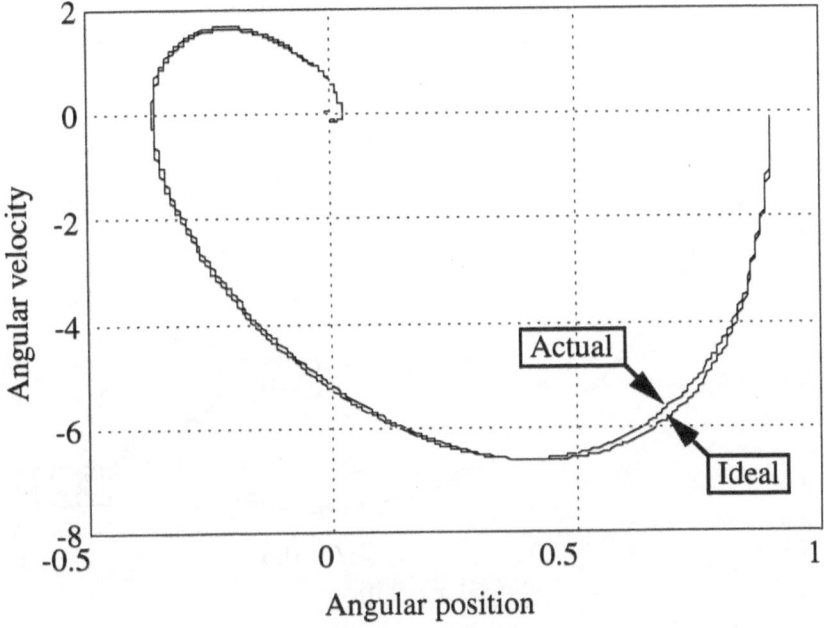

Fig. 3. Phase plane trajectories for $b = 1.1\bar{b}$

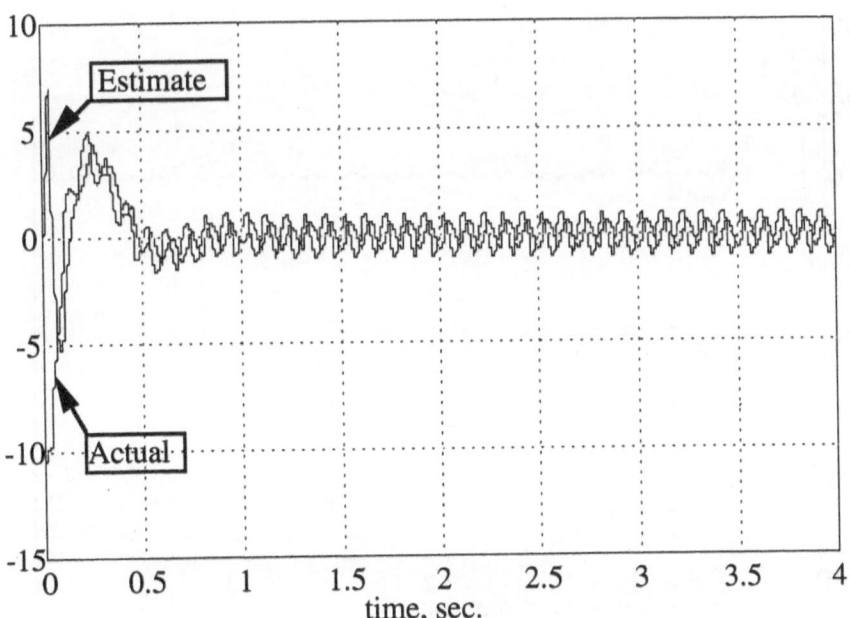

Fig. 4. Uncertainty estimate for $b = 1.1\bar{b}$ and external disturbance

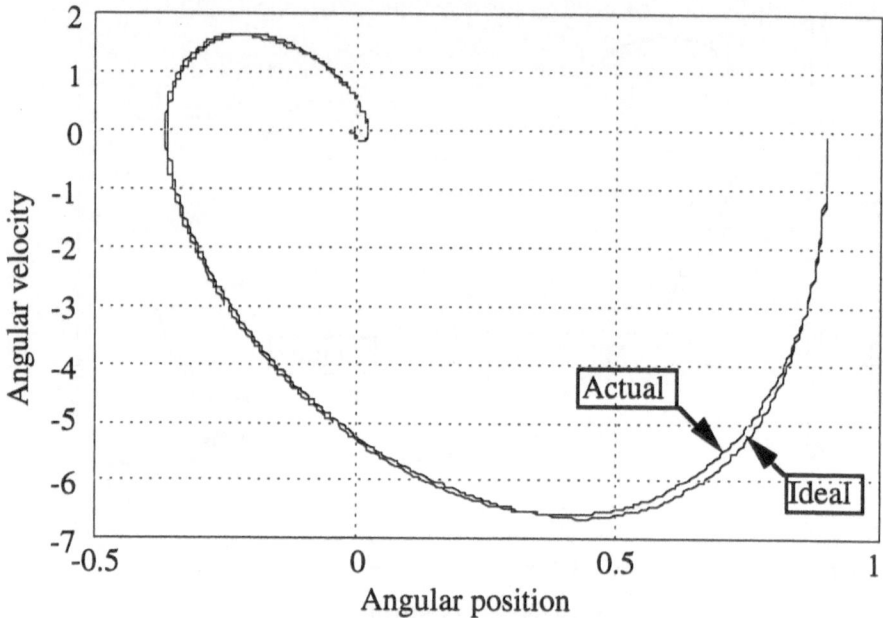

Fig. 5. Phase plane trajectories for $b = 1.1\bar{b}$ and external disturbance

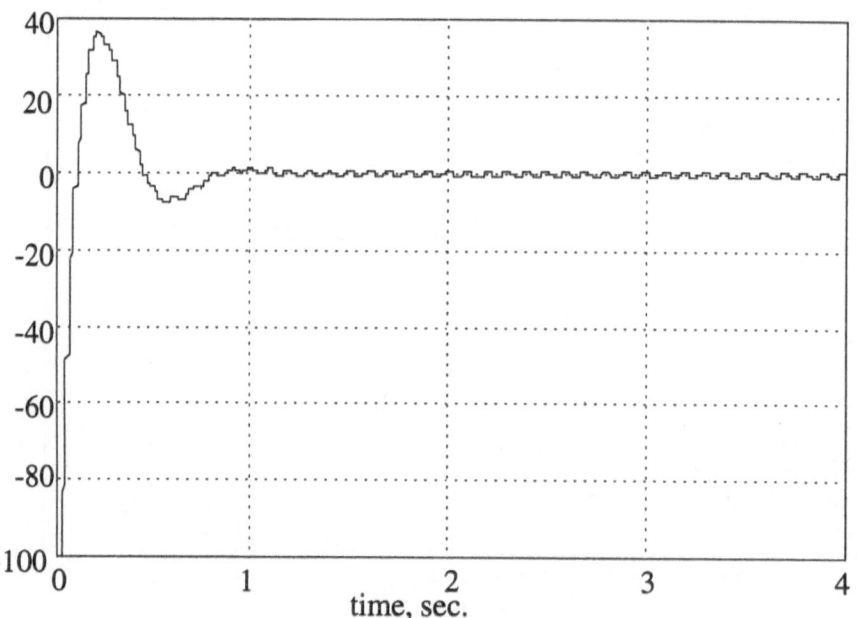

Fig. 6. Feedback control torque for $b = 1.1\bar{b}$ and external disturbance

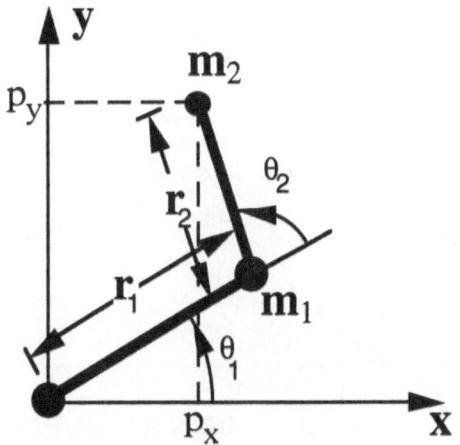

Fig. 7. Two link manipulator

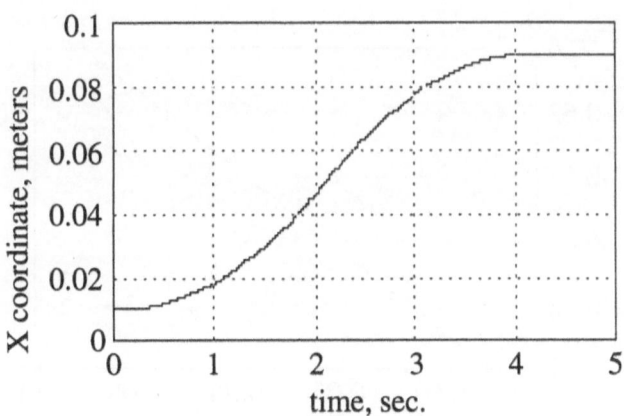

Fig. 8. X coordinate reference S-curve

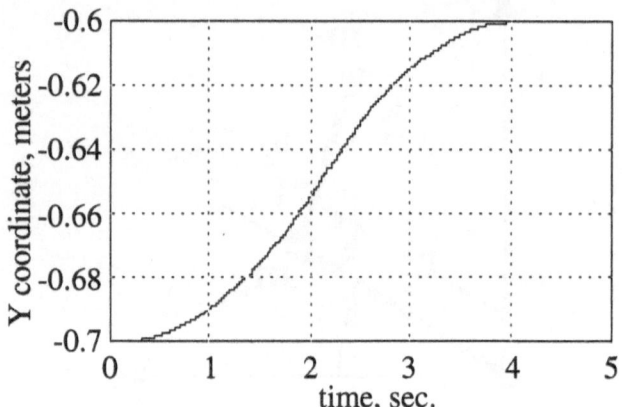

Fig. 9. Y coordinate reference S-curve

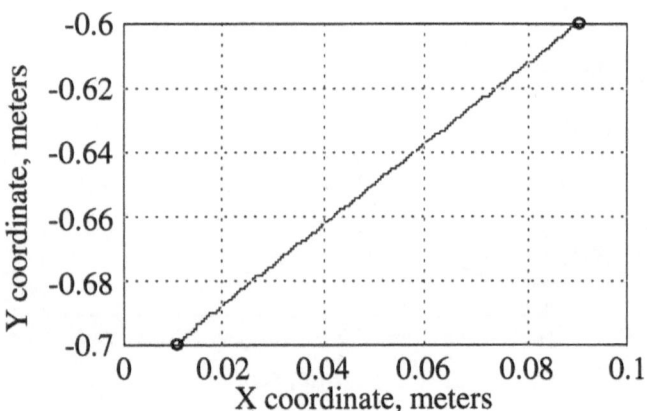

Fig. 10. Cartesian path reference in XY plane

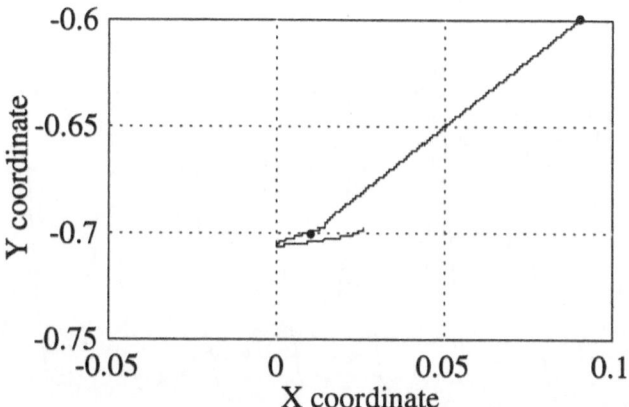

Fig. 11. Cartesian path tracking response and its reference in X-Y plane

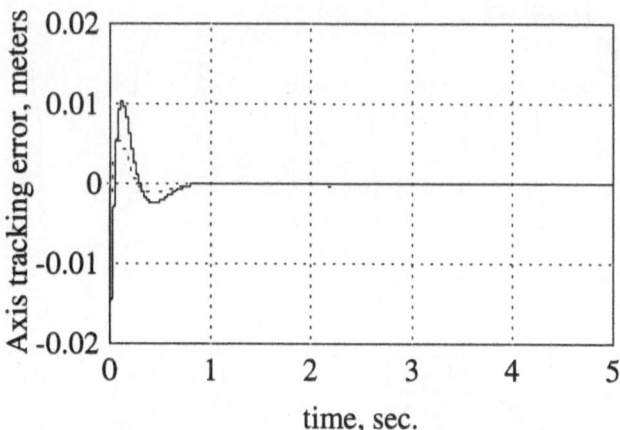

Fig. 12. Cartesian path tracking errors for X and Y axis

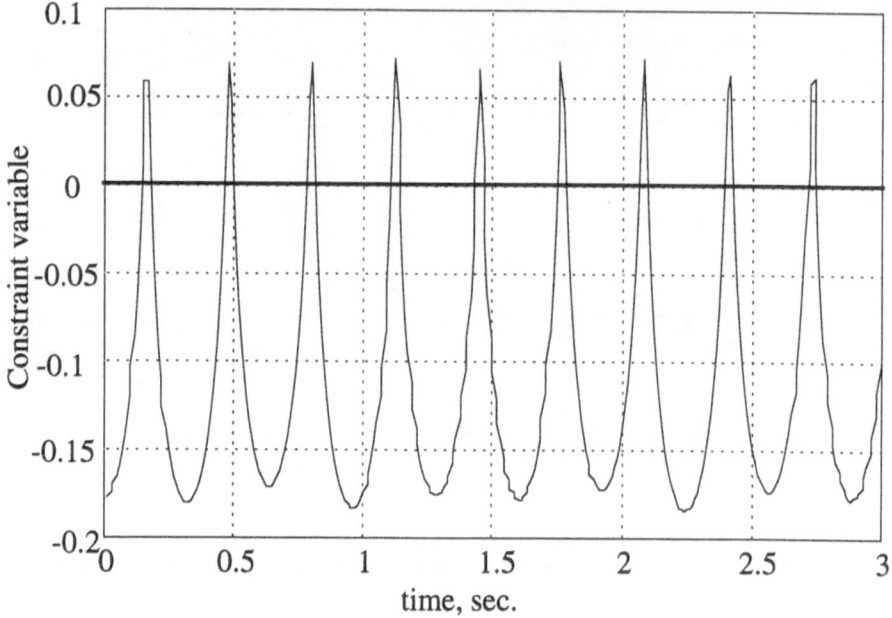

Fig. 13. Constraint dynamic state time response in free motion

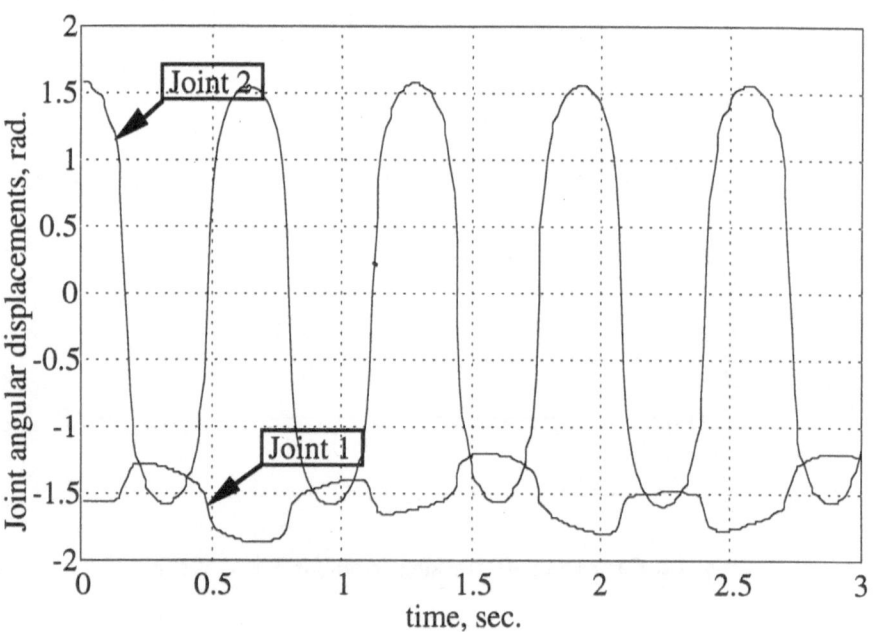

Fig. 14. Joint displacement time responses in free motion

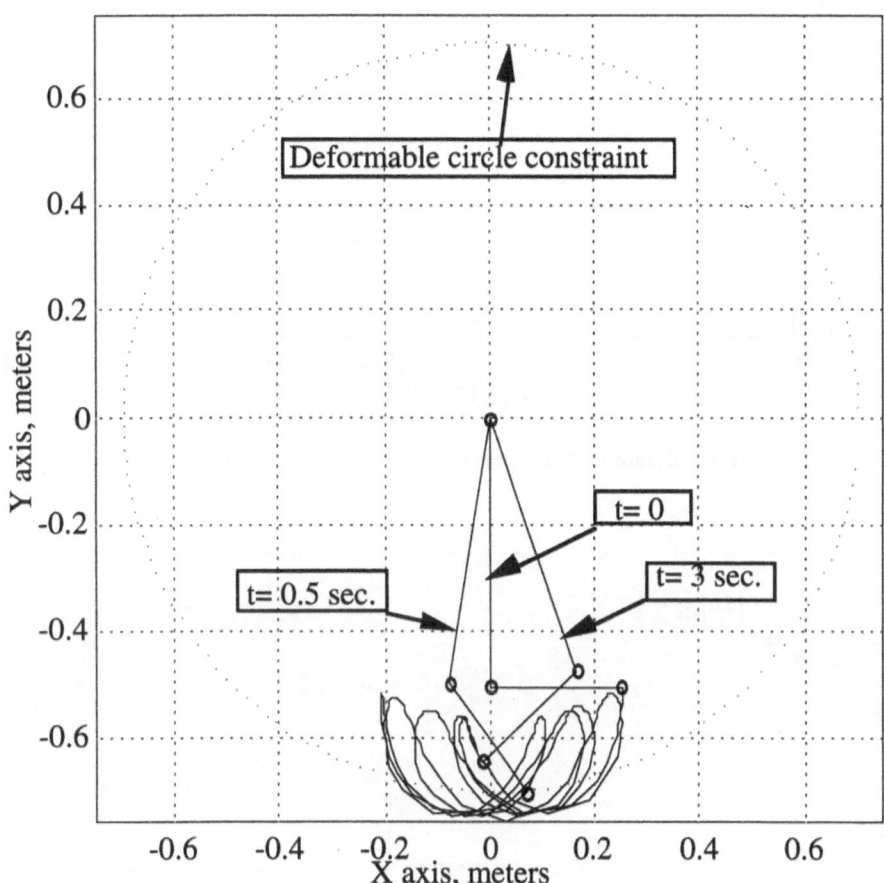

Fig. 15. Robotic manipulator's free motion in the configuration space with a deformable circle constraint

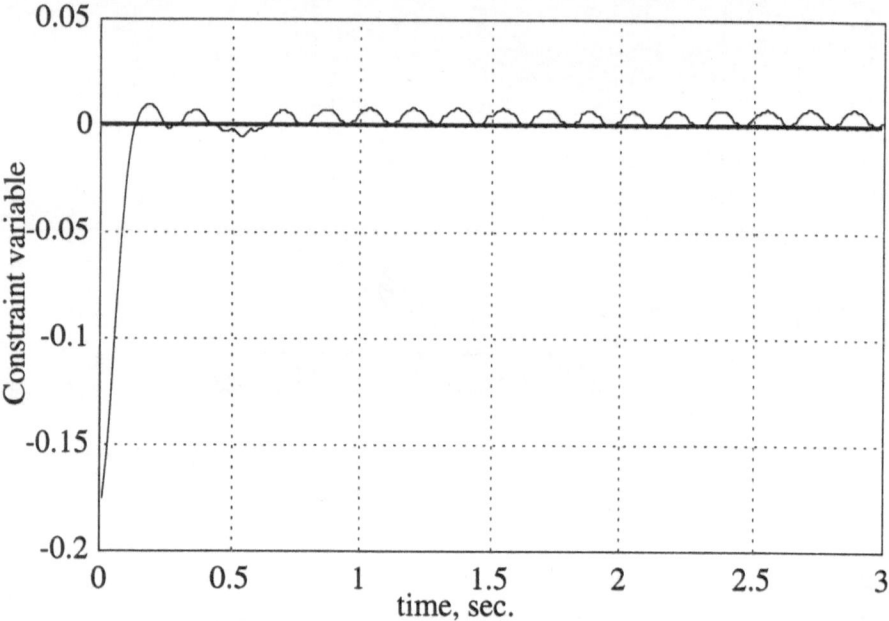

Fig. 16. Constraint dynamic state time response with constrained motion control

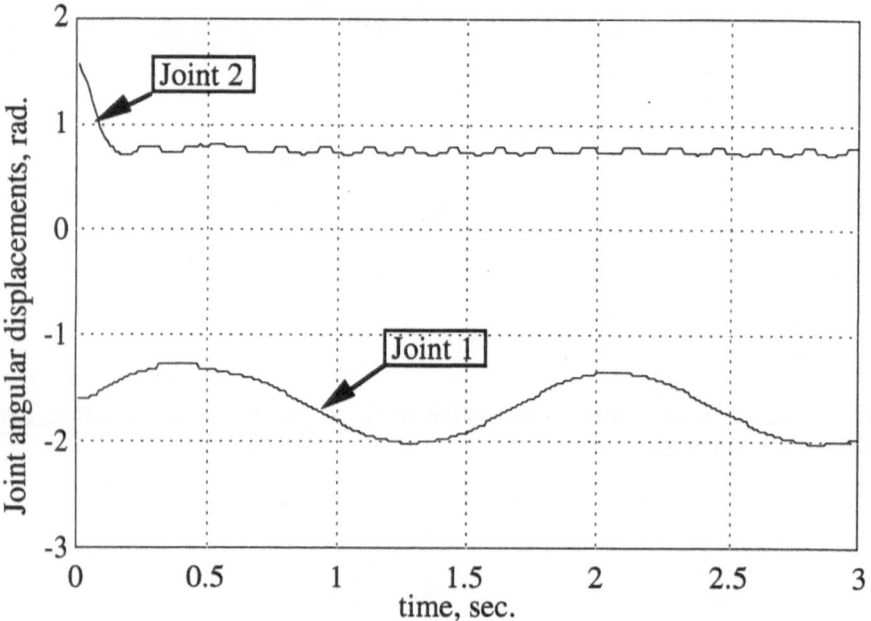

Fig. 17. Joint displacement time responses with constrained motion control

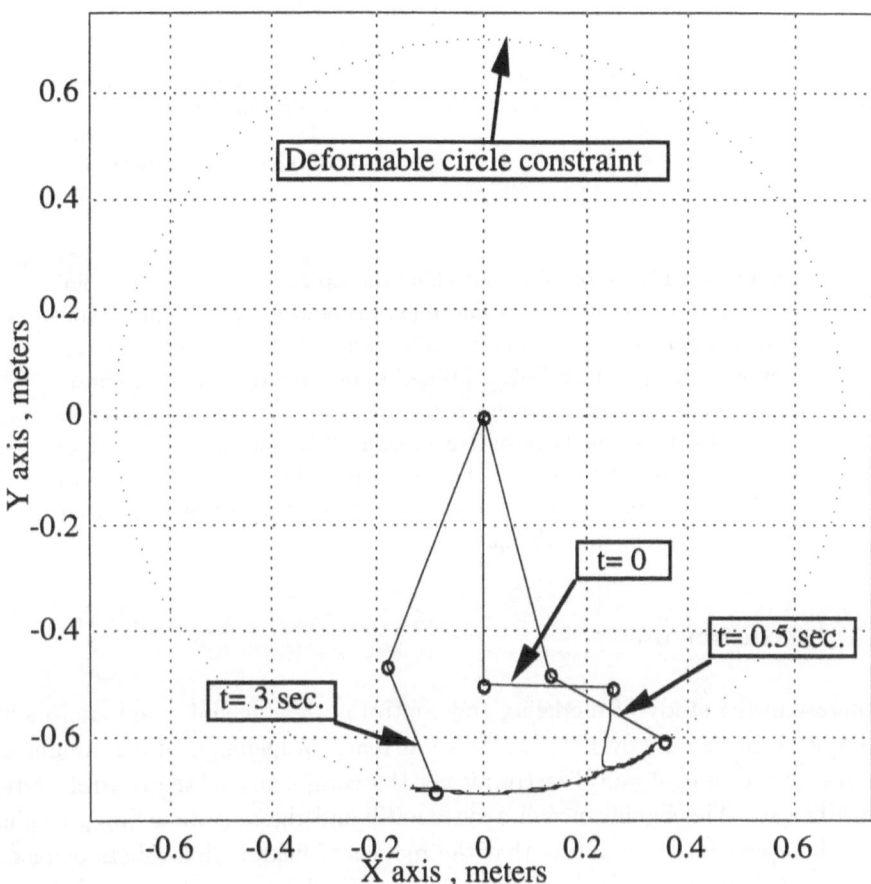

Fig. 18. Robotic manipulator's controlled motion in the configuration space with a deformable circle constraint

Control of Nonminimum Phase Singularly Perturbed Systems with Applications to Flexible Link Manipulators [*]

K. Hashtrudi Zaad[*] and K. Khorasani

[*]Department of Electrical Engineering
University of British Columbia (UBC)
2356 Main Mall
Vancouver, B.C., Canada V6T 1Z4

Department of Electrical and Computer Engineering
Concordia University
1455 De Maisonneuve Blvd. W.
Montreal, Quebec, Canada H3G 1M8

Abstract. The problem of controlling the tip position of a flexible–link manipulator is considered. A linear mathematical model of the flexible system is expressed in a standard singularly perturbed form. The concept of the integral manifolds is utilized to design a dynamical composite control strategy to guarantee a minimum phase closed–loop system restricted to the manifold, resulting in controlling the tip position to an arbitrary degree of accuracy by just measuring the hub angle and the tip position. Numerical simulations are included to demonstrate the advantages of the proposed technique.

1   Introduction

Interest in the study of modeling and control of flexible–link manipulators has been growing rapidly over the past several years. Modeling is of paramount importance for control and directly affects the complexity of the control system architecture. The significance of a systematic and rigorous modeling procedure may be appreciated by noting that the models of flexible link robots belong to a class of distributed parameter systems described by partial differential equations (PDE's). The assumed modes and finite element methods are two common approaches currently being used for approximating the PDE's by a system of ordinary differential equations. Once the lumped parameter approximation is obtained a further simplification may be made by linearizing the nonlinear equations of motion into a time–domain state–space representation or a frequency–domain transfer function representation. Generally speaking, these models are

[*] This work was supported in part by Natural Sciences and Engineering Research Council of Canada under Grant OGP0042515 and by the Fonds Pour la Formation de Chercheurs et l'Aide la Recherche, Programme Etablissement de Nouveaux Chercheurs under Grant NC0028.

acceptable for only "small" motion dynamics. The popularity of linear models for control of flexible–link manipulators is due to the availability of a variety of well–developed and practically tested time–domain and frequency–domain control techniques. However, as performance specifications such as higher speed and higher accuracy manipulations become more demanding, the control requirements become increasingly more crucial.

A common approach to compensate for the nonlinear dynamics of a rigid manipulator is the so–called "inverse dynamics" or "computed torque" strategy. By employing this scheme the manipulator dynamics is externally linearized and decoupled by the nonlinear controller introduced in the feedback loop. A servo controller is then constructed for the resulting decoupled linear model so that certain design specifications are satisfied. This scheme assumes exact cancellation of the nonlinear terms of the manipulator dynamics by the nonlinear controller. Extension of this approach to flexible–link manipulators has been hindered because of the nonminimum phase characteristics of the system. It is well–known that the zero dynamics associated with a flexible manipulator are unstable. The source of this problem may be traced to the noncolocated nature of the sensor and actuator used for controlling the tip position. Therefore, inverse dynamics type controllers result in an unstable closed–loop internal dynamics. There have been several methods proposed in the literature to resolve this issue. Wang and Vidyasagar [1] redefined the output of the nonlinear system by the "reflected" tip position, demonstrating that the zero dynamics associated with this new input–output map are stable. The drawback of this method is that it is not easy (if not impossible) to relate the desired tip position trajectories to the desired reflected tip position trajectories. Bayo and Moulin [2] proposed a noncausal torque solution which acts before the tip starts moving and after the tip stops moving. The drawback of this method is its heavy computational requirements for transforming the dynamic model and the input trajectories from time–domain into frequency–domain specifications. Kwon and Book [3] proposed to decompose the inverse dynamics (and the input torque) of the manipulator into a causal system and an anticausal system by using coordinate transformations. The causal part is integrated forward in time and the anticausal part integrated backward in time. This method is limited to only linear approximations of the flexible–link manipulator.

In this paper the linear model of the flexible system is expressed in a standard singularly perturbed form. The concept of the integral manifolds is utilized to design a dynamical composite control strategy to guarantee a minimum phase closed–loop system restricted to the manifold. Consequently, the tip position may be controlled to an arbitrary degree of accuracy. Numerical simulations are included to demonstrate the advantages of the proposed technique.

2 Model for the Flexible Manipulator

The position of any point along the link is denoted by $y(x, t) = xq(t) + w(x, t)$, as shown in Fig.1.

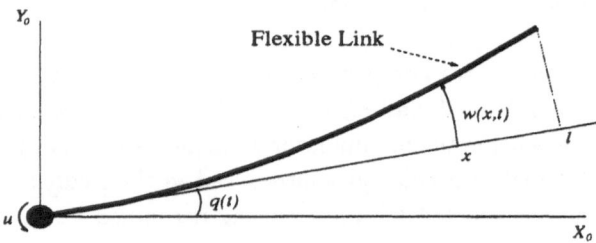

Fig. 1. Schematic of the flexible link

The effects of the rotary inertia and shear deformation are ignored by assuming that the cross–sectional area of the arm is small in comparison with its length l. Euler–Bernoulli beam theory and the assumed modes method are used to express the deflection of a point located at a distance x along the arm as

$$w(x, t) = \sum_{i=1}^{m} \sigma_i(x)\delta_i(t) \tag{1}$$

where $\delta_i(t)$ is the generalized coordinate of the ith mode, $\sigma_i(t)$ is the normalized, clamped–free eigenfunction of the ith mode as defined in [1], [2] and m is the number of modes taken into account. By accounting for the point mass at the tip and applying the Euler–Lagrange equations, the equations of motion are derived as

$$u = M_1(q, \delta)\ddot{q} + M_2(q, \delta)\ddot{\delta} + f_1(q, \dot{q}) + g_1(q, \dot{q}, \delta, \dot{\delta}) \tag{2}$$

$$0 = M_2(q, \delta)\ddot{q} + M_3(q, \delta)\ddot{\delta} + f_2(q, \dot{q}) + g_2(q, \dot{q}, \delta, \dot{\delta}) + K\delta \tag{3}$$

$$y = lq + \sum_{i=1}^{m} \sigma_i \delta_i \tag{4}$$

where $q \in \mathbb{R}$ is the hub angle, $\delta \in \mathbb{R}^m$ is the vector of m generalized coordinates, u is the input torque/force, f_1 and f_2 represent gravitational (only f_1), Coriolis and centrifugal torques/forces, g_1 and g_2 represent the interaction of joint variables and their derivatives with deflection variables and their derivatives, and K is a diagonal constant stiff matrix [4].

For a single link manipulator restricted to the horizontal plane the above system of equations may be simplified by observing that $f_i(q, \dot{q}) = 0$, and $g_i(q, \dot{q}, \delta, \dot{\delta}) = \mathcal{L}(\dot{q}^2\delta, \dot{q}\delta\dot{\delta})$, $i = 1, 2$ where $\mathcal{L}$ is a linear operator.

After linearizing equations (2)–(4) around the equilibrium point $\delta = \dot{\delta} = \ddot{\delta} = 0$, $\dot{q} = \ddot{q} = 0$, and $u = 0$, and using the state assignment $x_1 = q$, $x_2 = \dot{q}$, $\hat{z}_i = \frac{\delta_i}{\epsilon^2}$, $\hat{z}_{i+m} = \frac{\dot{\delta}_i}{\epsilon}$, $i = 1, \cdots, m$, $K = diag\{kk_i\}$, $i = 1, \cdots, m$, with $k_i \equiv 1$ and $\epsilon^2 = \frac{1}{k}$, we obtain

$$\dot{x}_1 = x_2$$
$$\dot{x}_2 = -m_{01}\hat{z}_1 - \cdots - m_{0m}\hat{z}_m + m_{00}u$$
$$\epsilon\dot{\hat{z}}_1 = \hat{z}_{m+1}$$
$$\vdots$$
$$\epsilon\dot{\hat{z}}_m = \hat{z}_{2m}$$
$$\epsilon\dot{\hat{z}}_{m+1} = -m_{11}\hat{z}_1 - \cdots - m_{1m}\hat{z}_m + m_{10}u \qquad (5)$$
$$\vdots$$
$$\epsilon\dot{\hat{z}}_{2m} = -m_{m1}\hat{z}_1 - \cdots - m_{mm}\hat{z}_m + m_{m0}u$$
$$y = lx_1 + \epsilon^2(\sigma_1(l)\hat{z}_1 + \cdots + \sigma_m(l)\hat{z}_m)$$

where it is assumed that the m_{ij}'s are the elements of the inverse of the positive definite inertia matrix $M(q_e, 0)$ of (2)–(3), with q_e an arbitrary constant at the equilibrium point. The above state–space representation is now expressed in a more compact form

$$\dot{x}_1 = x_2 \qquad (6)$$
$$\dot{x}_2 = Lz_1 + m_{00}u \quad , x_1, x_2 \in \mathbb{R}, \ u \in \mathbb{R} \qquad (7)$$
$$\epsilon\dot{z}_1 = z_2 \qquad (8)$$
$$\epsilon\dot{z}_2 = -Hz_1 + Bu \quad , z_1, z_2 \in \mathbb{R}^m \qquad (9)$$
$$y = lx_1 + \epsilon^2 Cz_1 \quad , y \in \mathbb{R} \qquad (10)$$

where

$$z_1 = \begin{bmatrix} \hat{z}_1 \\ \vdots \\ \hat{z}_m \end{bmatrix}, \quad z_2 = \begin{bmatrix} \hat{z}_{m+1} \\ \vdots \\ \hat{z}_{2m} \end{bmatrix}, \quad B = \begin{bmatrix} m_{10} \\ \vdots \\ m_{m0} \end{bmatrix}, \quad H = \begin{bmatrix} m_{11} & \cdots & m_{1m} \\ m_{21} & \cdots & m_{2m} \\ \vdots & \vdots & \vdots \\ m_{m1} & \cdots & m_{mm} \end{bmatrix},$$

$$L = \begin{bmatrix} -m_{01} & \cdots & -m_{0m} \end{bmatrix}, \quad C = \begin{bmatrix} \sigma_1(l) & \cdots & \sigma_m(l) \end{bmatrix}$$

The transfer function of the open–loop system given by

$$\frac{Y(s)}{U(s)} = \frac{(Ll + (\epsilon s)^2 C)[(\epsilon s)^2 I + H]^{-1}B + m_{00}l}{s^2} \qquad (11)$$

is expressed as

$$\frac{Y(s)}{U(s)} = \frac{l(\bar{c}_{m-1} + m_{00})(\varepsilon s)^{2m} + (l\bar{c}_{m-2} + \bar{a}_{m-1} + m_{00}l\bar{b}_{m-1})(\varepsilon s)^{2m-2} + \cdots}{s^2[(\varepsilon s)^{2m} + \bar{b}_{m-1}(\varepsilon s)^{2m-2} + \cdots + \bar{b}_1(\varepsilon s)^2 + \bar{b}_0]}$$

$$+ \frac{(l\bar{c}_0 + \bar{a}_1 + m_{00}l\bar{b}_1)\gamma_2)(\varepsilon s)^2 + (\bar{a}_0 + m_{00}l\bar{b}_0)}{s^2[(\varepsilon s)^{2m} + \bar{b}_{m-1}(\varepsilon s)^{2m-2} + \cdots + \bar{b}_1(\varepsilon s)^2 + \bar{b}_0]} \tag{12}$$

where $L\ adj((\varepsilon s)^2 I + H)B := \bar{a}_{m-1}(\varepsilon s)^{2(m-1)} + \cdots + \bar{a}_1(\varepsilon s)^2 + \bar{a}_0$, $C\ adj((\varepsilon s)^2 I + H)B := \bar{c}_{m-1}(\varepsilon s)^{2(m-1)} + \cdots + \bar{c}_1(\varepsilon s)^2 + \bar{c}_0$, and $det((\varepsilon s)^2 I + H) := (\varepsilon s)^{2m} + \bar{b}_{m-1}(\varepsilon s)^{2m-2} + \cdots + \bar{b}_1(\varepsilon s)^2 + \bar{b}_0$. The reduced–order "rigid" system obtained by neglecting ε has the transfer function

$$\frac{Y(s)}{U(s)} = \frac{\bar{a}_0 + m_{00}l\bar{b}_0}{\bar{b}_0 s^2}. \tag{13}$$

We are now in a position to state the following fact regarding the zero dynamics of system (6)–(10).

Property 1. *The full order flexible model (6)–(10) with its corresponding transfer function (12) is nonminimum phase for all $\varepsilon > 0$, and the reduced–order rigid system with its corresponding transfer function (13) is minimum phase.*

The objective pursued in the next section is to design dynamical composite corrected control strategies using *only* the tip position (output y) and the hub angle (state x_1) so that the output of the system tracks a desired trajectory. The composite control strategy consists of a fast control to stabilize the fast dynamics and a slow control to guarantee a minimum phase property for the slow dynamics as well as tracking for the slow output.

3   Design of Dynamical Composite Corrected Control

The composite control u in (6)–(9) is specified according to [5]

$$u = u(x, z, \varepsilon) = u_s(x; \varepsilon) + u_f(x, z) \tag{14}$$

where u_s denotes the slow control and u_f the fast control, respectively. The fast control is designed so that at $\varepsilon \equiv 0, u_f \equiv 0$. The slow control u_s is expressed in a power series expansion about $\varepsilon = 0$ according to

$$u_s(x, \varepsilon) = u_0(x) + \varepsilon u_1(x) + \varepsilon^2 u_2(x) + O(\varepsilon^3) \tag{15}$$

The uncorrected reduced–order fast subsystem is constructed by setting ε to zero in (8)–(9) and solving for the *uncorrected* quasi-steady-state (QSS) of the fast variables $\bar{z}_1 = H^{-1}Bu_0$ and $\bar{z}_2 = 0$. Then defining the deviation of the fast variables from the uncorrected QSS by $z_{1f0} := z_1 - \bar{z}_1$ and $z_{2f0} := z_2 - \bar{z}_2$, and using (8)–(10) which results in

$$\varepsilon \dot{z}_{1f0} = z_{2f0} - \varepsilon H^{-1} B \dot{u}_0 \tag{16}$$

$$\varepsilon \dot{z}_{2f0} = -H z_{1f0} + B u_f + \varepsilon B (u_1 + \varepsilon u_2 + O(\varepsilon^2)) \tag{17}$$

$$y_{f0} = \varepsilon^2 C z_{1f0} \tag{18}$$

where $y_{f0} = y - (lx_1 + \varepsilon^2 C \bar{z}_1)$ denotes the fast output. Setting $\varepsilon = 0$ in the right–hand–side of (16)–(18) yields the *uncorrected fast subsystem* [2]

$$\varepsilon \dot{\tilde{z}}_{1f0} = \tilde{z}_{2f0} \tag{19}$$

$$\varepsilon \dot{\tilde{z}}_{2f0} = -H \tilde{z}_{1f0} + B u_f \tag{20}$$

$$\tilde{y}_{f0} = 0 \tag{21}$$

Clearly, the equilibrium point of the uncorrected fast subsystem is unstable (the eigenvalues are either on the $j\omega$–axis or symmetric with respect to both $j\omega$ and real axes [3]). Consequently, a fast control is needed to guarantee asymptotic stability of this system. This objective has to be achieved by using only output feedback. However, since the uncorrected fast subsystem is not observable, we have to construct a more accurate fast subsystem. To this end, setting $\varepsilon^2 = 0$ in (16)–(17), results in the 1st–order–corrected QSS [4] of the fast variables $\bar{z}_{1f0} = \varepsilon H^{-1} B u_1$ and $\bar{z}_{2f0} = \varepsilon H^{-1} B \dot{u}_0$ [6]. Defining the deviation of the fast variables z_{1f0} and z_{2f0} from their QSS by $z_{1f1} := z_{1f0} - \bar{z}_{1f0}$ and $z_{2f1} := z_{2f0} - \bar{z}_{2f0}$, and using (16)–(18) results in

$$\varepsilon \dot{z}_{1f1} = z_{2f1} - \varepsilon^2 H^{-1} B \dot{u}_1 \tag{22}$$

$$\varepsilon \dot{z}_{2f1} = -H z_{1f1} + B u_f + \varepsilon^2 (B u_2 - H^{-1} B \ddot{u}_0) + \varepsilon^3 B (u_3 + O(\varepsilon)) \tag{23}$$

$$y_{f1} = \varepsilon^2 C z_{1f1} \tag{24}$$

from which the 1st–order–corrected fast subsystem is obtained by simply setting $\varepsilon^2 = 0$ in (22)–(24), that is

$$\varepsilon \dot{\tilde{z}}_{1f1} = \tilde{z}_{2f1} \tag{25}$$

$$\varepsilon \dot{\tilde{z}}_{2f1} = -H \tilde{z}_{1f1} + B u_f \tag{26}$$

$$\tilde{y}_{f1} = 0 \tag{27}$$

which is still unboservable up to $O(\varepsilon^2)$. Again if we find the 2nd–order–corrected QSS of the fast variables $\bar{z}_{1f1} = \varepsilon^2 H^{-1}(B u_2 - H^{-1} B \ddot{u}_0)$ and $\bar{z}_{2f1} = \varepsilon^2 H^{-1} B \dot{u}_1$

---

[2] $(\dot{\cdot})$ represents $(.)$ at $\varepsilon^{p+1} = 0$; in this case $p = 0$.

[3] The latter happens when $\exists k_i$ such that $k_i \neq k_j$, $j = 1, \cdots, m$, $i \neq j$, and as a result H becomes an asymmetric positive definite matrix.

[4] It is assumed that $u_f \equiv 0$ on $z_{f0} = \bar{z}_{f0}$ (1st–order–corrected slow manifold of z).

(by setting $\varepsilon^3 = 0$ in (22)–(23))[5] and then define the deviation of the fast variables z_{1f1} and z_{2f1} from their QSS by $z_{1f2} := z_{1f1} - \bar{z}_{1f1}$ and $z_{2f2} := z_{2f1} - \bar{z}_{2f1}$, equations (22)–(24) in the new coordinates become

$$\varepsilon \dot{z}_{1f2} = z_{2f2} - \varepsilon \dot{\bar{z}}_{1f1} \tag{28}$$

$$\varepsilon \dot{z}_{2f2} = -H z_{1f2} + B u_f + (\varepsilon^3 B u_3 - \varepsilon^2 \ddot{\bar{z}}_{1f0}) + \varepsilon^4 B(u_4 + O(\varepsilon)) \tag{29}$$

$$y_{f2} = \varepsilon^2 C z_{1f2} \tag{30}$$

and the resulting 2nd–order–corrected fast subsystem is (using (28)–(30) at $\varepsilon^3 = 0$)

$$\varepsilon \dot{\bar{z}}_{1f2} = \tilde{z}_{2f2} \tag{31}$$

$$\varepsilon \dot{\bar{z}}_{2f2} = -H \tilde{z}_{1f2} + B u_f \tag{32}$$

$$\tilde{y}_{f2} = \varepsilon^2 C \tilde{z}_{1f2} \tag{33}$$

The 2nd–order–corrected fast subsystem (31)–(33) may now be stabilized by using a *fast dynamical output* feedback startegy provided that the pairs $(\varepsilon^2 C, -H)$ and $(-H, B)$ are observable and controllable, respectively.

To achieve a more accurate control on the fast dynamics of the system, one may proceed in the same manner as above to find the pth–*order–corrected* fast subsystem and then design the fast dynamical output feedback. To this end, the pth–order–corrected QSS of the fast variables are obtained from[6]

$$\bar{z}_{1fp-1} = \varepsilon^p H^{-1} B u_p - \varepsilon^2 H^{-1} \ddot{\bar{z}}_{1fp-3} \tag{34}$$

$$\bar{z}_{2fp-1} = \varepsilon \dot{\bar{z}}_{1fp-2} \tag{35}$$

and the corresponding dynamics is governed by

$$\varepsilon \dot{z}_{1fp} = z_{2fp} - \varepsilon \dot{\bar{z}}_{1fp-1} \tag{36}$$

$$\varepsilon \dot{z}_{2fp} = -H z_{1fp} + B u_f + (\varepsilon^{p+1} B u_{p+1} - \varepsilon^2 \ddot{\bar{z}}_{1fp-2})$$
$$\qquad + \varepsilon^{p+2} B(u_{p+2} + O(\varepsilon)) \tag{37}$$

$$y_{fp} = \varepsilon^2 C z_{1fp} \tag{38}$$

where $z_{1fp} := z_{1fp-1} - \bar{z}_{1fp-1}$ and $z_{2fp} := z_{2fp-1} - \bar{z}_{2fp-1}$. The pth–order–corrected fast subsystem is now obtained as (using (36)–(38) at $\varepsilon^{p+1} = 0$)

$$\varepsilon \dot{\bar{z}}_{1fp} = \tilde{z}_{2fp} \tag{39}$$

$$\varepsilon \dot{\bar{z}}_{2fp} = -H \tilde{z}_{1fp} + B u_f \tag{40}$$

$$\tilde{y}_{fp} = \varepsilon^2 C \tilde{z}_{1fp} \tag{41}$$

---

[5] It is assumed that $u_f \equiv 0$ on $z_{f1} = \bar{z}_{f1}$ (2nd–*order–corrected slow manifold of* z).

[6] It is assumed that $u_f \equiv 0$ on $z_{fp-1} = \bar{z}_{fp-1}$ (p^{th}–*order–corrected slow manifold of* z).

Remark. For $p \geq 2$ the fast subsystems have an identical structure. The difference among these subsystems is that for higher values of p, $\tilde{z}_{fp}$ and $\tilde{y}_{fp}$ contain higher order fast components implying that a closer approximation to the exact fast dynamics is achieved.

Now, let the fast dynamical controller for the pth–order–corrected fast subsystem (39)–(41) be specified as

$$\varepsilon \dot{v} = Fv + Gy_{fp} \qquad , v \in \mathbb{R}^{2m} \tag{42}$$

$$u_f = Mv \tag{43}$$

By proper selection of F, G and M matrices, (42)–(43) may represent a Luenberger full order observer based controller for (36)–(38) where v is an estimate of z_{fp} up to $O(\varepsilon^{p+1})$. To verify this, let us write (36)–(38) as

$$\varepsilon \dot{z}_{fp} = A_p z_{fp} + B_p u_f + O(\varepsilon^{p+1}) \tag{44}$$

$$y_{fp} = \varepsilon^2 C_p z_{fp} \tag{45}$$

where

$$A_p = \begin{bmatrix} 0 & I \\ -H & 0 \end{bmatrix}, \quad B_p = \begin{bmatrix} 0 \\ B \end{bmatrix}, \quad C_p = \begin{bmatrix} C & 0 \end{bmatrix},$$

Now if F is taken to be

$$F = A_p + B_p M - \varepsilon^2 G C_p \tag{46}$$

then (42) may be expressed as

$$\varepsilon \dot{v} = A_p v + B_p u_f + G(y_{fp} - \hat{y}_{fp}) \tag{47}$$

where $\hat{y}_{fp}$, the output of the estimator, is defined as

$$\hat{y}_{fp} := \varepsilon^2 C_p v \tag{48}$$

Subtracting (44) from (47), yields

$$\varepsilon(\dot{v} - \dot{z}_{fp}) = (A_p - \varepsilon^2 G C_p)(v - z_{fp}) + O(\varepsilon^{p+1}) \tag{49}$$

which implies that there exists a small $T_{c1} > 0$ such that starting from $t = t_0$,

$$v = z_{fp} + O(\varepsilon^{p+1}) \qquad , \forall\, t \geq t_0 + T_{c1} \tag{50}$$

Since (49) represents the fast dynamics, T_{c1} is sufficiently small and is comparable to the boundary layer duration. Note that G and M are determined so as to place the poles of the observer and the fast subsystem, i.e. the eigenvalues of $A_p - \varepsilon^2 G C_p$ and $A_p + BM$ respectively, at some desired locations.

Remark. Since z_{1fp} is not available for feedback, y_{fp} is constructed from the tip position y in the following way

$$y_{fp} = y - y_{fs}^p \tag{51}$$

where [7]

$$y_{fs}^p := lx_1 + \varepsilon^2 C(\bar{z}_1 + \bar{z}_{1f0} + \cdots + \bar{z}_{1fp-1}) \tag{52}$$

Augmenting the fast dynamical controller to the full order system, the resulting exact closed–loop system becomes

$$\dot{x}_1 = x_2 \tag{53}$$

$$\dot{x}_2 = L(z_{1fp} + \bar{z}_1 + \bar{z}_{1f0} + \cdots + \bar{z}_{1fp-1}) + m_{00}(u_s + Mv) \tag{54}$$

$$\varepsilon \dot{z}_{1fp} = z_{2fp} - \varepsilon \dot{\bar{z}}_{1fp-1} \tag{55}$$

$$\varepsilon \dot{z}_{2fp} = -H z_{1fp} + BMv + (\varepsilon^{p+1} Bu_{p+1} - \varepsilon^2 \ddot{\bar{z}}_{1fp-2})$$
$$\qquad + \varepsilon^{p+2} B(u_{p+2} + O(\varepsilon)) \tag{56}$$

$$\varepsilon \dot{v} = \varepsilon^2 GC z_{1fp} + Fv \tag{57}$$

$$y_{fp} = \varepsilon^2 C z_{1fp} \tag{58}$$

Since the pth–order–corrected fast subsystem is stabilized by making the matrix

$$A_f := \begin{bmatrix} 0 & I & 0 \\ -H & 0 & BM \\ \varepsilon^2 GC & 0 & F \end{bmatrix}$$

Hurwitz, we are now in a position to construct the pth–order–corrected slow subsystem and design a slow control u_s to guarantee that the output of the pth–order–corrected slow subsystem, y_s^p (defined subsequently by (78)), tracks asymptotically a desired trajectory, $y_d(t)$. Towards this end, the slow control strategy is selected according to

$$u_s := u_0 + \varepsilon \beta_1 \dot{u}_0 + \cdots + \varepsilon^p \beta_p u_0^{(p)} \tag{59}$$

where $\beta_1, \cdots, \beta_p$ are chosen to ensure a minimum phase property from the uncorrected slow control u_0 to the slow output y_s^p. Using the above control strategy, $\bar{z}_{1f0}, \cdots, \bar{z}_{1fp-1}$ are found to be

$$\bar{z}_{1f0} = \varepsilon H^{-1} B\beta_1 \dot{u}_0 \tag{60}$$

$$\bar{z}_{1f1} = \varepsilon^2 H^{-1}(B\beta_2 - H^{-1}B)\ddot{u}_0 \tag{61}$$

$$\bar{z}_{1f2} = \varepsilon^3 H^{-1}(B\beta_3 - H^{-1}B\beta 1)u_0^{(3)} \tag{62}$$

$$\vdots$$

$$\bar{z}_{1fp-1} = \varepsilon^p H^{-1}(B\beta p - H^{-1}\Gamma_p)u_0^{(p)} \qquad , p \geq 2 \tag{63}$$

---

[7] It is called y_{fs}^p since it is a signal generated from the pth–order dynamical slow controller (cf. (59)–(63), (120)) including the fast components of x_1.

where for even p, Γ_p is defined as

$$\Gamma_p(\beta_{p-2}, \beta_{p-4}, \cdots, \beta_2) := B\beta_{p-2} - H^{-1}B\beta_{p-4} + \cdots$$
$$+ (-1)^{\frac{p-2}{2}} H^{-\frac{p-2}{2}} B \qquad , p \geq 2 \tag{64}$$

and for odd p, Γ_p is defined as

$$\Gamma_p(\beta_{p-2}, \beta_{p-4}, \cdots, \beta_1) := B\beta_{p-2} - H^{-1}B\beta_{p-4} + \cdots$$
$$+ (-1)^{\frac{p-3}{2}} H^{-\frac{p-3}{2}} B\beta_1 \qquad , p \geq 3 \tag{65}$$

The dynamical model (53)–(58) will be used as a basis for constructing reduced–order models by utilizing the integral manifold approach [6], [7]. The integral manifold for (53)–(57) is now defined as a surface

$$\mathcal{M}_\varepsilon : \begin{bmatrix} z_{fp} \\ v \end{bmatrix} = \Phi(x, u_s, \varepsilon) \tag{66}$$

in the (z_{fp}, v, t) space so that if $(z_{fp}(t), v(t)) \in \mathcal{M}_\varepsilon$ at $t = t_0$, then $(z_{fp}(t), v(t)) \in \mathcal{M}_\varepsilon$ for all $t \geq t_0$. The surface $\Phi := [\phi_1^T, \phi_2^T, \phi_3^T]^T$ satisfies the so–called manifold condition

$$\varepsilon\dot{\phi}_1 = \phi_2 - \varepsilon\dot{\bar{z}}_{1fp-1} \tag{67}$$
$$\varepsilon\dot{\phi}_2 = -H\phi_1 + BM\phi_3 - \varepsilon^2\ddot{\bar{z}}_{1fp-2} \tag{68}$$
$$\varepsilon\dot{\phi}_3 = \varepsilon^2 GC\phi_1 + F\phi_3 \tag{69}$$

from which by using a power series expansion of Φ in ε and equating terms with identical power in ε, we get

$$\phi_1 = -\varepsilon^2 H^{-1}\ddot{\bar{z}}_{1fp-2} - \varepsilon^2 H^{-1}\ddot{\bar{z}}_{1fp-1} + O(\varepsilon^{p+3}) \tag{70}$$
$$\phi_2 = \varepsilon\dot{\bar{z}}_{1fp-1} - \varepsilon^3 H^{-1}\bar{z}^{(3)}_{1fp-2} - \varepsilon^3 H^{-1}\bar{z}^{(3)}_{1fp-1} + O(\varepsilon^{p+4}) \tag{71}$$
$$\phi_3 = \varepsilon^4 F^{-1}GCH^{-1}\ddot{\bar{z}}_{1fp-2} + O(\varepsilon^{p+4}) \tag{72}$$

where the higher order terms in ϕ_1 and ϕ_2 depend also on F, G and M.

Remark. From (72) it is clear that on the pth–order–corrected slow manifold , v or u_f ($= Mv$) vanishes, which complies with our previous assumption that on $z_{fp-1} = \bar{z}_{fp-1}$ manifold $u_f \equiv 0$.

Dynamics of the full order system restricted to the exact manifold $\mathcal{M}_\varepsilon$ defines the exact reduced–order slow model

$$\dot{x}_1^e = x_2^e \tag{73}$$
$$\dot{x}_1^e = L(\phi_1(x_1^e, u_s, \varepsilon) + \bar{z}_1 + \bar{z}_{1f0} + \cdots + \bar{z}_{1fp-1})$$
$$+ m_{00}(u_s + M\phi_3(x_1^e, u_s, \varepsilon)) \tag{74}$$
$$y_s = lx_1^e + \varepsilon^2 C(\phi_1(x_1^e, u_s, \varepsilon) + \bar{z}_1 + \bar{z}_{1f0} + \cdots + \bar{z}_{1fp-1}) \tag{75}$$

In general computing the exact manifold and constructing the exact reduced–order slow model will be computationally involved. Instead one may be satisfied with an $O(\varepsilon^{p+1})$ approximation to the exact slow model. The pth–order–corrected slow subsystem is obtained by setting $\varepsilon^{p+1} \equiv 0$, to yield

$$\dot{\bar{x}}_1 = \bar{x}_2 \qquad (76)$$

$$\dot{\bar{x}}_2 = L(\bar{z}_1 + \bar{z}_{1f0} + \cdots + \bar{z}_{1fp-1}) + m_{00}(u_0 + \varepsilon\beta_1\dot{u}_0 + \cdots + \varepsilon^p\beta_p u_0^{(p)}) \quad (77)$$

$$y_s^p = l\bar{x}_1 + \varepsilon^2 C(\bar{z}_1 + \bar{z}_{1f0} + \cdots + \bar{z}_{1fp-3}) \qquad (78)$$

Alternatively, the input–output representation of the full order system (6)–(10) is obtained from

$$\ddot{y} = l\ddot{x}_1 + \varepsilon^2 C\ddot{z}_1 = l\dot{x}_2 + \varepsilon C\dot{z}_2$$
$$= l(Lz_1 + m_{00}(u_s + u_f)) + C(-Hz_1 + B(u_s + u_f))$$
$$= (Ll - CH)z_1 + (m_{00}l + CB)(u_s + u_f) \qquad (79)$$

Restricting the dynamics of the above system to the pth–order–corrected manifold of z (i.e. on $z_{fp-1} = \bar{z}_{fp-1}$), yields[8]

$$\ddot{y}_s^p = (Ll - CH)(\bar{z}_1 + \bar{z}_{1f0} + \cdots + \bar{z}_{1fp-1}) + (m_{00}l + CB)u_s \qquad (80)$$

Utilizing the expressions (60)–(65) in (80), the input–output representation of the pth–order–corrected slow model becomes

$$\ddot{y}_s^p = l(LH^{-1}B + m_{00})u_0 + \varepsilon l(LH^{-1}B + m_{00})\beta_1\dot{u}_0$$
$$+\varepsilon^2[l(LH^{-1}B + m_{00})\beta_2 - (Ll - CH)H^{-2}\Gamma_2]\ddot{u}_0$$
$$+\varepsilon^3[l(LH^{-1}B + m_{00})\beta_3 - (Ll - CH)H^{-2}\Gamma_3]u_0^{(3)}$$
$$\vdots$$
$$+\varepsilon^p[l(LH^{-1}B + m_{00})\beta_p - (Ll - CH)H^{-2}\Gamma_p]u_0^{(p)} \qquad (81)$$

Equation (81) is now written in a more compact form as

$$\ddot{y}_s^p = \alpha_0 u_0 + \varepsilon\alpha_0\beta_1\dot{u}_0 + \varepsilon^2(\alpha_0\beta_2 + \alpha_2)\ddot{u}_0 + \cdots + \varepsilon^p(\alpha_0\beta_p + \alpha_p)u_0^{(p)} \quad (82)$$

with obvious definitions for α_i's. Since $\alpha_i = \alpha_i(\beta_1(\beta_2), \cdots, \beta_{i-2}), i = 2, \cdots, p$, therefore it is always possible to stabilize the zero dynamics of (82) by proper selection of the parameters $\beta_j, j = 1, \cdots, p$, that is for any $\varepsilon \in (0, \varepsilon^*), \varepsilon^* > 0$, β_j may be designed so that the companion matrix

$$A_w = \begin{bmatrix} 0 & 1 & \cdots & 0 \\ 0 & 0 & \cdots & 0 \\ \vdots & \vdots & \vdots & \vdots \\ 0 & 0 & \cdots & 1 \\ -\dfrac{\alpha_0}{\varepsilon^p(\alpha_0\beta_p + \alpha_p)} & -\dfrac{\alpha_0\beta_1}{\varepsilon^{p-1}(\alpha_0\beta_p + \alpha_p)} & \cdots & -\dfrac{\alpha_0\beta_{p-1} + \alpha_{p-1}}{\varepsilon(\alpha_0\beta_p + \alpha_p)} \end{bmatrix}$$

---

[8] Note that on this manifold $u_f \equiv 0$.

is Hurwitz and the pth-order-corrected slow subsystem becomes minimum phase from u_0 to y_s^p. For the sake of convenience, from now on the following set of new parameters are used $a_0 := \alpha_0$, $a_1 := \alpha_0\beta_1$, $a_2 := \alpha_0\beta_2 + \alpha_2$, $\cdots$, $a_p := \alpha_0\beta_p + \alpha_p$.

Since ϕ_1 is $O(\varepsilon^{p+1})$, therefore the exact reduced-order slow model is an $O(\varepsilon^{p+1})$ perturbation of the pth-order-corrected slow subsystem. Hence y_s^p is an $O(\varepsilon^{p+1})$ approximation to y_s, that is, $y_s = y_s^p + O(\varepsilon^{p+1})$, therefore in principle it is possible to track the tip position y to a desired accuracy after the fast transient dies out. Towards this end, the *dynamic* control law is proposed as

$$a_0 u_0 + \varepsilon a_1 \dot{u}_0 + \varepsilon^2 a_2 \ddot{u}_0 + \cdots + \varepsilon^p a_p u_0^p = y^r(t) \tag{83}$$

where

$$y^r(t) := \ddot{y}_d(t) - \gamma_1(\dot{y}_s^p - \dot{y}_d(t)) - \gamma_0(y_s^p - y_d(t)) \tag{84}$$

and $y_d(t)$ denotes the desired tip position trajectory. The gains γ_0 and γ_1 are designed so that $s^2 + \gamma_1 s + \gamma_0$ is a Hurwitz polynomial. Applying (83)–(84) to (82), the closed-loop p^{th}-order slow subsystem becomes

$$\ddot{e} + \gamma_1 \dot{e} + \gamma_0 e = 0 \tag{85}$$

where $e := y_s^p - y_d$. The control law (83)–(84) requires feedback from y_s^p and $\dot{y}_s^p$; in other words, feedback from the hub angle and its velocity. As stated in Sect.2, we assume that only the hub angle, x_1 and the tip position, y are accessible for feedback; therefore, controller (83)–(84) is not implementable in its present form. But as we will see later, an $O(\varepsilon^{p+1})$ approximation to $\dot{y}_s^p$ is constructed by feeding an $O(\varepsilon^{p+1})$ approximation of y_s^p through a *slow* observer.

To implement y_s^p, $\bar{x}_1$ is needed. This can be obtained by solving (76)–(77), resulting in an increase of $(2p+2)$ in the order of the dynamic controller. Instead, one might be satisfied with an $O(\varepsilon^{p+1})$ steady state approximate of $\bar{x}_1$. To this end, let us use the following coordinate transformation

$$\hat{x} := x - \varepsilon J \begin{bmatrix} z_{fp} \\ v \end{bmatrix} \tag{86}$$

where

$$\hat{x} := \begin{bmatrix} \hat{x}_1 \\ \hat{x}_2 \end{bmatrix}, \quad J := \begin{bmatrix} J_1 & J_2 \\ J_3 & J_4 \end{bmatrix}, \quad J_1 := \begin{bmatrix} J_{11} & J_{12} \end{bmatrix}, \quad J_3 := \begin{bmatrix} J_{31} & J_{32} \end{bmatrix}$$

or equivalently

$$\hat{x}_1 = x_1 - J_{11}(\varepsilon z_{1fp}) - J_{12}(\varepsilon z_{2fp}) - J_2(\varepsilon v) \tag{87}$$

$$\hat{x}_2 = x_1 - J_{31}(\varepsilon z_{1fp}) - J_{32}(\varepsilon z_{2fp}) - J_4(\varepsilon v) \tag{88}$$

Combining (87)–(88) with (53)–(57), yields

$$\dot{\hat{x}}_1 = \hat{x}_2 + (\varepsilon J_{31} + J_{12}H - \varepsilon^2 J_2 GC)z_{1fp} + (\varepsilon J_{32} - J_{11})z_{2fp}$$
$$+(\varepsilon J_4 - J_{12}BM - J_2 F)v + \varepsilon J_{11}\dot{\bar{z}}_{1fp-1} + \varepsilon^2 J_{12}\ddot{\bar{z}}_{1fp-2} \qquad (89)$$
$$\dot{\hat{x}}_2 = L(\bar{z}_1 + \bar{z}_{1f0} + \cdots + \bar{z}_{1fp-1}) + m_{00}(u_0 + \varepsilon\beta_1\dot{u}_0 + \cdots + \varepsilon^p \beta_p u_0^{(p)})$$
$$+(L + J_{32}H - \varepsilon^2 J_4 GC)z_{1fp} + (-J_{31})z_{2fp} + (m_{00}M - J_{32}BM - J_4 F)v$$
$$+\varepsilon J_{31}\dot{\bar{z}}_{1fp-1} + \varepsilon^2 J_{32}\ddot{\bar{z}}_{1fp-2} \qquad (90)$$

To eliminate the fast components z_{1fp}, z_{2fp} and v in (89)–(90), the following set of equations have to be satisfied

$$-J_{31} = 0 \qquad (91)$$
$$\varepsilon J_{32} - J_{11} = 0 \qquad (92)$$
$$\varepsilon J_{31} + J_{12}H - \varepsilon^2 J_2 GC = 0 \qquad (93)$$
$$L + J_{32}H - \varepsilon^2 J_4 GC = 0 \qquad (94)$$
$$\varepsilon J_4 - J_{12}BM - J_2 F = 0 \qquad (95)$$
$$m_{00}M - J_{32}BM - J_4 F = 0 \qquad (96)$$

Solving the above equations for $J_{11}, \cdots, J_4$ in the proper order, yields

$$J_{31} = 0 \qquad (97)$$
$$J_4 = (LH^{-1}B + m_{00})M(F + \varepsilon^2 GCH^{-1}BM)^{-1}$$
$$= (LH^{-1}B + m_{00})MF^{-1} + O(\varepsilon^2) \qquad (98)$$
$$J_{32} = (\varepsilon^2 J_4 GC - L)H^{-1} = -LH^{-1} + O(\varepsilon^2) \qquad (99)$$
$$J_{11} = \varepsilon(\varepsilon^2 J_4 GC - L)H^{-1} = -\varepsilon LH^{-1} + O(\varepsilon^3) \qquad (100)$$
$$J_2 = \varepsilon J_4(F + \varepsilon^2 GCH^{-1}BM)^{-1} = \varepsilon(LH^{-1}B + m_{00})MF^{-2} + O(\varepsilon^3) \qquad (101)$$
$$J_{12} = \varepsilon^2 J_2 GCH^{-1} = \varepsilon^3(LH^{-1}B + m_{00})MF^{-2}GCH^{-1} + O(\varepsilon^5) \qquad (102)$$

Now that $J_{11}, \cdots, J_4$ satisfy (91)–(96), equations (89)–(90) are rewritten as

$$\dot{\hat{x}}_1 = \hat{x}_2 + \varepsilon J_{11}\dot{\bar{z}}_{1fp-1} + \varepsilon^2 J_{12}\ddot{\bar{z}}_{1fp-2} \qquad (103)$$
$$\dot{\hat{x}}_2 = L(\bar{z}_1 + \bar{z}_{1f0} + \cdots + \bar{z}_{1fp-1}) + m_{00}(u_0 + \varepsilon\beta_1\dot{u}_0 + \cdots + \varepsilon^p \beta_p u_0^{(p)})$$
$$+\varepsilon^2 J_{32}\ddot{\bar{z}}_{1fp-2} \qquad (104)$$

Since J_{11}, J_{12} and J_{32} are $O(\varepsilon), O(\varepsilon^3)$ and $O(1)$ terms respectively, therefore (103)–(104) may be represented as

$$\dot{\hat{x}}_1 = \hat{x}_2 + O(\varepsilon^{p+2}) \qquad (105)$$
$$\dot{\hat{x}}_2 = L(\bar{z}_1 + \bar{z}_{1f0} + \cdots + \bar{z}_{1fp-1}) + m_{00}(u_0 + \varepsilon\beta_1\dot{u}_0 + \cdots + \varepsilon^p \beta_p u_0^{(p)})$$
$$+O(\varepsilon^{p+1}) \qquad (106)$$

which if compared to (76)–(77) implies that

$$\hat{x} = \bar{x} + O(\varepsilon^{p+1}) \tag{107}$$

To use (86), z_{fp} is needed. But since z is not accessible, z_{fp} is not available either. However an estimate of z_{fp} is available and has already been constructed. Therefore, instead of (86), the following transformation is used

$$\tilde{x} := x - \varepsilon J \begin{bmatrix} v \\ v \end{bmatrix} = x - \varepsilon \begin{bmatrix} J_1 + J_2 \\ J_3 + J_4 \end{bmatrix} v \tag{108}$$

From (86) and (108), $\tilde{x}$ is expressed in terms of $\hat{x}$ as follows

$$\tilde{x} = \hat{x} + \varepsilon \begin{bmatrix} J_1 \\ J_3 \end{bmatrix} (z_{fp} - v) \tag{109}$$

which along with (50) and (108) implies that there exists a small $T_{c2} > 0$ such that

$$\tilde{x} = \hat{x} + O(\varepsilon^{p+2}) = x + O(\varepsilon^{p+1}) \quad , \forall\, t \geq t_0 + T_{c2} \tag{110}$$

As a result, an $O(\varepsilon^{p+1})$ approximation to y_s^p is constructed as

$$\tilde{y}_s^p = l\tilde{x}_1 + \varepsilon^2 C(\bar{z}_1 + \bar{z}_{1f0} + \cdots + \bar{z}_{1fp-3}) \tag{111}$$

which if compared to (78) and (110) shows that[9]

$$\tilde{y}_s^p = y_s^p + O(\varepsilon^{p+1}) \quad , \forall\, t \geq t_0 + T_{c2} \tag{112}$$

Now that $\tilde{y}_s^p$ is constructed (based on (108), (111)), we are in a position to estimate $\dot{\tilde{y}}_s^p$, which in steady state is an $O(\varepsilon^{p+1})$ approximation to $\dot{y}_s^p$ (cf. (132)), by the following *slow* observer

$$\dot{\zeta}_1 = \zeta_2 + k_2(\tilde{y}_s^p - \zeta_1) \tag{113}$$
$$\dot{\zeta}_2 = k_1(\tilde{y}_s^p - \zeta_1) + y_\zeta^r \tag{114}$$

where

$$y_\zeta^r := \ddot{y}_d(t) - \gamma_1(\zeta_2 - \dot{y}_d(t)) - \gamma_0(\zeta_1 - y_d(t)) \tag{115}$$

and the gains k_1 and k_2 are selected so that $s^2 + k_2 s + k_1$ is a Hurwitz polynomial. This guarantees the convergence of ζ_1 and ζ_2 to $\tilde{y}_s^p$ and $\dot{\tilde{y}}_s^p$ up to $O(\varepsilon^{p+1})$, respectively (It is assumed that v approaches z_{fp} quite rapidly). To see this,

---

[9] The details are found in Appendix.

let us define the new variables $\psi_1 := \tilde{y}_s^p - \zeta_1$ and $\psi_2 := \dot{\tilde{y}}_s^p - \zeta_2$. In the new coordiantes the observer dynamics are transformed into

$$\dot{\psi}_1 = \dot{\tilde{y}}_s^p - \dot{\zeta}_1 = \dot{\tilde{y}}_s^p - \zeta_2 - k_2(\tilde{y}_s^p - \zeta_1) = \psi_2 - k_2\psi_1 \tag{116}$$

$$\dot{\psi}_2 = \ddot{\tilde{y}}_s^p - \dot{\zeta}_2 = \ddot{\tilde{y}}_s^p - k_1(\tilde{y}_s^p - \zeta_1) - y_\zeta^r = (\ddot{\tilde{y}}_s^p - \ddot{y}_s^p) - k_1\psi_1 \tag{117}$$

The slow observer dynamics (116)–(117) and also (137) (in Appendix) show that there exists a $T_{c3} > 0, T_{c3} \gg T_{c1}, T_{c2}$, such that

$$\psi_1 = O(\varepsilon^{p+1}) \quad , \quad \psi_2 = O(\varepsilon^{p+1}) \quad , \forall\, t \geq t_0 + T_{c3} \tag{118}$$

or

$$\zeta_1 = \tilde{y}_s^p + O(\varepsilon^{p+1}) \quad , \quad \zeta_2 = \dot{\tilde{y}}_s^p + O(\varepsilon^{p+1}) \quad , \forall\, t \geq t_0 + T_{c3} \tag{119}$$

Using the observer dynamics (111)–(112) to construct y_ζ^r, and defining the state variables $w_0 := u_0, w_1 := \dot{u}_0, \cdots, w_{p-1} := u_0^{(p-1)}$, a canonical state-space representation of the proposed *dynamical* controller, becomes

$$\dot{w}_0 = w_1$$
$$\vdots \tag{120}$$
$$\dot{w}_{p-2} = w_{p-1}$$
$$\dot{w}_{p-1} = -\frac{1}{\varepsilon^p a_p}[a_0 w_0 + \varepsilon a_1 w_1 + \cdots + \varepsilon^{p-1} a_{p-1} w_{p-1} - y_\zeta^r]$$

The slow control u_s to be utilized in the full order system is now expressed as

$$u_s = u_0 + \varepsilon\beta_1 \dot{u}_0 + \cdots + \varepsilon^p \beta_p u_0^{(p)}$$
$$= (1 - \frac{a_0\beta_p}{a_p})w_0 + \varepsilon(\beta_1 - \frac{a_1\beta_p}{a_p})w_1 + \cdots + \varepsilon^{p-1}(\beta_{p-1} - \frac{a_{p-1}\beta_p}{a_p})w_{p-1}$$
$$+ \frac{\beta_p}{a_p}y_\zeta^r \tag{121}$$

From the Lyapunov stability analysis of the full order closed–loop system (details are omitted due to space limitations, cf. [8]), tracking error ($\hat{e} := y - y_d$), $\psi_1, \psi_2, w_0, \cdots, w_{p-1}, z_{1fp}, z_{2fp}$ and v are all proved to be uniformly ultimately bounded and moreover, in steady state $\hat{e}$ becomes $O(\varepsilon^p)$.

In summary, using x_1 and y measurements from the flexible manipulator along with (42)–(43), (51)–(52), (60)–(65), (108), (111), (113)–(115) , (120) and (121) regulates the fast dynamics and guarantees the tip position to track the desired trajectory $y_d(t)$ up to $O(\varepsilon^p)$.

4 Numerical Simulations

A numerical example is worked out to illustrate the proposed control strategy. The system under study has the following set of parameters

$$H = \begin{bmatrix} 2 & 1 \\ 1 & 4 \end{bmatrix}, \quad B = \begin{bmatrix} 1 \\ 1 \end{bmatrix}, \quad L = \begin{bmatrix} -1.2 & -2 \end{bmatrix}, \quad C = \begin{bmatrix} 2 & -2 \end{bmatrix}$$

$m_{00} = 1$, $l = 1$ and $m = 2$. The signal $y_d(t) = \sin(t)$ is applied as the desired output trajectory and the initial conditions are set at $x_1(0) = 2, x_2(0) = -1, \hat{z}_1(0) = 0.05, \hat{z}_2(0) = 0.01, \hat{z}_3(0) = \hat{z}_4(0) = 0, v(0) = 0, \zeta_1(0) = 0, \zeta_2(0) = 0$ and $w_0(0) = \cdots = w_{p-1}(0) = 0$. The dynamical composite control strategy is applied to the case with $\varepsilon = 0.1$ corresponding to $k_1 = k_2 = 100$ for $p = 2, 3, 4$ and 5. In the following simulations, $\lambda_e, \lambda_w, \lambda_\zeta, \lambda_v$ and λ_z denote the pole locations for the output error (e) dynamics, slow control dynamics, slow observer, fast observer and the closed-loop p^{th}–order–corrected fast subsystem, respectively. Moreover, for the sake of convenience, the poles of each set of dynamics are all assumed to be at a same location. For instance, $\lambda_e = -0.7$ implies that both of the poles of (85) are at -0.7.

Control inputs have been designed based on the $2^{nd}, 3^{rd}, 4^{th}$ and 5th–order–corrected slow and fast subsystems and are applied to the full order system with the following pole locations

- $p = 2$: $\lambda_e = -1.4, \lambda_w = -8.0, \lambda_\zeta = -6, \lambda_v = -30, \lambda_z = -90$
- $p = 3$: $\lambda_e = -0.7, \lambda_w = -4.2, \lambda_\zeta = -6, \lambda_v = -45, \lambda_z = -45$
- $p = 4$: $\lambda_e = -0.4, \lambda_w = -4.1, \lambda_\zeta = -6, \lambda_v = -30, \lambda_z = -90$
- $p = 5$: $\lambda_e = -0.38, \lambda_w = -8.9, \lambda_\zeta = -6, \lambda_v = -30, \lambda_z = -88$

Figures (2)–(5)[10] show the tracking error $\hat{e} := y - y_d$ for $p = 2, 3, 4, 5$ respectively. Clearly, there is a significant improvement as p increases from 2 to 5. Also, they confirm the results derived from the Lyapunov stability analysis [8], indicating that the steady state of $\hat{e}$ is $O(\varepsilon^p)$.

Figures (6)–(13) depict the slow control u_s and the fast control u_f. From these figures one may conclude that $w_0, \cdots, w_{p-1}, \psi_1, \psi_2, z_{1fp}, z_{2fp}$ and v are all bounded.

Remark. In all the above pole locations λ_e, λ_w and λ_ζ are slow and λ_v and λ_z are fast in nature. Also, from the simulation results we found that the transient performance is mostly affected by λ_e; the magnitude of the steady state of $\hat{e}$ is determined by λ_w, λ_ζ; and λ_v and λ_z along with the other eigenvalues affect the stability of the closed loop system.

---

[10] Figures (b) are rescaled version of Figs.(a), representing the steady state responses.

5 Conclusion

The problem of controlling the tip position of a flexible link manipulator, bearing non-minimum phase characteristics, is considered. The linearized model of the arm is represented in a singularly perturbed form. Utilizing the concept of integral manifolds, a dynamic composite corrected control is designed to first stabilize the fast dynamics of the system, second to guarantee the minimum phase property of the system restricted to the integral manifold and third to make the link tip position track a desired reference trajectory to an arbitrary degree of accuracy by just measureing the hub angle and the tip position. The proposed methodology may also be generalized to nonlinear systems. This problem is currently under investigation.

6 Appendix

From (49), (103)–(104) and (108), we obtain

$$\dot{\tilde{x}}_1 = \dot{\hat{x}}_1 + \varepsilon J_1(\dot{z}_{fp} - \dot{v})$$

$$= \hat{x}_2 + \varepsilon J_{11}\ddot{\tilde{z}}_{1fp-1} + \varepsilon^2 J_{12}\ddot{\tilde{z}}_{1fp-2}$$

$$+[J_{11}\ J_{12}]((A_p - \varepsilon^2 GC_p)(z_{fp} - v) - \begin{bmatrix} \varepsilon\ddot{\tilde{z}}_{1fp-1} \\ \varepsilon^2\ddot{\tilde{z}}_{1fp-2} \end{bmatrix})$$

$$= \hat{x}_2 + [J_{11}\ J_{12}](A_p - \varepsilon^2 GC_p)(z_{fp} - v) \tag{122}$$

$$= (\tilde{x}_2 - \varepsilon J_3(z_{fp} - v)) + J_1(A_p - \varepsilon^2 GC_p)(z_{fp} - v)$$

$$= \tilde{x}_2 + (J_1(A_p - \varepsilon^2 GC_p) - \varepsilon J_3)(z_{fp} - v) \tag{123}$$

and

$$\dot{\tilde{x}}_2 = \dot{\hat{x}}_2 + \varepsilon J_3(\dot{z}_{fp} - v)$$

$$= L(\bar{z}_1 + \bar{z}_{1f0} + \cdots + \bar{z}_{1fp-1}) + m_{00}(u_0 + \varepsilon\beta_1\dot{u}_0 + \cdots + \varepsilon^p\beta_p u_0^{(p)})$$

$$+\varepsilon^2 J_{32}\ddot{\tilde{z}}_{1fp-2} + [J_{31}\ J_{32}]((A_p - \varepsilon^2 GC_p)(z_{fp} - v) - \begin{bmatrix} \varepsilon\ddot{\tilde{z}}_{1fp-1} \\ \varepsilon^2\ddot{\tilde{z}}_{1fp-2} \end{bmatrix})$$

$$= L(\bar{z}_1 + \bar{z}_{1f0} + \cdots + \bar{z}_{1fp-1}) + m_{00}(u_0 + \varepsilon\beta_1\dot{u}_0 + \cdots + \varepsilon^p\beta_p u_0^{(p)})$$

$$+[0\ J_{32}](A_p - \varepsilon^2 GC_p)(z_{fp} - v) \tag{124}$$

Since from (50)

$$v = z_{fp} + O(\varepsilon^{p+1}) \qquad ,\forall\, t \geq t_0 + T_{c1} \tag{125}$$

therefore $\forall\, t \geq t_0 + T_{c1}$,

$$\dot{\tilde{x}}_1 = \tilde{x}_2 + O(\varepsilon^{p+2}) \tag{126}$$

$$\dot{\tilde{x}}_2 = L(\bar{z}_1 + \bar{z}_{1f0} + \cdots + \bar{z}_{1fp-1}) + m_{00}(u_0 + \varepsilon\beta_1\dot{u}_0 + \cdots + \varepsilon^p\beta_p u_0^{(p)})$$

$$+O(\varepsilon^{p+1}) \tag{127}$$

which compared with (76)–(77) proves that there exists a small $T_{c2} > 0$ such that

$$\tilde{x} = \bar{x} + O(\varepsilon^{p+1}) \quad , \forall \, t \geq t_0 + T_{c2} \tag{128}$$

and

$$\tilde{y}_s^p = y_s^p + l(\tilde{x} - \bar{x}) = y_s^p + O(\varepsilon^{p+1}) \quad , \forall \, t \geq t_0 + T_{c2} \tag{129}$$

As far as $\dot{\tilde{y}}_s^p$ is concerned, from (111), (122) and (123) we may write

$$\dot{\tilde{y}}_s^p = l\dot{\tilde{x}}_1 + \varepsilon^2 C(\dot{\bar{z}}_1 + \dot{\bar{z}}_{1f0} + \cdots + \dot{\bar{z}}_{1fp-3})$$
$$= l\hat{x}_2 + lJ_1(A_p - \varepsilon^2 GC_p)(z_{fp} - v) + \varepsilon C(\bar{z}_2 + \bar{z}_{2f0} + \cdots + \bar{z}_{2fp-2}) \tag{130}$$

or equivalently

$$\dot{\tilde{y}}_s^p = l\tilde{x}_2 + \varepsilon C(\bar{z}_2 + \bar{z}_{2f0} + \cdots + \bar{z}_{2fp-2})$$
$$+ l(J_1(A_p - \varepsilon^2 GC_p) - \varepsilon J_3)(z_{fp} - v) \tag{131}$$

which shows that there exists a small $T_{c4} > 0$, such that

$$\dot{\tilde{y}}_s^p = \dot{y}_s^p + O(\varepsilon^{p+1}) \quad , \forall \, t \geq t_0 + T_{c4} \tag{132}$$

where

$$\dot{y}_s^p = l\dot{\tilde{x}}_1 + \varepsilon^2 C(\dot{\bar{z}}_1 + \dot{\bar{z}}_{1f0} + \cdots + \dot{\bar{z}}_{1fp-3})$$
$$= l\bar{x}_2 + \varepsilon C(\bar{z}_2 + \bar{z}_{2f0} + \cdots + \bar{z}_{2fp-2}) \tag{133}$$

Also, $\ddot{\tilde{y}}_s^p$ is found from (130) according to

$$\ddot{\tilde{y}}_s^p = l\dot{\hat{x}}_2 + \varepsilon C(\dot{\bar{z}}_2 + \dot{\bar{z}}_{2f0} + \cdots + \dot{\bar{z}}_{2fp-2}) + lJ_1(A_p - \varepsilon^2 GC_p)(\dot{z}_{fp} - \dot{v}) \tag{134}$$

The QSS terms $\bar{z}_1, \cdots, \bar{z}_{1fp-1}$ and $\bar{z}_2, \cdots, \bar{z}_{2fp-1}$ are the first p terms of z_1 and z_2 expansions when z is restricted to the integral manifold of (6)–(9) found from the manifold condition

$$\varepsilon(\dot{\bar{z}}_2 + \dot{\bar{z}}_{2f0} + \cdots + \dot{\bar{z}}_{2fp-2} + O(\varepsilon^p)) = -H(\bar{z}_1 + \bar{z}_{1f0} + \cdots + \bar{z}_{1fp-1} + O(\varepsilon^{p+1}))$$
$$+ Bu_s \tag{135}$$

Thus if we apply (92)–(93), (104) and (135) to (134), we get

$$\ddot{\tilde{y}}_s^p = l(L(\bar{z}_1 + \bar{z}_{1f0} + \cdots + \bar{z}_{1fp-1}) + m_{00}u_s + \varepsilon^2 J_{32}\ddot{\bar{z}}_{1fp-2})$$
$$+ C(-H(\bar{z}_1 + \bar{z}_{1f0} + \cdots + \bar{z}_{1fp-1}) + Bu_s)$$
$$+ l[\varepsilon J_{32} \ \ \varepsilon^2 J_2 GCH^{-1}](A_p - \varepsilon^2 GC_p)(\dot{z}_{fp} - \dot{v})$$
$$= (Ll - CH)(\bar{z}_1 + \bar{z}_{1f0} + \cdots + \bar{z}_{1fp-1}) + (m_{00}l + CB)u_s + \varepsilon^2 lJ_{32}\ddot{\bar{z}}_{1fp-2}$$
$$+ l[J_{32} \ \ \varepsilon J_2 GCH^{-1}](A_p - \varepsilon^2 GC_p)((A_p - \varepsilon^2 GC_p)(z_{fp} - v) - \begin{bmatrix} \varepsilon \dot{\bar{z}}_{1fp-1} \\ \varepsilon^2 \ddot{\bar{z}}_{1fp-2} \end{bmatrix})$$
$$= (Ll - CH)(\bar{z}_1 + \bar{z}_{1f0} + \cdots + \bar{z}_{1fp-1}) + (m_{00}l + CB)u_s$$
$$+ l[J_{32} \ \ \varepsilon J_2 GCH^{-1}](A_p - \varepsilon^2 GC_p)^2(z_{fp} - v) + O(\varepsilon^{p+1})$$
$$= \ddot{y}_s^p + l[J_{32} \ \ \varepsilon J_2 GCH^{-1}](A_p - \varepsilon^2 GC_p)^2(z_{fp} - v) + O(\varepsilon^{p+1}) \tag{136}$$

which if compared to (75) shows that there exists a small $T_{c5} > 0$, such that

$$\ddot{\tilde{y}}_s^p = \ddot{y}_s^p + O(\varepsilon^{p+1}) \qquad , \forall\, t \geq t_0 + T_{c5} \; . \tag{137}$$

References

[1] Wang D., Vidyasagar M.: Transfer functions for a single flexible link. Proceedings of IEEE International Conference on Robotics and Automation, (1989)

[2] Bayo E., Moulin H.: An efficient computation of the inverse dynamics of flexible manipulators in the time domain. Proceedings of IEEE International Conference on Robotics and Automation, (1989)

[3] Kwon D.S., Book W.J.: An inverse dynamics method yielding flexible manipulator state trajectories. Proceedings of American Control Conference, (1990)

[4] Book W.J., Siciliano B.: A Singular Perturbation Approach to Control of Lightweight Flexible Manipulators. The International Journal of Robotics Research, vol.7, No.4 (1989)

[5] Kokotovic P.V., Khalil H.K. and O'Reilly J.: Singular Perturbation Methods in Control; Analysis and Design. Academic Press, (1986)

[6] Khorasani K., Spong M.W.: Invariant Manifolds and their application to Robot Manipulators with Flexible Joints. Proceedings of IEEE International Conference on Robotics and Automation (St. Louis, MO), (1985)

[7] Khorasani K., Kokotovic P.V.: A corrective design for nonlinear systems with fast actuators. IEEE Transactions on Automatic Control, Vol. 31, No. 1, (1986)

[8] Hashtrudi Zaad K., Khorasani K.: Control of nonminimum phase singularly perturbed systems with applications to flexible link manipulators. Submitted to International Journal of Control, 1994.

This article was processed using the LaTeX macro package with LLNCS style

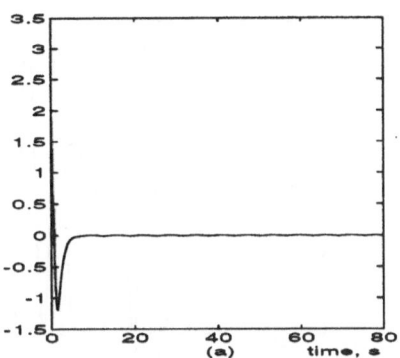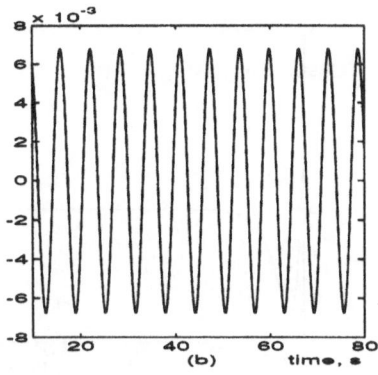

Fig. 2. Output tracking error ($\hat{e} = y - y_d$) for $p = 2, \varepsilon = 0.1$

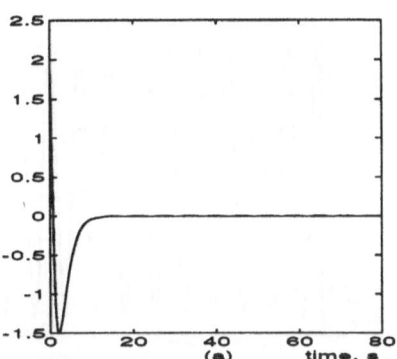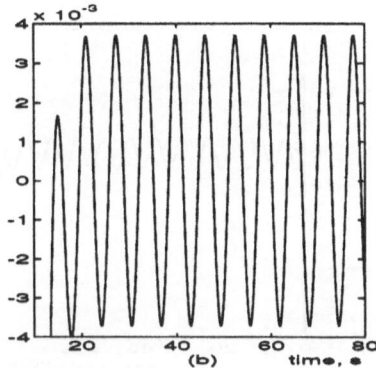

Fig. 3. Output tracking error ($\hat{e} = y - y_d$) for $p = 3, \varepsilon = 0.1$

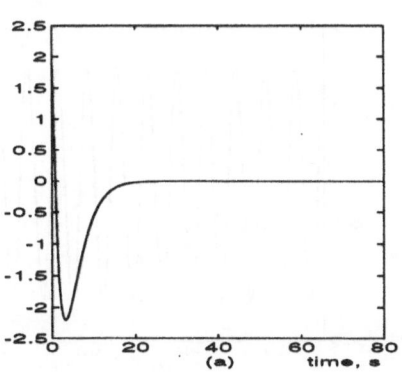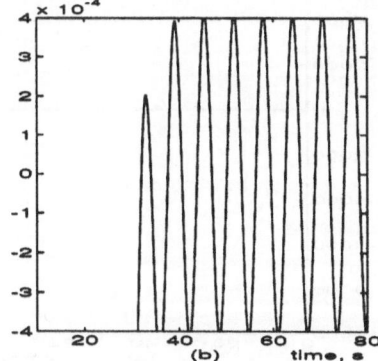

Fig. 4. Output tracking error ($\hat{e} = y - y_d$) for $p = 4, \varepsilon = 0.1$

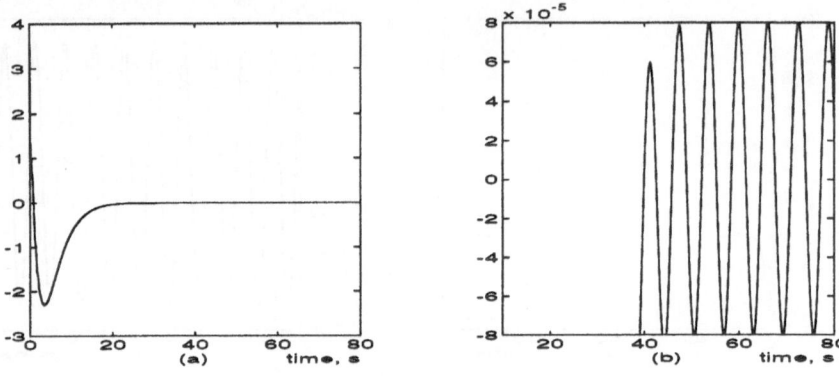

Fig. 5. Output tracking error $(\hat{e} = y - y_d)$ for $p = 5, \varepsilon = 0.1$

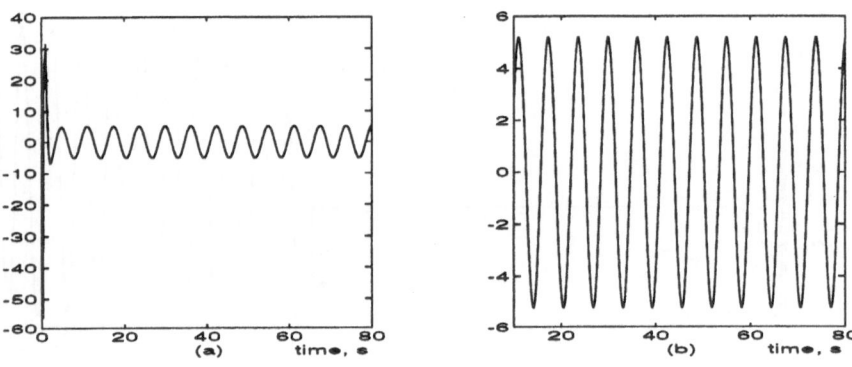

Fig. 6. Slow control (u_s) for $p = 2, \varepsilon = 0.1$

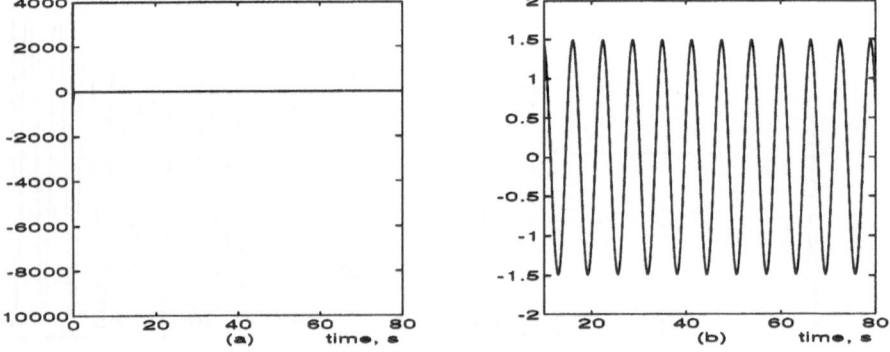

Fig. 7. Fast control (u_f) for $p = 2, \varepsilon = 0.1$

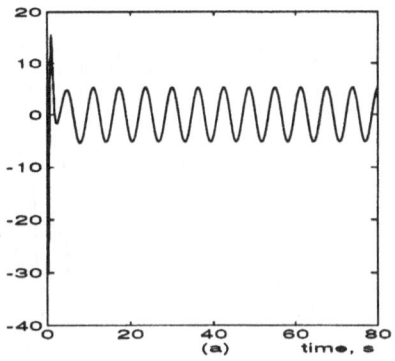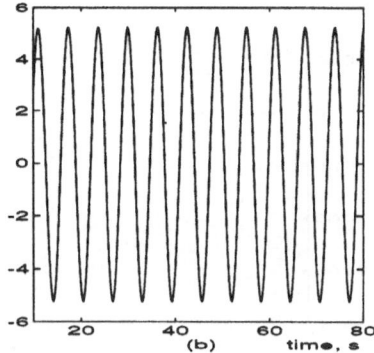

Fig. 8. Slow control (u_s) for $p = 3, \varepsilon = 0.1$

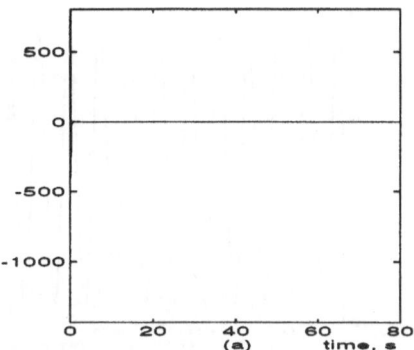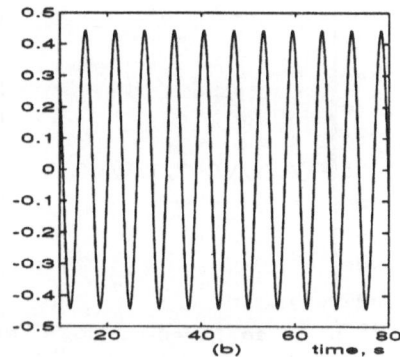

Fig. 9. Fast control (u_f) for $p = 3, \varepsilon = 0.1$

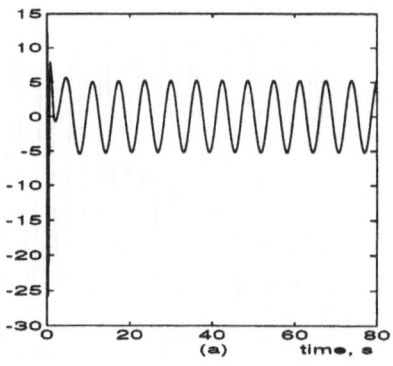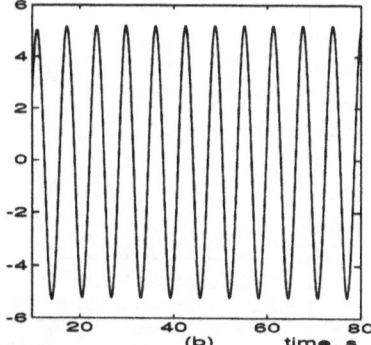

Fig. 10. Slow control (u_s) for $p = 4, \varepsilon = 0.1$

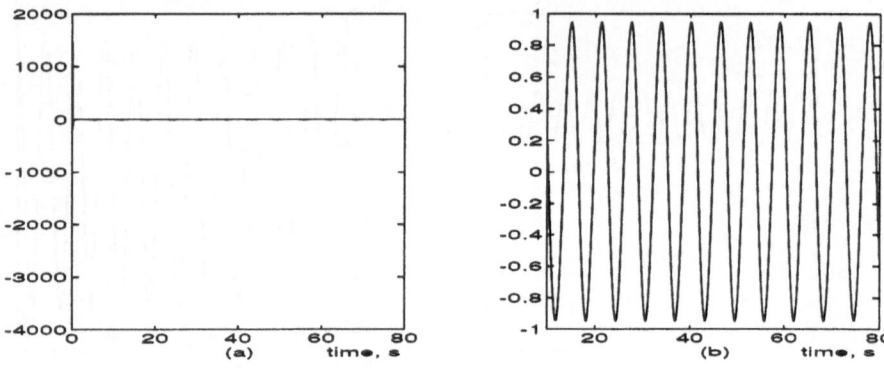

Fig. 11. Fast control (u_f) for $p = 4, \varepsilon = 0.1$

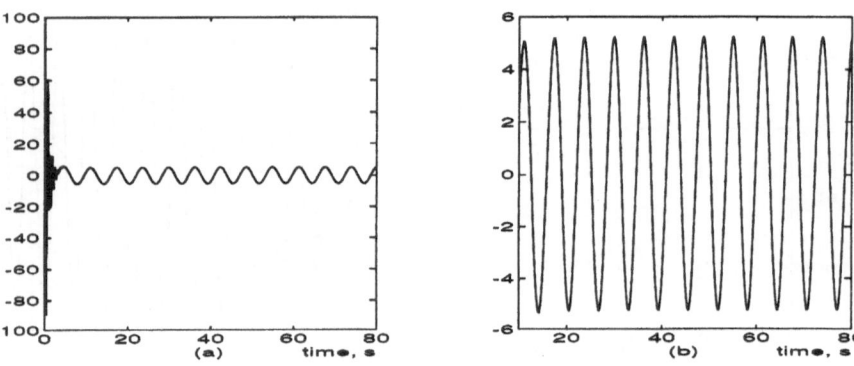

Fig. 12. Slow control (u_s) for $p = 5, \varepsilon = 0.1$

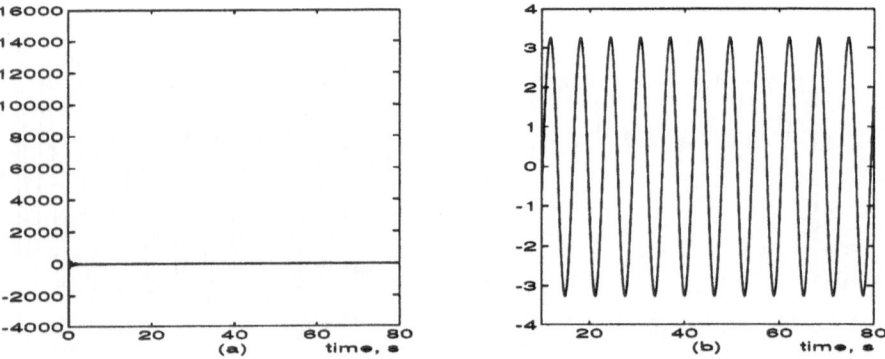

Fig. 13. Fast control (u_f) for $p = 5, \varepsilon = 0.1$

Digital Reduced-Order Modeling and Feedback Design for Nonlinear Systems*

David G. Taylor

Georgia Institute of Technology, School of Electrical and Computer Engineering,
Atlanta, Georgia 30332-0250

Abstract. Digital processors provide the most convenient means to perform the real-time computations required for the implementation of complex nonlinear feedback systems. In this paper, sampled-data techniques which accurately model the effects of digital implementation are applied to nonlinear singularly perturbed systems. Using the permanent-magnet synchronous motor as an example, the digital reduced-order design method is illustrated and auxiliary issues, such as state estimation and parameter identification, are also addressed.

1 Introduction

Singularly perturbed systems, i.e. systems which exhibit two or more distinct time-scales, arise frequently in a variety of engineering disciplines. For instance, electromechanical systems typically have electrical and mechanical time constants which differ in magnitude, leading to two-time-scale behavior. In order to relieve the numerical stiffness and lessen the excessive implementation requirements often associated with such systems, reduced-order modeling approaches are desired. Many reduced-order analysis and design results, for both linear and nonlinear singularly perturbed systems, are available for the continuous-time and discrete-time cases [4].

For the sampled-data case, however, in which the control signal is computed in discrete-time by a digital processor and enters the continuous-time plant through a zero-order hold, the modeling problem is somewhat more involved. As first illustrated in [2], reduced-order feedback design for sampled-data systems may lack robustness if the reduced-order design model is obtained simply by discretizing the "standard" reduced-order continuous-time model which ignores the piecewise-constant nature of the control signal. Without proper modeling, a controller which stabilizes the system when the singular perturbation parameter is zero may actually destabilize the system for nonzero but arbitrarily small values of the singular perturbation parameter. This robustness issue was clarified and resolved for linear two-time-scale systems in [2]. One purpose of the present paper is to extend the notion of sampled-data reduced-order modeling to a class

* This work was supported in part by the National Science Foundation under Grant ECS-9158037 and by the Air Force Office of Scientific Research under Grant F49620-93-1-0147.

of nonlinear two-time-scale systems. Although the reduced-order state equation is quite straightforward, the corresponding output equation typically possesses a direct transmission term which should be modeled with a delay.

The main part of the paper is concerned with applications of digital reduced-order modeling to velocity control design for permanent-magnet synchronous motors. The velocity control algorithm itself is quite intuitive, since it needs only the state equation modeling result. However, two supplementary technologies are developed which feature the less intuitive output equation modeling result. In the first case, an asymptotically valid discrete-time approximation to the back-emf is obtained by declaring the stator current as the sampled output, and this signal is used to implement state estimation. In the second case, again by declaring the stator current to be the sampled output, an asymptotically valid discrete-time approximation for the combined resistive and back-emf losses is obtained, and this signal is shown to be useful for parameter identification. These two supplementary technologies may be used to implement digital reduced-order velocity control without shaft-mounted sensors or without pre-calibration, respectively.

The organization of this paper, which summarizes some of the material originally appearing in [6], is as follows. First, a class of digitally controlled continuous-time nonlinear singularly perturbed systems is introduced and discussed. Discrete-time design models for such systems are obtained by constructing asymptotically valid reduced-order continuous-time state and output equations which are subsequently discretized using Taylor series methods. Second, applications of reduced-order discrete-time modeling and approximate input-output linearization to permanent-magnet synchronous motors are developed. The main emphasis for this case study is the presentation of two supplementary technologies, namely state estimation for shaft-sensorless control and parameter identification for self-tuning control, featuring novel uses of piecewise-linear approximations of the motor nonlinearities. Third, experimental results for the permanent-magnet synchronous motor application are shown. The experimental results illustrate several advantages of the new approaches over existing technology.

2 Discrete-Time Modeling

2.1 Order Reduction

Consider the singularly perturbed system

$$\dot{x}(t) = a_1(x(t)) + A_1(x(t))z(t) + B_1(x(t))u(t) \tag{1}$$
$$\epsilon\dot{z}(t) = a_2(x(t)) + A_2(x(t))z(t) + B_2(x(t))u(t) \tag{2}$$
$$y(t) = c(x(t)) + C(x(t))z(t) \tag{3}$$

where $x \in R^N$, $z \in R^M$, $u \in R^P$, $y \in R^Q$ and $\epsilon > 0$ is a small parameter. Attention is restricted to the case of *stable* parasitics, i.e. it is assumed that there exists a $\kappa > 0$ such that $Re\{\lambda(A_2(x))\} \leq -\kappa$ for all $x \in R^N$. This system is digitally controlled with an input

$$u(t) = u[n] \quad \forall\, t \in [nT, (n+1)T) \tag{4}$$

which is piecewise-constant over half-open sampling intervals of length $T > 0$, selected to be commensurate with the slow (dominant) dynamics of the system. Throughout this paper, it will be assumed that ϵ is sufficiently small relative to T. This assumption implies that the fast transients induced by the control input at each sampling instant will decay during the sampling interval. For the more complicated case when this assumption is violated, more refined approximation techniques would be required (see [6] for details).

Due to the discontinuity in $u(t)$ at each sampling instant, traditional approaches to reduced-order modeling cannot be directly applied. Instead, one must account for the recurring boundary layer effects. It is conceptually useful to consider a sequence of modeling problems, one for each sampling interval. Formally setting $\epsilon = 0$ in (2) and solving for $z(t)$ yields a family of slow manifolds

$$z_{s(n)}(t) := -A_2^{-1}(x(t))\left(a_2(x(t)) + B_2(x(t))u[n]\right) \quad \forall\, t \in [nT, (n+1)T) \tag{5}$$

Using singular perturbation theory [4], it can be shown that if ϵ is sufficiently small relative to T, then there exists $0 < \delta \ll T$ such that

$$z(t) = z_{s(n)}(t) + O(\epsilon) \quad \forall\, t \in [nT + \delta, (n+1)T) \tag{6}$$

This means that $z_{s(n)}(t)$ is a good approximation of $z(t)$ everywhere except near the sampling instants. The approximation (6) is the point of departure for the development of an asymptotically valid *state equation* for the reduced-order discrete-time design model.

It is clear that (5) cannot be used directly to approximate $z(t)$ at $t = nT$. The step change in $u(t)$ at $t = nT$ is not transmitted instantaneously through the $z(t)$ dynamics. However, $z(t)$ is the solution of a differential equation and is thus continuous. Assuming again that ϵ is sufficiently small relative to T, this continuity allows (5) and (6) to be used indirectly as

$$z(nT) = \lim_{\mu \to 0} z(nT - \mu) \tag{7}$$

$$= \lim_{\mu \to 0} z_{s(n-1)}(nT - \mu) + O(\epsilon) \tag{8}$$

$$= z_{s(n-1)}(nT) + O(\epsilon) \tag{9}$$

The algebraic constraint (9), which relates samples of z to samples of x and u up to an $O(\epsilon)$ approximation error, is the point of departure for the development of an asymptotically valid *output equation* for the reduced-order discrete-time design model.

2.2 Discretization

Consider a generic autonomous nonlinear system

$$\dot{x}(t) = f(x(t)) \tag{10}$$

where $x \in R^N$ and f is analytic. The solution of (10), $x(t)$, given an initial condition at $t = t_0$, can be approximated using a formal Taylor series expansion

$$x(t) = x^{(0)}(t_0) + (t - t_0)x^{(1)}(t_0) + \frac{1}{2}(t - t_0)^2 x^{(2)}(t_0) + \cdots \tag{11}$$

where the superscript $^{(i)}$ denotes the ith time derivative. The series (11) will be convergent if $t - t_0 > 0$ is sufficiently small. Using a convenient Lie derivative notation

$$L_f^i x \big|_{x(t)} := x^{(i)}(t) \tag{12}$$

such that $L_f^0 x = x$, $L_f^1 x = f(x)$, $L_f^2 x = \frac{\partial f}{\partial x} f(x)$, etc., this approximation technique is compactly expressed by

$$x(t) = e^{(t-t_0)L_f} x \big|_{x(t_0)} \tag{13}$$

where it is understood that

$$e^{\tau L_f} x := \sum_{i=0}^{\infty} \frac{\tau^i}{i!} L_f^i x \tag{14}$$

denotes a formal exponential series involving Lie derivatives. The series (13) (or, equivalently, (11)) forms the basis for a more useful result, as shown below.

Now consider a digitally controlled nonlinear system

$$\dot{x}(t) = f(x(t)) + \sum_{i=1}^{P} g_i(x(t))u_i(t) \tag{15}$$

where $x \in R^N$, $u_i \in R$ for $i = 1, \ldots, P$, $f, g_1, \ldots, g_P$ are analytic, and

$$u_i(t) = u_i[n] \quad \forall\, t \in [nT, (n+1)T) \tag{16}$$

where $T > 0$ is the sampling period. Since each control is constant over any sampling interval, it follows from (13) that

$$x(t) = e^{(t-nT)L_{(f+\sum_{i=1}^{P} u_i[n]g_i)}} x \big|_{x(nT)} \quad \forall\, t \in [nT, (n+1)T) \tag{17}$$

Due to the continuity of $x(t)$ and the property $L_{\alpha+\beta}(\cdot) = L_\alpha(\cdot) + L_\beta(\cdot)$, this intermediate result leads to the discrete-time model

$$x[n+1] = e^{T(L_f + \sum_{i=1}^{P} u_i[n]L_{g_i})} x \big|_{x[n]} \tag{18}$$

where $x[n] := x(nT)$. Not only is the series in (18) formally correct, it is absolutely and uniformly convergent if T is sufficiently small. If the $O(T^2)$ terms are dropped, then simple Euler discretization is achieved. Greater accuracy can be achieved by including higher order terms, though this yields a nonlinear dependence on the $u_i[n]$.

Example 1. The discrete-time model (18) covers the zero-order hold discrete-time equivalent model for linear systems as a special case. To see this, define $f(x)$ and $g(x)$ according to

$$f(x) = Ax \quad g(x) = b$$

For this case, $L_f^i x = A^i x$, $L_g L_f^i x = A^i b$, and all other Lie derivatives are zero, leading to

$$x[n+1] = x[n] + T(Ax[n] + bu[n]) + \frac{T^2}{2}(A^2 x[n] + Abu[n]) + \cdots$$

$$= e^{AT} x[n] + \int_0^T e^{A\eta} d\eta b\, u[n]$$

which is the expected result. Note that in this example, $e^{(\cdot)}$ denotes the matrix exponential.

Example 2. As a nonlinear example, consider the system defined by

$$f(x) = \begin{bmatrix} 0 \\ x_1 + x_1^2 \end{bmatrix} \quad g(x) = \begin{bmatrix} 1 \\ 1 \end{bmatrix}$$

For this case,

$$L_f x = \begin{bmatrix} 0 \\ x_1 + x_1^2 \end{bmatrix} \quad L_g x = \begin{bmatrix} 1 \\ 1 \end{bmatrix} \quad L_g L_f x = \begin{bmatrix} 0 \\ 1 + 2x_1 \end{bmatrix} \quad L_f^2 L_f x = \begin{bmatrix} 0 \\ 2 \end{bmatrix}$$

and all other Lie derivatives are identically zero. Hence, from (18)

$$x[n+1] = x[n] + T \begin{bmatrix} u[n] \\ x_1[n] + x_1^2[n] + u[n] \end{bmatrix} + \frac{T^2}{2} \begin{bmatrix} 0 \\ u[n](1 + 2x_1[n]) \end{bmatrix} + \frac{T^3}{6} \begin{bmatrix} 0 \\ 2u^2[n] \end{bmatrix}$$

In this example, the exact discrete-time model may be expressed using a finite number of terms. Such is not the case in general.

2.3 Design Model

The model (18) representing the discrete-time dynamics of the generic digitally controlled system (15)–(16) may be used to formulate an approximate reduced-order discrete-time state equation for the original system (1)–(2). To proceed, it will now be necessary to refer back to the development (5)–(9).

Although the discrepancy between $z_{s(n)}(t)$ and $z(t)$ over the presumably short boundary layer interval $[nT, nT + \delta)$ may be large, the effect of this discrepancy is insignificant when considering the $x(t)$ dynamics. This is so because, for sufficiently small ϵ relative to T, the time interval of width $0 < \delta \ll T$ will be so small that the error $z(t) - z_{s(n)}(t)$ will not significantly contribute to the integration of the right-hand-side of (1). Consequently, from (6) it follows that an asymptotically valid state equation useful for approximating $x(t)$ is

$$\dot{\bar{x}}(t) = a_s(\bar{x}(t)) + B_s(\bar{x}(t))u[n] \tag{19}$$

where

$$a_s(\cdot) := a_1(\cdot) - A_1(\cdot)A_2^{-1}(\cdot)a_2(\cdot) \tag{20}$$
$$B_s(\cdot) := B_1(\cdot) - A_1(\cdot)A_2^{-1}(\cdot)B_2(\cdot) \tag{21}$$

and the overbar is used to distinguish solutions of (19) from solutions of the original system (1)–(2). According to (18), the associated discrete-time representation of (19) would be

$$\bar{x}[n+1] = F(\bar{x}[n], u[n]) \tag{22}$$

where

$$F(x, u) := e^{T\left(L_{a_s} + \sum_{i=1}^{P} u_i L_{b_{s,i}}\right)}x \tag{23}$$

and $b_{s,i}$ denotes the ith column of B_s.

As for an asymptotically valid output equation to approximate $y(t)$ at sampling instants, the algebraic constraint (9) provides the key. Indeed, direct substitution of (9) into (3) suggests

$$\bar{y}[n] = h(\bar{x}[n], u[n-1]) \tag{24}$$

where

$$h(x, u) := c_s(x) + D_s(x)u \tag{25}$$

and

$$c_s(\cdot) := c(\cdot) - C(\cdot)A_2^{-1}(\cdot)a_2(\cdot) \tag{26}$$
$$D_s(\cdot) := -C(\cdot)A_2^{-1}(\cdot)B_2(\cdot) \tag{27}$$

The delay in the input term, $u[n-1]$, appearing in (24) but not in (22), is as predicted in [2] for linear systems.

In summary, the approximate design model is

$$\bar{x}[n+1] = F(\bar{x}[n], u[n]) \tag{28}$$
$$\bar{y}[n] = h(\bar{x}[n], u[n-1]) \tag{29}$$

where F and h are defined in (23) and (25). It should now be clear that for all ϵ sufficiently small relative to T

$$x[n] = \bar{x}[n] + O(\epsilon) \tag{30}$$
$$y[n] = \bar{y}[n] + O(\epsilon) \tag{31}$$

for all $n \geq 0$ such that the state trajectories remain $O(1)$ bounded, where $x[n]$ and $y[n]$ denote the sample values $x(nT)$ and $y(nT)$ of the original full-order system (1)–(3).

3 Applications

3.1 PM Synchronous Motors

Some physical systems of interest are modeled by state equations of the form (1)–(2). For instance, consider a three-phase permanent-magnet synchronous motor with smooth rotor (but perhaps non-smooth stator) modeled by

$$\frac{d\theta}{dt} = \omega(t) \tag{32}$$

$$J\frac{d\omega}{dt} = K'(\theta(t))i(t) - \tau_m(\theta(t), \omega(t)) \tag{33}$$

$$L\frac{di}{dt} = v(t) - Ri(t) - K(\theta(t))\omega(t) \tag{34}$$

where the (slow) mechanical states are angular position θ and angular velocity ω, the (fast) electrical states are the three components of the phase current vector i, and the inputs are the three components of the phase voltage vector v. The only nonlinearities are the periodic vector function $K(\theta)$, which represents the angle-dependent shape of torque and back-emf, and the load torque function $\tau_m(\theta, \omega)$, which includes magnetic cogging torque (if present). The period of the typically non-sinusoidal function $K(\theta)$ is $2\pi/N_p$, where N_p is the number of magnetic pole pairs. Also appearing above are the total rotor and load inertia J, the symmetric positive-definite matrix of phase inductances L, and the diagonal matrix of phase resistances R. The motor is driven by an amplifier with linear current loops, which in turn is fed by a piecewise-constant control vector u, i.e.

$$v(t) = K_a(u(t) - i(t)) \tag{35}$$
$$u(t) = u[n] \quad \forall\, t \in [nT, (n+1)T) \tag{36}$$

where K_a is a diagonal matrix of strictly positive amplifier gains, T is the sampling period selected on the basis of the mechanical dynamics, and n is the discrete-time index.

The state equations of this composite system are indeed of the form (1)–(2), and it is a simple exercise to convert the model into the notation used in previous sections. However, in the remainder of this paper, the more intuitive notation introduced above will be used for added clarity. Note that the singular perturbation parameter ϵ may be taken as the scaling coefficient for matrix L, i.e. $L := \epsilon\tilde{L}$ where $\tilde{L} = O(1)$. Without loss of generality, it will be assumed that $\tilde{L} = I$ since, after order-reduction, this term naturally drops out anyway.

Assuming that ϵ is sufficiently small relative to T, the key step in order reduction is to formulate the family of slow manifolds by setting $\epsilon = 0$ and solving for $i(t)$, yielding

$$i_{s(n)}(t) := R_e^{-1}(K_a u[n] - K(\theta(t))\omega(t)) \quad \forall\, t \in [nT, (n+1)T) \tag{37}$$

where $R_e := R + K_a$ is the effective resistance matrix. From (37), it is straightforward to synthesize approximate discrete-time design models of the form (28)–(29). The discrete-time reduced-order state equation analogous to (28) using Euler (first-order Taylor series) discretization is expressed by

$$\theta[n+1] = \theta[n] + T\omega[n] \tag{38}$$

$$\omega[n+1] = \omega[n] + T\frac{1}{J}(\tau_e(\theta[n], \omega[n], u[n]) - \tau_m(\theta[n], \omega[n])) \tag{39}$$

where $\tau_e(\theta, \omega, u)$ is an approximation to the electrical torque produced in response to the control u, i.e.

$$\tau_e(\theta[n], \omega[n], u[n]) := K'(\theta[n])R_e^{-1}K_a u[n] - K'(\theta[n])R_e^{-1}K(\theta[n])\omega[n] \tag{40}$$

and where $\theta[n] := \theta(nT)$ and $\omega[n] := \omega(nT)$. On the other hand, the discrete-time reduced-order output equation analogous to (29) is

$$i[n] = R_e^{-1}\left(K_a u[n-1] - K(\theta[n])\omega[n]\right) \tag{41}$$

for the case when the sampled output is taken to be the stator current.

3.2 Velocity Control

The problem of designing a velocity tracking control can be decomposed into three sub-problems: determining a commutation of the three components of $u[n]$ consistent with instantaneous torque control; determining a torque command which linearizes the mechanical reduced-order design model; determining a linear feedback for the linearized mechanical dynamics.

The commutation problem arises from the fact that there are multiple torque-producing phases, yet just a scalar torque to control. An optimal approach to the commutation problem would be to find the unique control vector which simultaneously produces the desired instantaneous torque and minimizes the power dissipated in the phase resistances. Alternative means for eliminating the extra degrees of freedom could be to minimize the voltage feed requirement or the amplifier bandwidth requirement. With the objective of minimizing copper losses, however, the optimal commutation problem may be formulated as

$$\text{minimize } i'i \tag{42}$$

$$\text{subject to } K'(\theta)i = \tau_d \tag{43}$$

where τ_d is the desired electric torque signal, to be supplied by the outer loops of the velocity controller. Using the method of Lagrange multipliers, it is easy to show that the solution to this constrained optimization problem is

$$i = K(\theta)\left(K'(\theta)K(\theta)\right)^{-1}\tau_d \tag{44}$$

which, according to (37), suggests the commutation

$$u[n] = K_a^{-1}\left(K(\theta[n])\omega[n] + R_e\frac{K(\theta[n])}{\|K(\theta[n])\|^2}\tau_d[n]\right) \tag{45}$$

Note that it is safe to assume $\|K(\theta)\| \neq 0$ for all θ since, otherwise, there would be rotor angles at which no torque could be produced by any phase.

The velocity control design is completed by augmenting the commutation scheme with a torque command

$$\tau_d[n] = \tau_m(\theta[n], \omega[n]) + J\alpha_d[n] \tag{46}$$

which linearizes the mechanical dynamics, and with an acceleration command

$$\alpha_d[n] = \frac{1}{T}(\gamma(\omega[n] - \omega_d[n]) - \omega[n] + \omega_d[n+1]), \quad |\gamma| < 1 \tag{47}$$

which stabilizes the velocity tracking error dynamics. The controller is thus ultimately commanded by $\{\omega_d[n] : n \geq 0\}$, which represents the desired velocity trajectory.

In order to implement digital reduced-order velocity tracking control as defined above, the following requirements must be met.

- Sensor Requirements:
  - Current sensors are needed to implement the analog linear feedback within the amplifier.
  - A high-resolution shaft-mounted position sensor is needed to implement optimal commutation and load nonlinearity compensation.
  - A velocity signal (derived from position measurements) is needed to implement back-emf compensation and the velocity tracking loop.
- Calibration Requirements:
  - The electrical drive model, $K(\theta)$, R and K_a, must be accurately known to perform commutation (45).
  - The mechanical load model, $\tau_m(\theta, \omega)$ and J, must be accurately known to perform load compensation (46).
  - The inductance matrix L is neglected by the controller and hence need not be accurately known.

Clearly, it would be beneficial to develop the means to relax some of these restrictive requirements. The following two sections illustrate how the digital reduced-order design models may be used to (i) estimate the rotor motion from stator current measurements, thus eliminating the need for a shaft-mounted position sensor, or (ii) identify the model parameters using recursive algorithms on-line, thus eliminating the need for pre-calibration of the controller. In both cases, piecewise-linear approximations of the various nonlinearities are used to achieve computationally efficient numerical implementations.

3.3 State Estimation

Low performance motor drives often use internally-mounted Hall-effect position sensors, since they provide sufficient resolution for simple commutation and ensure that synchronicity is maintained. However, these sensors only detect the polarity of the rotor magnets and, hence, permit only six distinct excitation states per electrical cycle for a three-phase motor, leading to significant torque ripple. The velocity controller derived in the previous section has the potential to achieve accurate motion control without torque ripple, but it requires a high-resolution shaft-mounted position sensor. More refined excitation of the phases is possible in this case, because the resolution of such a sensor is not limited by the motor's magnetic geometry. However, use of any shaft-mounted sensor, whether it be an optical encoder or magnetic resolver, is a disadvantage in certain applications for a variety of reasons: increased size, weight and cost of the drive; decreased reliability of the drive, due in part to additional cabling; components which are prone to failure in harse environments; and the need for mechanical couplers.

Because of the desire for refined commutation and accurate motion control without stringent position sensor requirements, several methods for shaft-sensorless control have previously been developed. For permanent-magnet synchronous motors, many of the existing methods rely on the back-emf waveform, and typically the zero-crossing angles of this waveform are used for commutation purposes. Existing schemes for back-emf reconstruction use analog circuits which are prone to drift. Existing schemes for sense coil measurement of back-emf require a non-standard stator. Existing schemes for direct-detection of back-emf require excitation methods which assure sequentially open-circuited phase windings. All of the above methods provide limited resolution position sensing. On the other hand, recursive state observers using on-line simulation of the motor model augmented with a current prediction error term can provide high-resolution motion estimates. Unfortunately, this approach requires an accurate model of the mechanical load and velocity-scheduled innovation gains.

A new method for position sensor elimination will be described below. The new method involves digital reduced-order computation of back-emf to obtain the entire three-phase back-emf waveform. Position and velocity are estimated using a nonlinear least-squares formulation, which is implemented numerically using piecewise-linear approximations of the relevant nonlinearities. No mechanical load model is needed for motion estimation, and there is no need for sequentially open-circuited phase windings, meaning that the method is consistent with minimum copper loss optimal commutation. The main disadvantage of the new method, shared by all back-emf based methods, is that motion estimation is not possible at (or near) zero velocity. Hence, shaft-sensorless controllers which incorporate the new method must transition in and out of an open-loop mode at low velocities.

From the digital reduced-order output equation (41), if K_a and R_e are accurately known it is possible to define a computable signal

$$y_{emf}[n] := K_a u[n-1] - R_e i[n] \tag{48}$$

which is an asymptotically valid approximation of the back-emf, i.e.

$$y_{emf}[n] \approx K(\theta[n])\omega[n] \tag{49}$$

Since the back-emf depends on rotor motion, but in a highly nonlinear way, it is reasonable to formulate the state estimation problem as a map inversion problem

[5] using the state-to-measurement map

$$H(\theta, \omega) := K(\theta)\omega \qquad (50)$$

In other words, given an accurate model for $K(\theta)$ and hence for $H(\theta, \omega)$, the objective is to seek an approximate solution to the overdetermined system of three constraint equations

$$H(\theta[n], \omega[n]) = y_{emf}[n] \qquad (51)$$

for the two unknowns $\theta[n]$ and $\omega[n]$. More formally, the nonlinear least squares estimation problem may be stated as

$$\begin{bmatrix} \hat{\theta}[n] \\ \hat{\omega}[n] \end{bmatrix} = \arg \min_{\theta, \omega} \|H(\theta, \omega) - y_{emf}[n]\| \qquad (52)$$

This formulation is advantageous from the point of view that it seeks a best fit to the possibly noisy measurement data. The Jacobian matrix of the state-to-measurement map will typically have full rank except at zero velocity, i.e.

$$\text{rank} \left[\frac{dK(\theta)}{d\theta} \omega \ K(\theta) \right] = \begin{cases} 2, & \omega \neq 0 \\ 1, & \omega = 0 \end{cases} \qquad (53)$$

thereby guaranteeing existence of minima, provided that the rotor is turning with sufficient velocity. On the other hand, these minima are necessarily non-unique. Due to periodicity, $K(\theta) = K(\theta + 2\pi/N_p)$, and hence all electrical cycles are indistinguishable. For velocity control applications with load torque $\tau_m(\theta, \omega)$ being θ-independent, this ambiguity does not present a problem since the controller requires only the "electrical angle" for feedback. Due to symmetry, $K(\theta) = -K(\theta + \pi/N_p)$, and hence the two rotation directions are indistinguishable. Again, this ambiguity does not present a problem if heuristic procedures are used to ensure that $\text{sgn}(\omega) = \text{sgn}(\omega_d)$.

Instead of using traditional iterative methods to solve the nonlinear least squares problem, such as the Gauss-Newton or Levenberg-Marquardt algorithms, a simple and computationally efficient iteration method has been developed which is based on piecewise-linear approximation of $K(\theta)$. Specifically, by dividing each electrical cycle into N_s equal width segments, $K(\theta)$ may be approximated by

$$K(\theta) \approx \alpha_j \vartheta + \beta_j, \quad \forall \vartheta \in [j\Theta, (j+1)\Theta) \qquad (54)$$

where ϑ is the electrical angle and Θ is the segment width, i.e.

$$\vartheta := \theta \bmod \frac{2\pi}{N_p} \quad \Theta := \frac{2\pi}{N_p N_s} \quad j = 0, \ldots, N_s - 1 \qquad (55)$$

The parameter vectors α_j and β_j associated with the jth segment are related to $K(\theta)$ by

$$\alpha_j := (K((j+1)\Theta) - K(j\Theta)) / \Theta \qquad (56)$$
$$\beta_j := (j+1)K(j\Theta) - jK((j+1)\Theta) \qquad (57)$$

so that the approximation of $K(\theta)$ is continuous. If an integer j were known, such that the unknown ϑ was located within the jth segment, then the nonlinear least squares problem reduces to the linear least squares problem

$$\begin{bmatrix} x_1[n] \\ x_2[n] \end{bmatrix} = \arg \min_{x_1,x_2} \|\alpha_j x_1 + \beta_j x_2 - y_{emf}[n]\| \tag{58}$$

$$\begin{bmatrix} \hat{\vartheta}_j[n] \\ \hat{\omega}_j[n] \end{bmatrix} = \begin{bmatrix} x_1[n]/x_2[n] \\ x_2[n] \end{bmatrix} \tag{59}$$

which can be explicitly solved to obtain

$$\hat{\vartheta}_j[n] = \frac{(\beta'_j \beta_j \alpha'_j - \alpha'_j \beta_j \beta'_j) y_{emf}[n]}{(\alpha'_j \alpha_j \beta'_j - \alpha'_j \beta_j \alpha'_j) y_{emf}[n]} \tag{60}$$

$$\hat{\omega}_j[n] = \frac{\alpha'_j \alpha_j \beta'_j - \alpha'_j \beta_j \alpha'_j}{\alpha'_j \alpha_j \beta'_j \beta_j - (\alpha'_j \beta_j)^2} y_{emf}[n] \tag{61}$$

Of course, initially it is not possible to know which segment is the correct one, so this information is determined recursively as follows. Given an initial guess for the correct segment, one first checks to see if the guess is either a consistent segment or a boundary segment. Segment j is called a consistent segment if

$$\hat{\vartheta}_j[n] \in [j\Theta, (j+1)\Theta) \tag{62}$$

or a boundary segment if

$$\hat{\vartheta}_{j-1}[n] \geq j\Theta \quad \hat{\vartheta}_j[n] < j\Theta \tag{63}$$

where the electrical angle solution is obtained from (60). If the initial guess is neither a consistent segment nor a boundary segment, then a new candidate segment is determined, by incrementing or decrementing according to the direction suggested by the failed consistency check, and the process repeats. Eventually, after at most N_s iterations this recursion will lead to either a consistent segment or a boundary segment. If the algorithm converges to a consistent segment, then the appropriate solution is provided by (60)–(61). If the algorithm converges to a boundary segment, then the solution is provided by the reduced-order linear least squares problem

$$\hat{\omega}_j[n] = \arg \min_{\omega} \|(\alpha_j(j\Theta) + \beta_j)\omega - y_{emf}[n]\| \tag{64}$$

yielding the explicit solution

$$\hat{\vartheta}_j[n] = j\Theta \tag{65}$$

$$\hat{\omega}_j[n] = \frac{j\Theta\alpha'_j + \beta'_j}{j^2\Theta^2\alpha'_j\alpha_j + 2j\Theta\alpha'_j\beta_j + \beta'_j\beta_j} y_{emf}[n] \tag{66}$$

For further details, please see [8].

3.4 Parameter Identification

Many applications of electric motors involve mechanical loads which vary substantially and are not known a priori. Even some of the electrical parameters may
drift due to changes in operating conditions. If a fixed-parameter model-based
control system is used without accurate calibration, then poor performance (and
perhaps even instability) can be expected. The benefits of optimal commutation,
instantaneous torque control, and linearization-based velocity control are available only to the extent that the parameters of the motor drive and load are
accurately known.

To help relax the requirement of accurate parameter knowledge for controller
calibration, a number of methods for adaptive or self-tuning control have been
developed. These self-tuning systems have an on-line parameter identification
scheme which provides recursively computed parameter estimates to a parameterized model-based control scheme. Most of the existing parameter identification methods are based on an assumption that the periodic function $K(\theta)$ is
sinusoidal without harmonics. This assumption simplies the problem, because
it allows a change of variables into the rotor frame of reference and eliminates
all periodic motor nonlinearities. Self-tuning controllers based on these methods
will suffer from residual torque-ripple when applied to the more general class
of non-sinusoidal motors. Existing parameter identification methods for motors
with non-sinusoidal magnetics account for higher-order harmonics using a truncated Fourier series parameterization [1]. With these methods, consideration of
additional harmonics requires a larger number of parameters and, hence, greater
computational effort.

A new method for on-line parameter identification will be described below.
Although there may be applications in which only several motor or load parameters are uncertain, the new method handles the case of total parametric
uncertainty (with the exception of the amplifier gain). In other words, every parameter needed to implement linearization-based velocity control with optimal
commutation can be included in the identification algorithm. The identifier has
a nested structure, in order to avoid a nonlinear dependence on parameters or
over-parameterization, with an inner-loop for the electrical parameters and an
outer-loop for the mechanical parameters. All periodic functions, namely $K(\theta)$
and any periodic components of the load torque $\tau_m(\theta, \omega)$ such as magnetic cogging, are parameterized using piecewise-linear basis functions rather than truncated Fourier series. The result is that the computational requirements of the
new method are quite low, and in fact do not depend at all on the harmonic content of the functions being identified (although the memory size needed to store
the complete set of identified parameters does scale with harmonic content).

From the digital reduced-order output equation (41), it is possible to define
a computable signal, namely the combined resistive and back-emf voltage, which
permits parameterization of the uncertainty related to the electrical parameters
in $K(\theta)$ and R. Given accurate knowledge of gain K_a, an asymptotically valid
electrical parameterization for the jth phase is

$$y_{e_j}[n] := K_{a_j} u_j[n-1] \approx w'_{e_j}[n] \Theta_{e_j} \tag{67}$$

where the electrical regressor vector $w_{e_j}[n]$ and the unknown electrical parameter vector Θ_{e_j} for phase j are defined by

$$w_{e_j}[n] := \left[i_j[n] \; f'_j[n] \right]' \tag{68}$$

$$\Theta_{e_j} := \left[R_{e_j} \; \Theta'_{f_j} \right]' \tag{69}$$

and where any appropriate set of basis functions is used to define the periodic back-emf function, i.e.

$$f'_j[n]\Theta_{f_j} \equiv K_j(\theta[n])\omega[n] \tag{70}$$

A typical choice might be a truncated Fourier series.

On the other hand, the digital reduced-order state equation (39) leads to an approximately computable signal representing the electric torque, which permits parameterization of the uncertainty related to the mechanical parameters in $\tau_m(\theta, \omega)$ and J. Given accurate knowledge of the (unknown) electrical parameter vector Θ_e, an asymptotically valid mechanical parameterization is

$$y_m[n] := \tau_e(\theta[n-1], \omega[n-1], u[n-1])|_{\Theta_e} \approx w'_m[n]\Theta_m \tag{71}$$

where the mechanical regressor vector $w_m[n]$ and the unknown mechanical parameter vector Θ_m are defined by

$$w_m[n] := \left[\tfrac{1}{T} \left(\omega[n] - \omega[n-1] \right) \; g'[n-1] \right]' \tag{72}$$

$$\Theta_m := \left[J \; \Theta'_g \right]' \tag{73}$$

and where any appropriate basis functions are used to account for periodic components of load torque, i.e.

$$g'[n]\Theta_g \equiv \tau_m(\theta[n], \omega[n]) \tag{74}$$

Again, a typical choice would be a truncated Fourier series. Note that calculation of $y_m[n]$ requires accurate knowledge of K_a and measurement of θ and ω, according to the definition of the electric torque function $\tau_e(\theta, \omega, u)$ given in (40). The implicit dependence of $y_m[n]$ on the unknown Θ_e calls for a nested identifier structure, so that the mechanical parameter identification will be driven by an approximation

$$\hat{y}_m[n] := \tau_e(\theta[n-1], \omega[n-1], u[n-1])|_{\hat{\Theta}_e[n]} \tag{75}$$

where $\hat{\Theta}_e[n]$ denotes the estimated value of Θ_e.

It is significant that most of the complex nonlinearities in the motor and load model are simply periodic functions of the rotor angle θ. Rather than expressing this complex dependence on θ through a truncated Fourier series or other common basis functions, it is advantageous to represent this dependence using a piecewise-linear approximation to $K(\theta)$ and any periodic components of $\tau_m(\theta, \omega)$. By doing so, just two parameters per scalar periodic function will need to be updated in the identification scheme at any one time. In order to guarantee continuity of the periodic functions estimated from the identification scheme, linear interpolation formulas are used to parameterize the piecewise-linear approximations. Specifically, by dividing each electrical cycle into N_s equal width segments, $K(\theta)$ may be approximated by

$$K(\theta) \approx (1 - \phi_k)\alpha_k + \phi_k\beta_k, \quad \forall \vartheta \in [k\Theta, (k+1)\Theta) \tag{76}$$

where, as before, ϑ denotes the electrical angle and Θ denotes the segment width, i.e.

$$\vartheta := \theta \bmod \frac{2\pi}{N_p} \quad \Theta := \frac{2\pi}{N_p N_s} \quad k = 0, \ldots, N_s - 1 \tag{77}$$

The scalar function ϕ_k is a normalized measure of electrical angle, scaled to be between 0 and 1, and the vectors α_k and β_k now represent the value of $K(\theta)$ at the boundary points of the kth segment, i.e.

$$\phi_k := \frac{\vartheta - k\Theta}{\Theta} \quad \alpha_k := K(k\Theta) \quad \beta_k := K((k+1)\Theta) \tag{78}$$

Hence, it is clear that the back-emf of the jth phase may be rewritten (and redefined from (70)) in the linearly parameterized form

$$f_j'[n]\Theta_{f_j} \equiv \begin{bmatrix} \omega(1 - \phi_k) & \omega(\phi_k) \end{bmatrix} \begin{bmatrix} (\alpha_k)_j \\ (\beta_k)_j \end{bmatrix} \quad \forall \vartheta \in [k\Theta, (k+1)\Theta) \tag{79}$$

This re-parameterization illustrates the two-parameter method of identification which reduces the computational effort, by permitting the regressor vector and parameter vector to be determined by the segment index k. A piecewise-linear approximation of periodic components in load torque $\tau_m(\theta, \omega)$ could be obtained in a completely analogous fashion.

On-line identification of the parameter vectors is achieved by using standard algorithms, such as normalized gradient updates, first for the electrical parameters

$$\hat{\Theta}_{e_j}[n+1] = \hat{\Theta}_{e_j}[n] + \frac{\gamma_e w_{e_j}[n]}{\kappa_e + w_{e_j}'[n]w_{e_j}[n]}(y_{e_j}[n] - w_{e_j}'[n]\hat{\Theta}_{e_j}[n]) \tag{80}$$

and, subsequently, for the mechanical parameters

$$\hat{\Theta}_m[n+1] = \hat{\Theta}_m[n] + \frac{\gamma_m w_m[n]}{\kappa_m + w_m'[n]w_m[n]}(\hat{y}_m[n] - w_m'[n]\hat{\Theta}_m[n]) \tag{81}$$

where $\hat{y}_m[n]$ is defined in (75). A deadzone on the prediction errors, or some other modification to the update laws, may be used for robustness. For further details, please see [6].

4 Experimental Results

4.1 Experimental Apparatus

The experimental implementation of both the shaft-sensorless and the self-tuning systems is now described. The computational engine was a Texas Instruments TMS320C30 floating point digital signal processor. Position was measured by a 1000 line optical encoder, providing a 1.57 mrad resolution after decoding. Velocity was computed from encoder measurements using a linear least-squares curve fit to four data points. Currents were measured by Hall-effect current transducers. Power was supplied to the motor by a linear amplifier, with limits of ± 25 V and ± 12 A. All code associated with the control system was implemented in C with a sampling period of $T = 1$ ms. The torque-angle characteristic functions and cogging are shown in Fig. 1. Nominal parameter values for the motor (unloaded) are given in Table 1.

4.2 Shaft-Sensorless Control

To explore the shaft-sensorless controller's ability to operate at widely varying rotor velocities, a bidirectional, speed-reversing test was run. In this test, the motor was commanded to smoothly ramp from rest at $t = 0$ to 60 rad/s at $t = 0.375$ s. Thereafter, the motor was commanded to smoothly transition between ± 60 rad/s every 5 s, with the transitions taking 0.75 s. This requires the controller to shift in and out of an open-loop mode near zero velocity due to the state estimator singularity. This is done smoothly, such that the controller operates purely open-loop below 10 rad/s, purely closed-loop above 20 rad/s, and linearly transitions between these thresholds. The amplifier gain K_a was 17, and the velocity gain γ was 0.95.

The results of running the motor under three different schemes are shown in Figs. 2–4, in order of increasing implementation cost (and performance). The open-loop controller (constant amplitude, variable frequency ac excitation) does a poor job of tracking the desired velocity trajectory, exhibiting underdamped ripple, and requires excessive energy, even when the desired velocity is constant. This is because the simple constant excitation must be sufficiently large to achieve the highest acceleration of the trajectory. The shaft-sensorless controller yields a significant performance improvement over the open-loop controller. The only significant tracking error is during the transitions, and this is because the controller is effectively open-loop when the velocity is less than 15 rad/s in magnitude. Furthermore, the instantaneous energy required is decreased to a near-optimal level. Even further improvements in performance are witnessed using the closed-loop control. However, the performance advantages of the closed-loop controller over the shaft-sensorless controller are limited to the low velocity transitions. From an implementation viewpoint, the shaft-sensorless controller is likely to be advantageous compared to the closed-loop controller in certain applications.

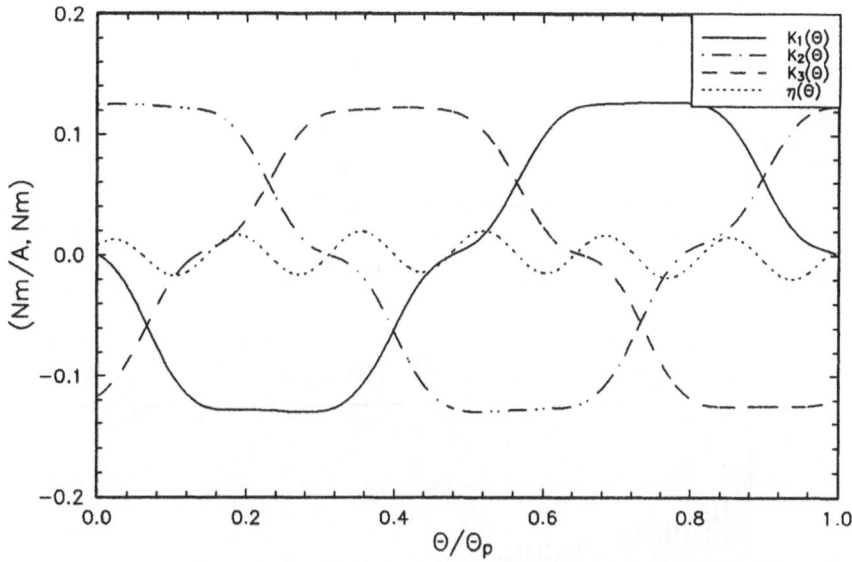

Fig. 1. Nominal functions for experimental motor.

Parameter	Value	Units
M	3	phases
N_p	4	pole pairs
L	diag$\{2.0, 2.0, 2.0\}$	mH
R	diag$\{1.36, 1.35, 1.33\}$	Ω
J	0.128	g m^2
B	0.250	g m^2/s

Table 1. Nominal parameters for experimental motor.

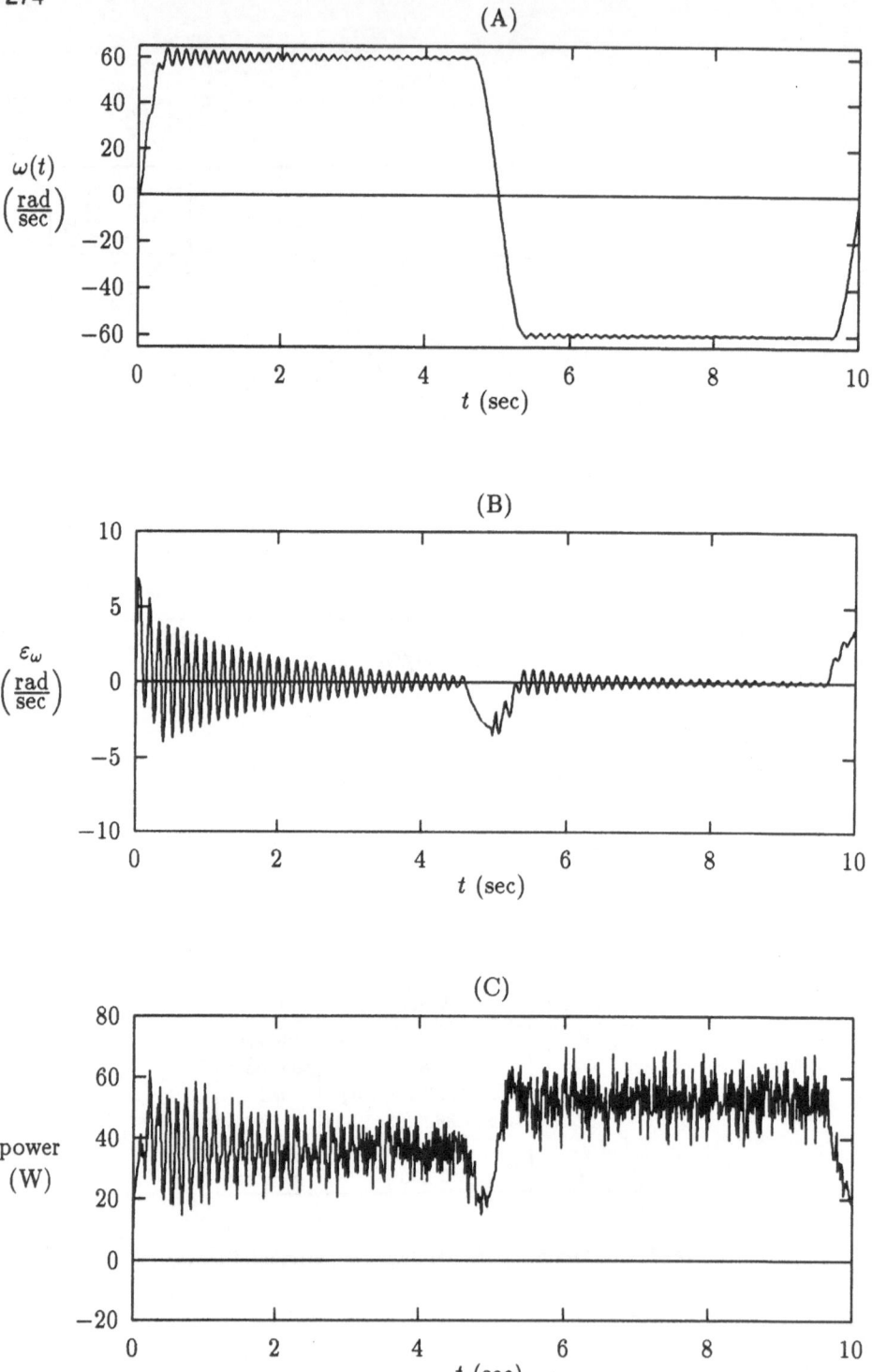

Fig. 2. Open-loop velocity control response.

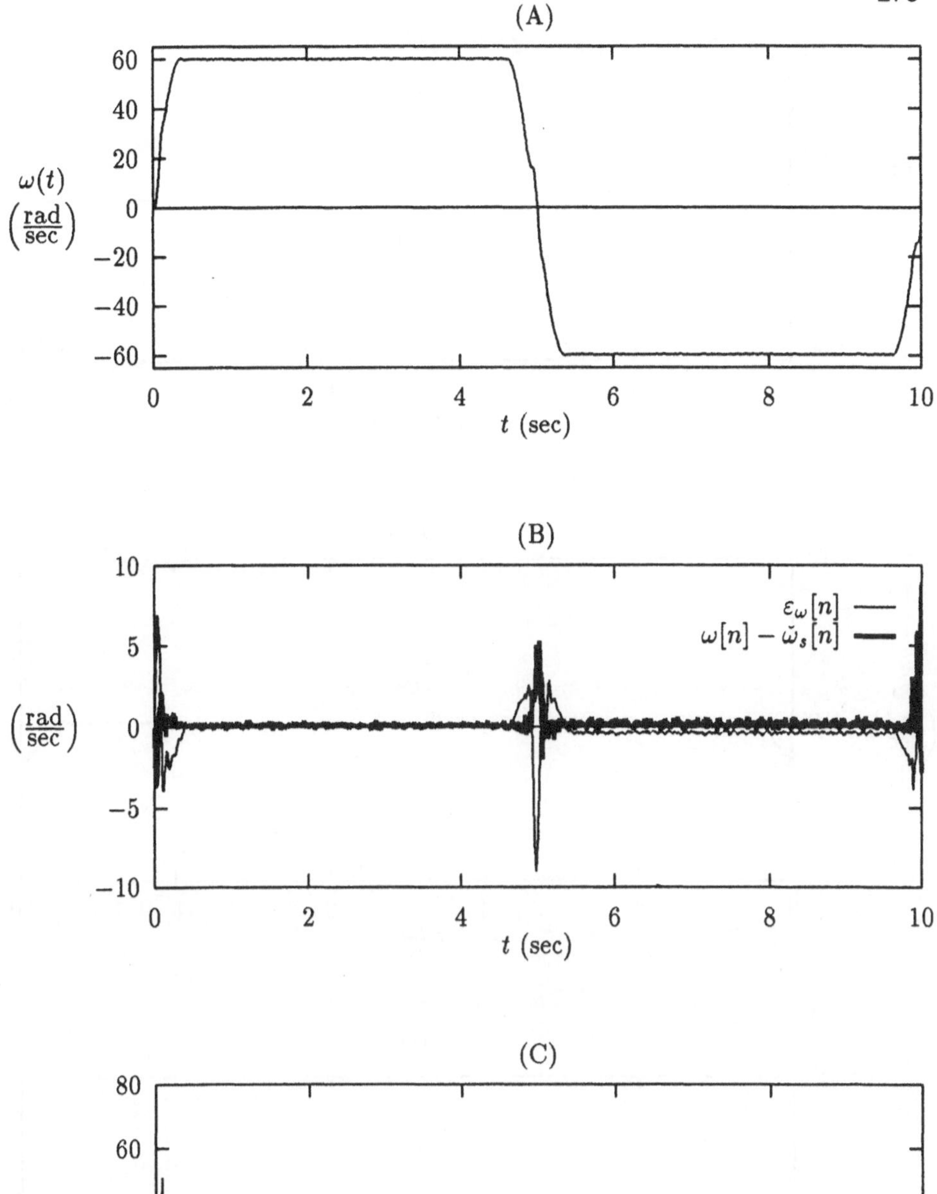

(A)

$\omega(t)$
$\left(\dfrac{\text{rad}}{\text{sec}}\right)$

t (sec)

(B)

$\left(\dfrac{\text{rad}}{\text{sec}}\right)$

$\varepsilon_\omega[n]$ ⎯⎯
$\omega[n] - \tilde{\omega}_s[n]$ ▬▬

t (sec)

(C)

power
(W)

t (sec)

Fig. 3. Shaft-sensorless velocity control response.

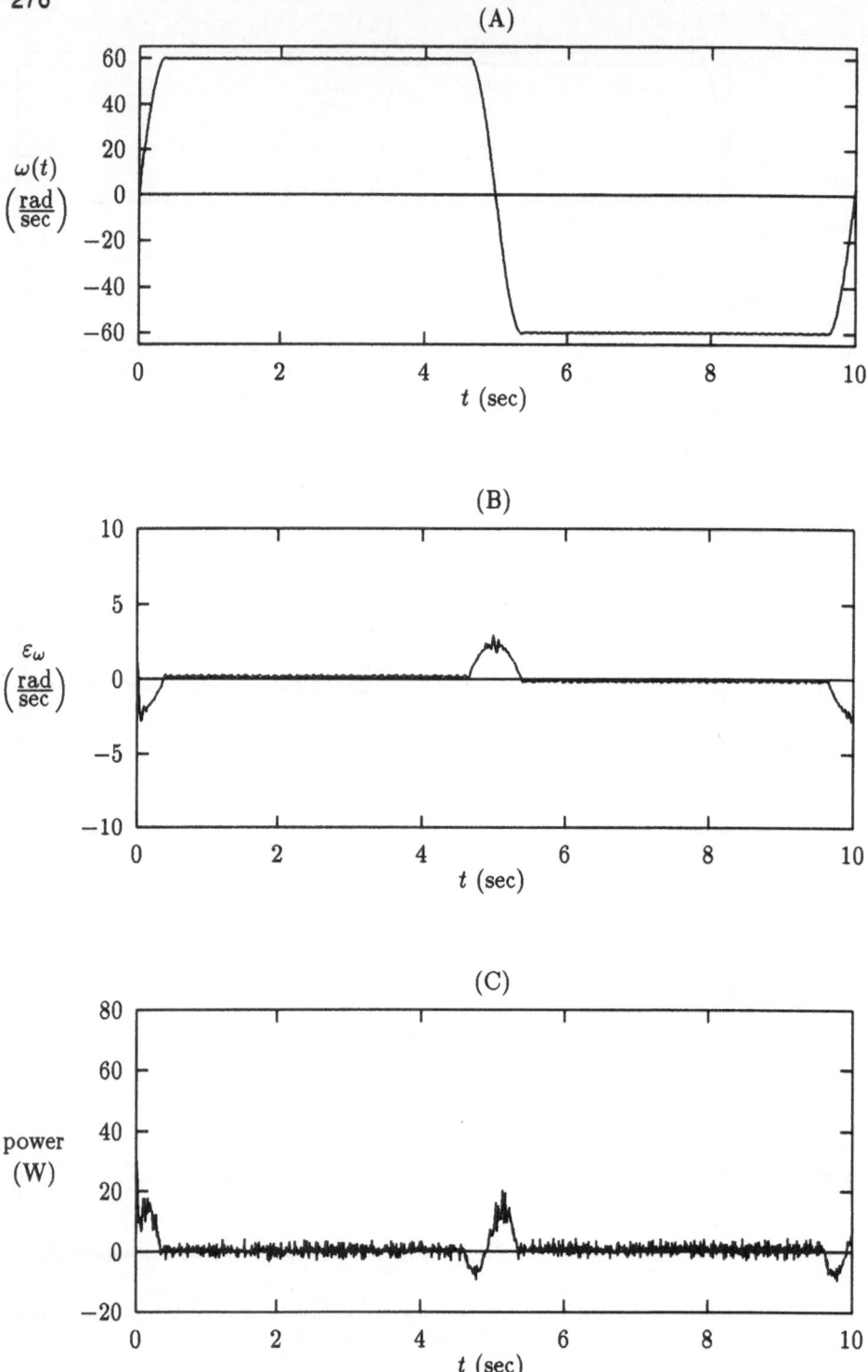

Fig. 4. Closed-loop velocity control response.

4.3 Self-Tuning Control

For this experimental work, the unknown mechanical parameters are inertia J, friction B, and the coefficients describing magnetic cogging. The unknown electrical parameters are the effective resistances R_e, and the coefficients describing the back-emf characteristics. The piecewise-linear approximations of $K(\theta)$ and cogging used 25 segments and 50 segments, respectively. To test the self-tuning scheme, the motor was commanded to track a smooth trajectory with alternating intervals of ± 40 rad/s and transition times of 0.2 s. The velocity gain γ was 0.8, and the amplifier gain K_a was 19. Adaptive tuning was enabled only after 5 s.

The results are shown in Fig. 5. The controller does a poor job when adaptation is disabled, with tracking errors of as much as 50 rad/s. When adaptation is enabled the response improves considerably, with the tracking error decreasing after only about 2 s to about 0.25 rad/s during the constant velocity intervals, and about 4 rad/s during the transients. The instantaneous power gives further evidence of the improvement resulting from the adaptive tuning. The assumed initial parameter estimates are shown in Fig. 6 and Table 2. The identified final parameter estimates obtained after 100 s of self-tuning are shown in Fig. 7 and Table 3. Also shown in Fig. 5 are the tracking error and instantaneous power when the controller is run with the fixed parameter values shown in Fig. 7 and Table 3. After approximately 4 s of tuning, the self-tuning controller performs as well as the fixed parameter controller.

5 Conclusions

In this paper, the notion of asymptotically valid reduced-order modeling for sampled-data nonlinear two-time-scale systems was explored. In particular, applications of this modeling concept in velocity control design for permanent-magnet synchronous motors were developed. While the velocity control algorithm itself was quite intuitive, needing only the state equation modeling result, two supplementary technologies were developed which featured the less intuitive output equation modeling result. In the first case, an asymptotically valid discrete-time approximation to the back-emf was obtained by declaring the stator current as the sampled output, and this signal was used to implement state estimation. The shaft-sensorless control experiments based on this state estimation illustrated the possibilities of model-based nonlinear control without the need for shaft-mounted sensors. In the second case, again by declaring the stator current to be the sampled output, an asymptotically valid discrete-time approximation for the combined resistive and back-emf losses was obtained, and this signal was shown to be useful for parameter identification. The self-tuning control experiments based on this parameter identification demonstrated the possibilities of model-based nonlinear control without the need for pre-calibration.

Interested readers are encouraged to consult [5] for additional details and applications of digital reduced-order modeling and feedback design for nonlinear

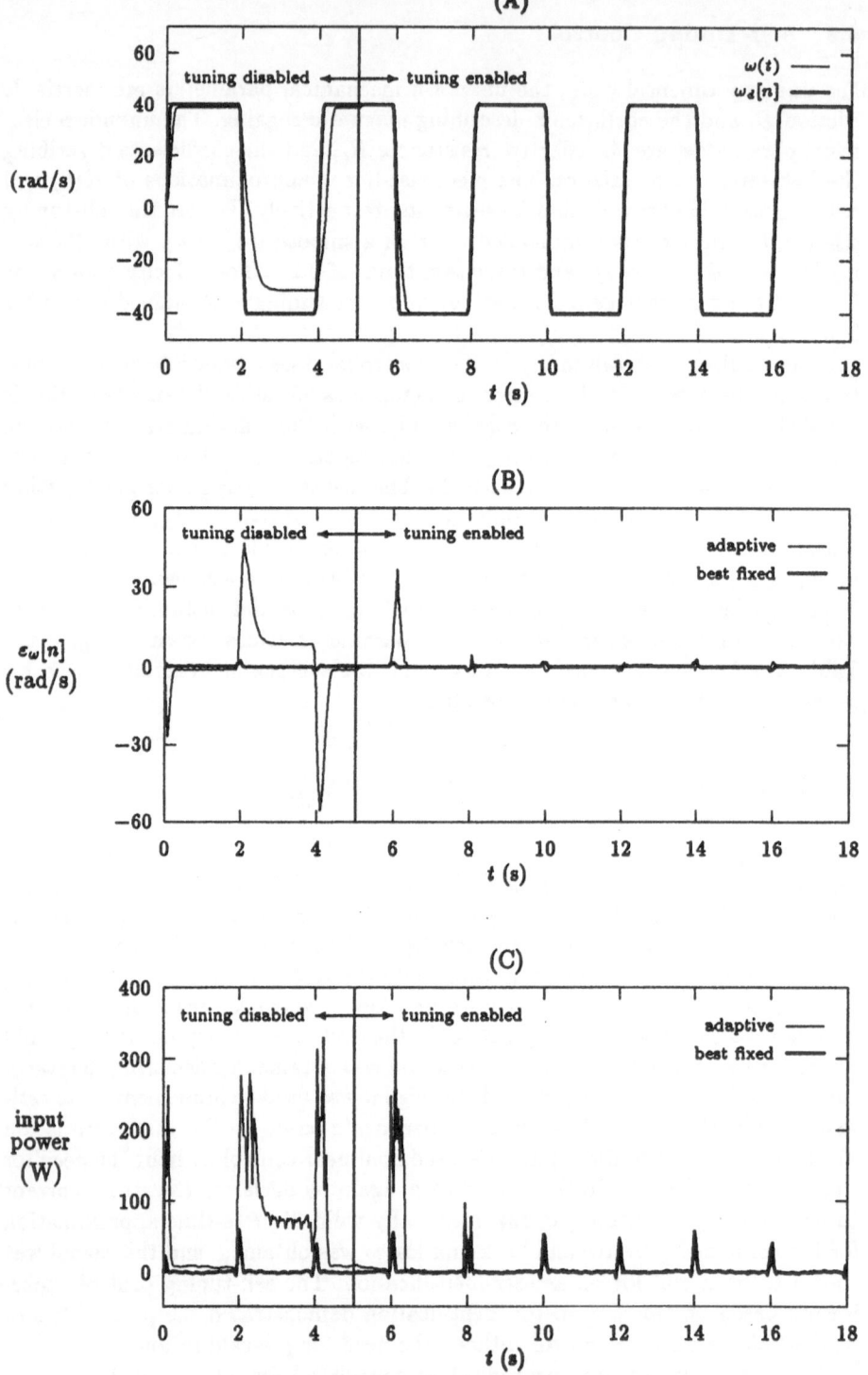

Fig. 5. Self-tuning velocity control response.

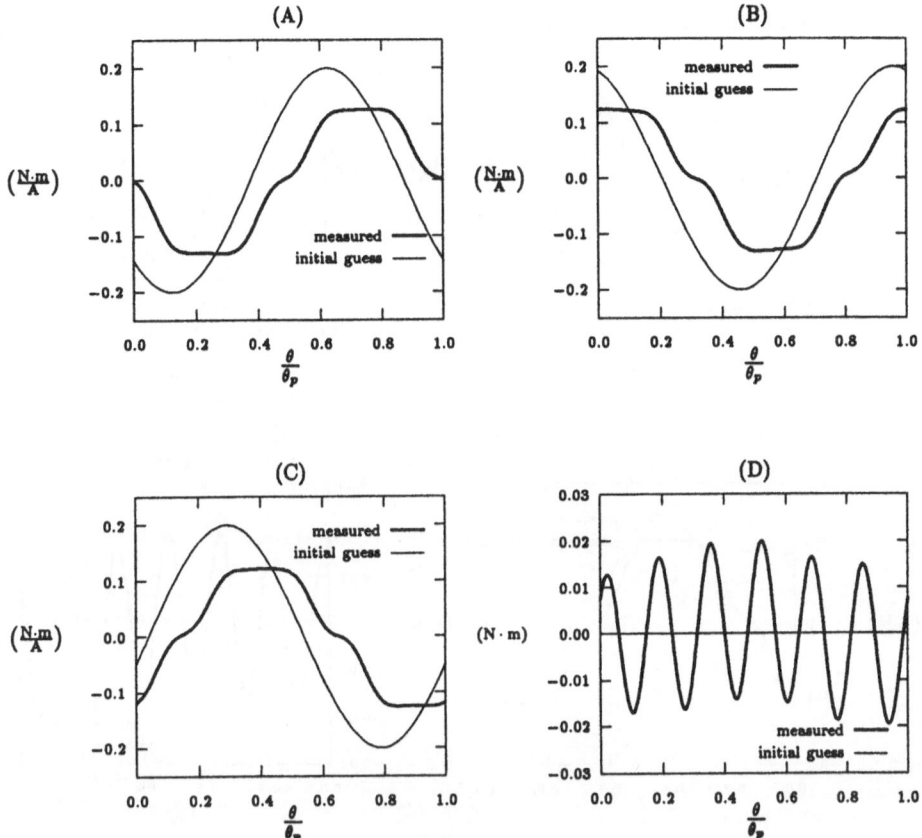

Fig. 6. Initial PL function estimates.

Parameter	Value	Units
$\hat{J}[0]$	1.0	g m^2
$\hat{B}[0]$	1.0	g m^2/s
$\hat{R}_{e_1}[0]$	15.0	Ω
$\hat{R}_{e_2}[0]$	15.0	Ω
$\hat{R}_{e_3}[0]$	15.0	Ω

Table 2. Initial parameter estimates.

280

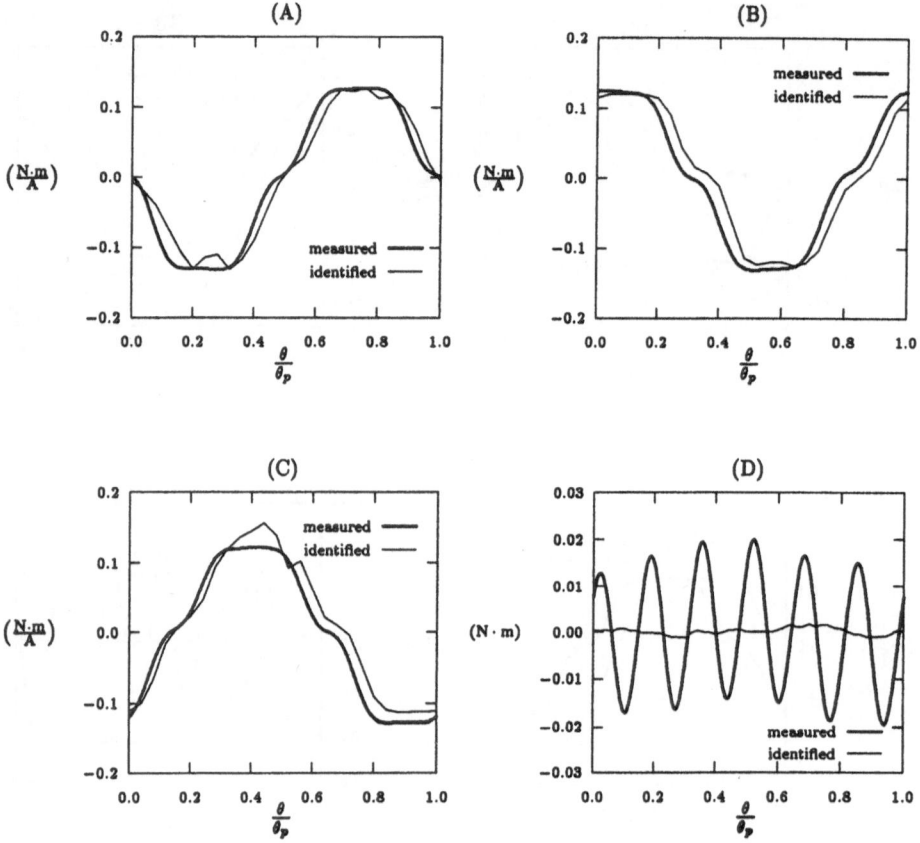

Fig. 7. Final PL function estimates.

Parameter	Value	Units
$\hat{J}[10^5]$	0.6595	g m^2
$\hat{B}[10^5]$	0.1883	g m^2/s
$\hat{R}_{e_1}[10^5]$	15.7578	Ω
$\hat{R}_{e_2}[10^5]$	15.2045	Ω
$\hat{R}_{e_3}[10^5]$	14.3248	Ω

Table 3. Final parameter estimates.

two-time-scale systems in general, and [6] and [7] for additional information regarding the applications to shaft-sensorless and self-tuning control of permanent-magnet synchronous motors in particular. Further work is underway on the extension of the methods in [6] and [7] to switched reluctance motor drives, which have additional nonlinearities due to doubly-salient construction and magnetic saturation.

References

1. D. Chen and B. Paden, "Adaptive linearization of hybrid step-motors: Stability analysis," *IEEE Transactions on Automatic Control*, vol. 38, no. 6, pp. 874–887, 1993.
2. F. Esfandiari and H. K. Khalil, "On the robustness of sampled-data control to unmodeled high-frequency dynamics," *IEEE Transactions on Automatic Control*, vol. 34, no. 8, pp. 900–903, 1989.
3. P. V. Kokotovic, H. K. Khalil and J. O'Reilly, *Singular Perturbation Methods in Control: Analysis and Design*. New York: Academic Press, 1986.
4. P. E. Moraal and J. W. Grizzle, "Observer design for nonlinear systems with discrete-time measurements," *IEEE Transactions on Automatic Control*, vol. 40, no. 3, pp. 395–404, 1995.
5. K. R. Shouse, *Reduced-Order Block Techniques for Singularly Perturbed Systems with Application to Permanent-Magnet Synchronous Motors*. Ph.D. Thesis, Georgia Institute of Technology, Atlanta, Georgia, August 1993.
6. K. R. Shouse and D. G. Taylor, "A digital self-tuning tracking controller for permanent-magnet synchronous motors," *Proceedings of the 32nd IEEE Conference on Decision and Control*, San Antonio, Texas, pp. 3397–3402, December 1993.
7. K. R. Shouse and D. G. Taylor, "Sensorless velocity control of permanent-magnet synchronous motors," *Proceedings of the 33rd IEEE Conference on Decision and Control*, Orlando, Florida, pp. 1844–1849, December 1994.

This article was processed using the LaTeX macro package with LLNCS style

A Nonlinear Model Reduction Formulation for Power System Slow Coherency and Aggregation

Joe H. Chow

Electrical, Computer, and Systems Engineering Department
Rensselaer Polytechnic Institute
Troy, NY 12180-3590, USA

Abstract. This paper presents a new nonlinear model reduction formulation for obtaining reduced models of large power systems. The new formulation is based on identifying the slow-coherent areas and the external network. For each slow-coherent area, the network and the dynamics of the reduced model are aggregated separately. The network aggregation of a coherent area is performed based on only its parameters and is independent of the other coherent areas. The dynamic model aggregation is formulated as the approximation of a nonlinear input-output model. The formulation would allow the development of new approaches to obtain aggregate generator and control device models.

1 Introduction

In the economic operation of a power system involving a large amount of power transfer, utility engineers constantly have to perform dynamic security assessment of the current and imminent operating conditions. The security assessment may involve transient stability and small-signal stability investigations. In an interconnected power pool it is essential that in these investigations, a large data set be reduced to a size consistent with the available computing resources. Thus methodologies and software for obtaining reduced-order power system models suitable for the various tasks in dynamic security assessment are crucial for reliable system operation.

The use of slow-coherency ideas [1] to identify the coherent areas of generators has been gaining popularity [2, 3, 4]. Some recent advances in using slow-coherency ideas for coherent-generator identification and for network aggregation are discussed in [5, 6, 7]. However, the aggregation of generator models and control devices such as exciters and turbine-governors has not been treated as thoroughly.

The purpose of this paper is to develop a new nonlinear model reduction formulation that encompasses all the aspects of power system slow coherency and aggregation. The new elements in the formulation are the external network and the current injections at the boundary buses, which serve as the connections to the other coherent areas. The benefit is that the new formulation provides a more tractable setting for performing nonlinear input-output model reduction.

To establish the new nonlinear model reduction formulation, this paper also extends the established slow-coherency and network aggregation algorithms to the new formulation. Some ideas on how to use the new formulation to obtain aggregate generator and control device models will be discussed. Algorithms and insights to obtain aggregate generator and control device models will be discussed in a future paper.

The algorithms found in this paper are being implemented in the EPRI (Electric Power Research Institute) program DYNRED [8]. They have also been implemented in the Power System Toolbox [9], a MATLAB-based power system simulation software.

The paper is organized as follows. The nonlinear model reduction formulation is proposed in Section 2. Section 3 establishes the link between weak connections and time scales in a large power system. Section 4 examines a tolerance-based coherent-area identification algorithm. The aggregation of slow-coherent areas is discussed in Section 5.

2 Problem Statement

In a large power system, groups of strongly connected generators tend to exhibit slow coherent inter-area swings as a result of electrically distant disturbances. A strongly connected group of generators usually belongs to the same utility which tends to have more connections within its service area and fewer connections to neighboring utilities. We call a strongly connected group of generators as a *slow-coherent* area. To be more concise, we define a slow-coherent area to consist of

1. the terminal buses of the generators in a strongly connected group,
2. the load buses that are strongly connected to these generator buses,
3. the transmission lines interconnecting the generator and load buses, and
4. the generators and their control devices.

In the sequel, we often call a slow-coherent area simply as a coherent area, with the understanding that coherency is always defined with respect to the slow inter-area modes.

To be consistent, all the buses in a coherent area are interconnected; that is, there are no disconnected buses or groups of buses within the area. The buses that have connections to those buses not in the same coherent area are called the boundary buses. For convenience, we allow single-generator coherent areas. The buses and transmission lines that are not included in any coherent areas from the *external* network. This external network consists of inter-tie lines connecting neighboring as well as far away utilities, which may be loaded with a significant amount of power transfer. It may also include HVDC systems, static var systems (SVC), and thyristor-controlled series compensations (TCSC). However, the external network will not contain any generators.

Consider a power system having n_g generators and n_b buses. Assume that there are r coherent areas and an external network in the system. In Area i there are n_{gi} generators and n_{ci} buses, of which n_{bbi} are boundary buses. Including the control devices, Area i has n_{si} state variables which are contained in the state vector $x_i \in \mathcal{R}^{n_{si}}$. These state variables include the generator angle δ, the generator speed ω, the generator fluxes, the exciter states, the power system stabilizer states, and the turbine-governor states. We denote the real and imaginary parts of the bus voltage phasors by the vector $V_{ci} \in \mathcal{R}^{2n_{ci}}$ and the real and imaginary parts of the boundary bus voltage phasors by the vector $V_{bbi} \in \mathcal{R}^{2n_{bbi}}$. Note that V_{bbi} is a sub-vector of V_{ci}.

The nonlinear dynamics of Area i are modeled as

$$\dot{x}_i = f_i(x_i, V_{ci}, I_i) \tag{1}$$
$$0 = g_i(x_i, V_{ci}, I_i) \tag{2}$$

where the current injection vector $I_i \in \mathcal{R}^{2n_{bbi}}$ consists of the real and imaginary parts of the line current flowing into Area i at the boundary buses. The nonlinear vector function $f_i \in \mathcal{R}^{n_{si}}$ represents the dynamics of the generators, exciters, stabilizers, and turbine-governors present in Area i, and the nonlinear vector function $g_i \in \mathcal{R}^{2n_{ci}}$ represents the network equations. The current injection I_i acts as the forcing function to this area. If there are no internal disturbances in Area i, then the states in the area will respond solely to the current injection.

Assembling the vectors

$$x = \begin{bmatrix} x_1 \\ x_2 \\ \vdots \\ x_r \end{bmatrix}, \ V_c = \begin{bmatrix} V_{c1} \\ V_{c2} \\ \vdots \\ V_{cr} \end{bmatrix}, \ I = \begin{bmatrix} I_1 \\ I_2 \\ \vdots \\ I_r \end{bmatrix}, \ f = \begin{bmatrix} f_1 \\ f_2 \\ \vdots \\ f_r \end{bmatrix}, \ g = \begin{bmatrix} g_1 \\ g_2 \\ \vdots \\ g_r \end{bmatrix} \tag{3}$$

where $x \in \mathcal{R}^{n_s}$, $n_s = \sum_{i=1}^{r} n_{si}$, $V_c \in \mathcal{R}^{2n_c}$, $n_c = \sum_{i=1}^{r} n_{ci}$, $I \in \mathcal{R}^{2n_{bb}}$, $n_{bb} = \sum_{i=1}^{r} n_{bbi}$, $f \in \mathcal{R}^{n_s}$, and $g \in \mathcal{R}^{2n_c}$, the models (1) and (2) for all the coherent areas can be combined to form

$$\dot{x} = f(x, V_c, I) \tag{4}$$
$$0 = g(x, V_c, I) \tag{5}$$

We denote the number of the buses in the external network by n_e and the real and imaginary parts of the bus voltage phasors by the vector $V_e \in \mathcal{R}^{2n_e}$. Assuming that the external network has no dynamic components, its model can be written as

$$0 = h(V_e, V_{bb}) \tag{6}$$
$$I = h_I(V_e, V_{bb}) \tag{7}$$

where

$$V_{bb} = \begin{bmatrix} V_{bb1} \\ V_{bb2} \\ \vdots \\ V_{bbr} \end{bmatrix} \in \mathcal{R}^{2n_{bb}}$$

The nonlinear vector function $h \in \mathcal{R}^{2n_e}$ represents the network equations and the nonlinear vector function $h_I \in \mathcal{R}^{2n_{bb}}$ relates the current injections to the bus voltages. Dynamic equations can be readily added to (6) if any dynamic components such as HVDC, SVC, and TCSC are contained in the external network. The nonlinear equations for the coherent areas (4) and (5) and the external network (6) and (7) form the model of the overall power system. Note that the algebraic equations (5), (6) and (7) are coupled and have to be solved together.

When the generators in a coherent area are subject to an electrically distant fault, their rotor angles exhibit similar swings. In other words, the intra-generator dynamics that are local to the coherent area are not stimulated significantly such that the dynamic model of the coherent area may be greatly reduced. It is desired to approximate the model (1) and (2) for Area i by a set of reduced equations

$$\bar{x}_i = \bar{f}_i(\bar{x}_i, \bar{V}_{ci}, I_i) \qquad (8)$$

$$0 = \bar{g}_i(\bar{x}_i, \bar{V}_{ci}, I_i) \qquad (9)$$

where the reduced state vector $\bar{x}_i \in \mathcal{R}^{\bar{n}_{si}}$ has a dimension $\bar{n}_{si} < n_{si}$, the voltage vector $\bar{V}_{ci} \in \mathcal{R}^{2\bar{n}_{ci}}$ may have a dimension different from that of V_{ci}, and the current injection I_i remains unchanged. For practical applications, it is essential that the vector function $\bar{f}_i \in \mathcal{R}^{\bar{n}_{si}}$ represents the dynamics of the equivalent generators, exciters, stabilizers, and turbine-governors, and the vector function $\bar{g}_i \in \mathcal{R}^{2\bar{n}_{ci}}$ models the network equations of an equivalent power network. In addition, all the boundary bus voltages are included in $\bar{V}_{ci}$ such that the aggregate coherent area can be connected to the external network.

If a particular security assessment does not concern the internal intra-generator dynamics of some coherent areas, then the models of those coherent areas can be reduced to obtain (8) and (9). These models when combined with the detailed models (1) and (2) of the remaining coherent areas and the external network (6) and (7) will form a reduced power system model that can be analyzed using existing power system analysis tools.

Based on the preceeding discussion, dynamic equivalencing can be formulated as two distinct problems of

1. finding a partition of a power system into coherent areas, and
2. obtaining the reduced coherent-area models (8) and (9).

The first problem is commonly known as the coherency grouping and the second problem as the dynamic aggregation. They will be discussed in Sections 4 and 5.

3 Slow Coherency and Weak Connections

The reduction of the coherent areas is based on the fact that the slow inter-area oscillations do not significantly disturb the fast intra-generator dynamics in the coherent areas that are electrically distant from the transient disturbance. This two-time-scale phenomenon arises because the connections between the generators within a coherent area are more dense or stiff than the connections between the coherent areas. The slow-coherency approach using singular perturbation techniques [10] has been developed to analytically establish the link between time scales and weak connections.

The two-time-scale behavior of a large power system having weak and strong connections can be captured using only the electromechanical models of the generators, without the generator flux variations and the control devices. The slow-coherency analysis can be shown for both the nonlinear and the linearized electromechanical models [1, 11, 12]. Here, we will present only the linearized analysis, because most practical techniques proposed to identify the slow-coherent areas are based on the linearized system models.

Restricting (4) to the electromechanical model and combining with (5) and (6), we obtain the nonlinear model

$$M\ddot{\delta} = f_\delta(\delta, V_c, V_e)$$
$$0 = g_\delta(\delta, V_c, V_e)$$
$$0 = h(V_c, V_e) \tag{10}$$

where $\delta \in R^{n_g}$ is the vector of the generator angles and M is the diagonal inertia matrix. The current injection I has been eliminated in the electromechanical model (10) using (7). Note that the damping is neglected in the model (10) because its impact on the coherency of the generators is negligible. We linearize (10) about the stable load-flow equilibrium $(\delta_o, V_{co}, V_{eo})$ to obtain the linear model

$$M\Delta\ddot{\delta} = \frac{\partial f(\delta, V_c, V_e)}{\partial \delta}\bigg|_o \Delta\delta + \frac{\partial f(\delta, V_c, V_e)}{\partial V_c}\bigg|_o \Delta V_c + \frac{\partial f(\delta, V_c, V_e)}{\partial V_e}\bigg|_o \Delta V_e$$
$$= K_1\Delta\delta + K_2\Delta V_c + K_3\Delta V_e$$
$$0 = \frac{\partial g(\delta, V_c, V_e)}{\partial \delta}\bigg|_o \Delta\delta + \frac{\partial g(\delta, V_c, V_e)}{\partial V_c}\bigg|_o \Delta V_c + \frac{\partial g(\delta, V_c, V_e)}{\partial V_e}\bigg|_o \Delta V_e$$
$$= K_4\Delta\delta + K_5\Delta V_c + K_6\Delta V_e$$
$$0 = \frac{\partial h(\delta, V_c, V_e)}{\partial V_c}\bigg|_o \Delta V_c + \frac{\partial h(\delta, V_c, V_e)}{\partial V_e}\bigg|_o \Delta V_e$$
$$= K_7\Delta V_c + K_8\Delta V_e \tag{11}$$

where $\Delta\delta \in \mathcal{R}^{n_g}$ is an n_g-vector of the generator angle deviations from δ_o, and $\Delta V_c \in \mathcal{R}^{2n_c}$ and $\Delta V_e \in \mathcal{R}^{2n_e}$ are vectors of the real and imaginary parts of the load bus voltage deviations from V_{co} and V_{eo}, respectively. The subscript "o" denotes that the partial derivatives are evaluated at the equilibrium $(\delta_o, V_{co}, V_{eo})$. In addition,

$$K_n = \begin{bmatrix} K_5 & K_6 \\ K_7 & K_8 \end{bmatrix} \tag{12}$$

is the network admittance matrix.

To model the strong connection between the generators within a coherent area and the weak connection between the coherent areas, we separate the admittance matrix K_n into

$$K_n = K_n^I + \epsilon K_n^E \tag{13}$$

where K_n^I is the matrix of internal connections between the buses in the same coherent areas, K_n^E is the matrix of connections in the external network, and ϵ is a parameter representing the ratio of the external connection strength to the internal connection strength. The parameter ϵ is small if the external connections are weak or sparse [13]. For large power systems, it is more likely that ϵ represents the sparseness of the connections between the coherent areas. Thus the partition of an interconnected power system into slow-coherent areas is often along utility company boundaries, since most utilities have more connections within the company service areas than those connecting to the neighboring utilities.

For the jth generator in Area i, we define $\Delta\delta_{ij}$ to be its rotor angle and m_{ij} to be its inertia. We also order the generators such that $\Delta\delta_{ij}$ from the same coherent areas appears consecutively in $\Delta\delta$. In this case the matrices $K_1,...,K_7$ are block-diagonal. To describe the slow motion, we define for each area an inertia weighted *aggregate variable*

$$y_i = \sum_{j=1}^{n_{gi}} m_{ij} \Delta\delta_{ij}/m_i, \quad i = 1, 2, ..., r \tag{14}$$

where the aggregate inertia is

$$m_i = \sum_{j=1}^{n_{gi}} m_{ij}, \quad i = 1, 2, ..., r \tag{15}$$

Denoting by y the r-vector whose ith entry is y_i, the matrix form of (14) is

$$y = C\Delta\delta = M_a^{-1} U^T M \Delta\delta \tag{16}$$

where

$$U = \text{diag}\,(u_1, u_2, \ldots, u_r) \tag{17}$$

is the grouping matrix consisting of the $n_i \times 1$ vectors

$$u_i = \begin{bmatrix} 1\,1\ldots1 \end{bmatrix}^T, \quad i = 1, 2, ..., r \tag{18}$$

and

$$M_a = \text{diag}\,(m_1, m_2, \ldots, m_r) = U^T M U \tag{19}$$

is the $r \times r$ aggregate inertia matrix.

To separate the fast dynamics, we select the first generator in each area to be the reference generator and define the motions of the other generators in the same area relative to this reference generator by the *local variables*

$$z_{i(j-1)} = \Delta\delta_{ij} - \Delta\delta_{i1}, \quad j = 2, 3, \ldots, n_{gi}, \quad i = 1, 2, ..., r \tag{20}$$

Denoting by z_i the $(n_i - 1)$-vector whose entries are $z_{i(j-1)}$ and considering z_i as the ith subvector of the $(n_g - r)$-vector z, we rewrite (20) as

$$z = G\Delta\delta = \text{diag}\,(G_1, G_2, \ldots, G_r)\,\Delta\delta \tag{21}$$

where G_i is the $(n_i - 1) \times n_i$ matrix

$$G_i = \begin{bmatrix} -1 & 1 & 0 & \cdot & 0 \\ -1 & 0 & 1 & \cdot & 0 \\ \cdot & \cdot & \cdot & \cdot & \cdot \\ -1 & 0 & 0 & \cdot & 1 \end{bmatrix} \tag{22}$$

If Area i has only a single generator, then z_i and G_i are empty.

Combining (16) and (21), the transformation of the original state $\Delta\delta$ into the aggregate and local variables is

$$\begin{bmatrix} y \\ z \end{bmatrix} = \begin{bmatrix} C \\ G \end{bmatrix} \Delta\delta \tag{23}$$

whose inverse is

$$\Delta\delta = [U\ G^+] \begin{bmatrix} y \\ z \end{bmatrix} \tag{24}$$

where

$$G^+ = G^T(GG^T)^{-1} \tag{25}$$

is also block-diagonal.

Applying the transformation (23) to the model (11), we obtain

$$M_a\ddot{y} = K_{11}y + K_{12}z + K_{13}\Delta V_c + K_{14}\Delta V_e$$
$$M_d\ddot{z} = K_{21}y + K_{22}z + K_{23}\Delta V_c + K_{24}\Delta V_e$$
$$0 = K_{31}y + K_{32}z + K_5\Delta V_c + K_6\Delta V_e$$
$$0 = K_7\Delta V_c + K_8\Delta V_e \tag{26}$$

where

$$M_d = GMG^+$$
$$K_{11} = U^T K_1 U,\ K_{12} = U^T K_1 G^+,\ K_{13} = U^T K_2,\ K_{14} = U^T K_3$$
$$K_{21} = (G^+)^T K_1 U,\ K_{22} = (G^+)^T K_1 G^+,\ K_{23} = (G^+)^T K_2,\ K_{23} = (G^+)^T K_3$$
$$K_{31} = K_4 U,\ K_{32} = K_4 G^+ \tag{27}$$

Following the derivations in [6], the elimination of the ΔV_c and ΔV_e variables in (26) will result in a singularly perturbed system with y containing the slow variables and z containing the fast variables. In addition, the fast variables in a coherent area are weakly coupled to the fast variables in the other coherent areas. The small parameter ϵ serves as the singular perturbation parameter as well as the weak-coupling parameter.

The ΔV_c and ΔV_e variables will be eliminated from (26) for determining coherency in the next section. However, it is important that they be retained in performing network aggregation to preserve the structure of a network in the reduction process.

4 IDENTIFICATION OF SLOW-COHERENT AREAS

The first step in the dynamic equivalencing of a power system is to identify its coherent areas. Although we define a coherent area to consist of the coherent generators and the load buses that are strongly connected to them, it suffices to first identify only the coherent generators and leave the identification of the buses to later. On the basis of the analysis in the previous section, the identification of the coherent generators can be achieved based on either the weak coupling property or the two-time-scale property.

For clarity we eliminate the ΔV_c and ΔV_e variables from (26) to obtain the model

$$\Delta \ddot{\delta} = M^{-1} K \, \Delta \delta = A \, \Delta \delta$$
$$K = K_1 - \begin{bmatrix} K_2 & K_3 \end{bmatrix} K_n^{-1} \begin{bmatrix} K_4 \\ 0 \end{bmatrix} \tag{28}$$

Since the generators from different coherent groups are weakly coupled, the state matrix A can be written as

$$A = A^I + \epsilon A^E \tag{29}$$

If the coherent generators are ordered consecutively in $\Delta \delta$, then A^I is block-diagonal. However, the generators are initially ordered in $\Delta \delta$ according to the sequence they appear in the data set. As a result, A^I is not necessary block-diagonal. Thus a search algorithm has to be used to locate the ϵ parameters in A and identify its weak-coupling structure. These A-matrix coherency algorithms include the clustering algorithm in [14], the RMS coherency technique in [15], and the weak-link algorithm in [16].

An alternative method to identify the slow-coherent generators is to use the two-time-scale property. The mode shapes of the slow-coherent generators with respect to the low-frequency modes must be similar. In other words, if W is the matrix of the eigenvectors corresponding to the small eigenvalues of the A matrix (29), then a slow-coherent group of generators must have similar row vectors in W. This slow-coherency condition is independent of the power system initial conditions which may vary according to the applied disturbance. Algorithms to find the slow-coherent generators using the slow-eigenvector matrix W include the slow-coherency grouping algorithm in [1, 5] and the signed-coherency grouping algorithm in [17]. For very large power systems, the slow-eigenvector matrix W can be computed from selective eigenvalue computation packages such as those discussed in [18, 19].

One of the key steps in the identification of the coherent generators is selecting the desired number of coherent areas in a power system. If an A-matrix-based algorithm is used for the identification, the number of the coherent areas will be dependent on the threshold used to define weak coupling. For the W-matrix-based algorithm in [1, 5], the number of areas is equal to the number of slow eigenvectors included in W. In this algorithm, the small eigenvalues of A are ordered and the largest gap between successive eigenvalues, that is, the largest time-scale separation, is determined. In some real systems the low-frequency modes tend to differ only slightly in frequencies and a sufficiently large gap may not be obvious. However, if only the lowest frequency inter-area modes are of interest, we can include at least 3 or 4 times as many the smallest eigenvalues to arrive at the matrix W, thus ensuring that the modes of interest will be adequately approximated after the aggregation.

4.1 Tolerance-based Slow-Coherency Algorithm

In a real power system the separation of the generators into slow-coherent areas is not always a clear-cut process. A generator may be partially coherent with two separate groups of generators. For the W-matrix-based algorithm [1] that results in a fixed number of coherent areas, this generator has to be assigned to one of these coherent groups, which can produce large errors during the model aggregation process. Thus some additional considerations have to be included to take into account the relative degree of slow coherency.

The tolerance-based coherency grouping algorithm in [6] includes a measure of the slow coherency between the generators in a coherent area. Let the columns of slow-eigenvector matrix W be normalized to unit vectors and denote its jth row by w_j. Define the slow-coherency measure

$$c_{jk} = w_j w_k^T / (|w_j||w_k|) \tag{30}$$

as the cosine of the angle between w_j and w_k. If the generators j and k are perfectly coherent with respect to the slow modes, then $w_j = w_k$ and the cosine of the angle between w_j and w_k is 1. A tolerance value γ, typically in the range of 0.9 to 0.95, can be selected such that two generators are said to be coherent if $c_{jk} > \gamma$. We define a coherency measure matrix C to be a matrix whose (j, k) entry is given by

$$C_{jk} = c_{jk} - \gamma \tag{31}$$

Based on the coherency measure, a set of rules can be formulated to incorporate the tolerance γ to determine the coherent areas:

1. The generators j and k are coherent if $C_{jk} > 0$.
2. If the generators j and k are coherent and the generators k and l are coherent, then the generators j and l are also coherent.
3. A *loose-coherent* area J_i is formed by the generators that are coherent under Rules 1 and 2. Let C_i be a submatrix of C corresponding to J_i.
4. If the column sums of C_i excluding the diagonal entries are all positive, then J_i is a *tight-coherent* area.
5. If any of the column sums of C_i excluding the diagonal entries is negative, then J_i should be decomposed into tight-coherent areas.
6. The least coherent generator in J_i corresponds to the columns of C_i without the diagonal entries having the smallest sum.
7. The coherency of J_i may be improved by removing its least coherent generator and reassigning it to a different area.

8. Given two partitions J_{i1} and J_{i2} of J_i, J_{i1} is tighter than J_{i2} if the sum of the off-diagonal entries of C_i corresponding to J_{i1} is smaller than that of J_{i2}.

The coherency rules can be used to construct the following algorithm.

Tolerance-based Coherent-Generator Identification Algorithm

1. Find the loose-coherent areas using Rules 1 - 3.
2. For each coherent area J_i,
    (a) Use Rule 4 to determine whether J_i is a tight-coherent area, which requires no further decomposition.
    (b) If the area is not tightly coherent, then decompose the area into tight-coherent areas. Start the decomposition by identifying the least coherent generator using Rule 6 and reassigning it using Rule 8. Continue until the area has been decomposed into tight-coherent areas and no further decrease in the sum of the off-diagonal entries is possible under Rule 8.

When the tolerance-based algorithm was applied to a 48-generator 140-bus system, the resulting coherent areas contained generators that were electrically close to each other [6]. In addition, the low-frequency modes of the reduced-order model using the coherent areas obtained from this tolerance-based algorithm were better approximations than those of the reduced-order model obtained from the two-time-scale algorithm without using a coherency measure.

5 AGGREGATION

Once the coherent areas have been identified, we can proceed with the aggregation of the coherent areas to obtain the nonlinear reduced models (8) and (9). Because one of the requirements is that the reduced model for each coherent area should consist of a network with an equivalent generator, the aggregation problem is usually decomposed into two parts. In the first part, the network of a coherent area is aggregated such that a new equivalent generator terminal bus is created. In the second part, the generators and their control devices in a coherent area are aggregated to obtain the equivalent generator and its control devices. This aggregate generator will be connected to the new equivalent generator bus. These two aggregation sub-problems will be discussed in the following subsections.

5.1 Network Aggregation

In a coherent area each generator is connected to its generator bus. When these generators are aggregated into a single equivalent generator, their generator buses also need to be aggregated. This process involving only the network is called the *network aggregation*. In this section, we will discuss three network aggregation techniques. The aggregation process in each technique will be illustrated with figures. To simplify the discussion, these figures will assume the coherent area to have only two generators, which will be labeled as **A** and **B** and are connected to the generator terminal buses **a** and **b**, respectively.

The *terminal-bus aggregation* technique [20] for aggregating the network in a coherent area is shown in Fig. 1. In this technique, it is assumed that the voltage phasors at the generator terminal buses **a** and **b** (Fig. 1-i) are coherent. As a result these buses are tied to a common bus **q** with infinite admittances (Fig. 1-ii). The voltage at Bus **q** can be set either to the average of the voltages at Buses **a** and **b** or to the weighted average with respect to the active and reactive power generation. To preserve the power flow, ideal transformers with complex turns ratios $\alpha_a \arg(\phi_a)$ and $\alpha_b \arg(\phi_b)$ and zero impedances link Buses **a** and **b** to Bus **q**. The phase angles ϕ_a and ϕ_b can be represented separately as phase shifters. At this point the network aggregation is completed. If the aggregate generator is to be represented by an electromechanical model, then its inertia m_{eq} and transient reactance $(x'_d)_{eq}$ are set to

$$m_{eq} = m_A + m_B, \quad (x'_d)_{eq} = (1/(x'_d)_A + 1/(x'_d)_B)^{-1} \tag{32}$$

where m_A and m_B are the inertias and $(x'_d)_A$ and $(x'_d)_B$ are the transient reactances of the generators **A** and **B**.

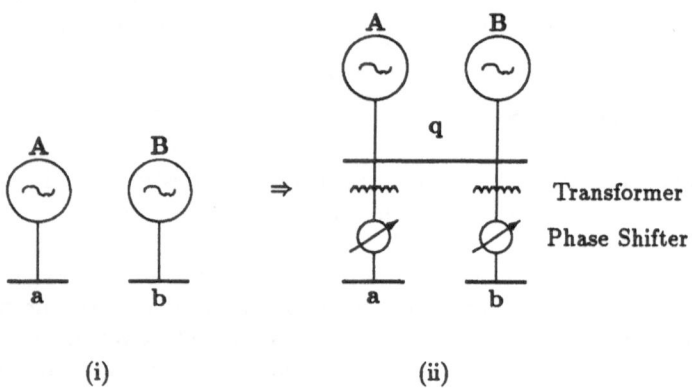

Fig. 1. Terminal-Bus Aggregation

By connecting the generator terminal buses, the terminal-bus aggregation technique stiffens the resulting aggregate network [1, 21]. This stiffening effect will manifest as the frequencies of the inter-area modes in the aggregate network being higher than those in the full-order network. The development in the last section will be to derive improved aggregation techniques.

5.1.1   Internal-node Aggregation

From the singular perturbation theory [10], the slow subsystem of (26) can be obtained as an asymptotic series expansion in the small parameter ϵ. We will consider only the zero- and first-order terms in the series expansion. Because the slow variable y is coupled to the fast variable z through ϵ, z can be considered to be a constant equal to zero as a zero-order approximation. Consequently (26) reduces to

$$M_a \ddot{y}_{in} = K_{11} y_{in} + K_{13} \Delta V_c + K_{14} \Delta V_e$$
$$0 = K_{31} y_{in} + K_5 \Delta V_c + K_6 \Delta V_e \tag{33}$$

where y_{in} approximates y. We can also obtain this model by linking the internal nodes of the coherent generators by infinite admittances [1, 21]. We call this method the *internal-node aggregation* technique. The internal-node aggregate model (33) is different from that obtained using the terminal-bus aggregation technique in [20], which aggregates the coherent generators at the terminal buses.

The model (33) is a linearized model. Furthermore, y is the rotor angle of the aggregate generator. However, there is no need to generate the linearized model. We can directly manipulate the network as illustrated in Fig. 2. In the internal-node aggregation technique, the generator internal-node voltages E'_A and E'_B are computed (Fig. 2-i). From (33), the generator internal nodes are tied to a common bus p with infinite admittances (Fig. 2-ii). The voltage at Bus p, E'_{eq}, can be set either to the average of E'_A and E'_B or to the weighted average with respect to the active and reactive power generation. To preserve the power flow, ideal transformers with complex turns ratios $\alpha_a \arg(\phi_a)$ and $\alpha_b \arg(\phi_b)$ and zero impedances are used to link Buses a and b to Bus p.

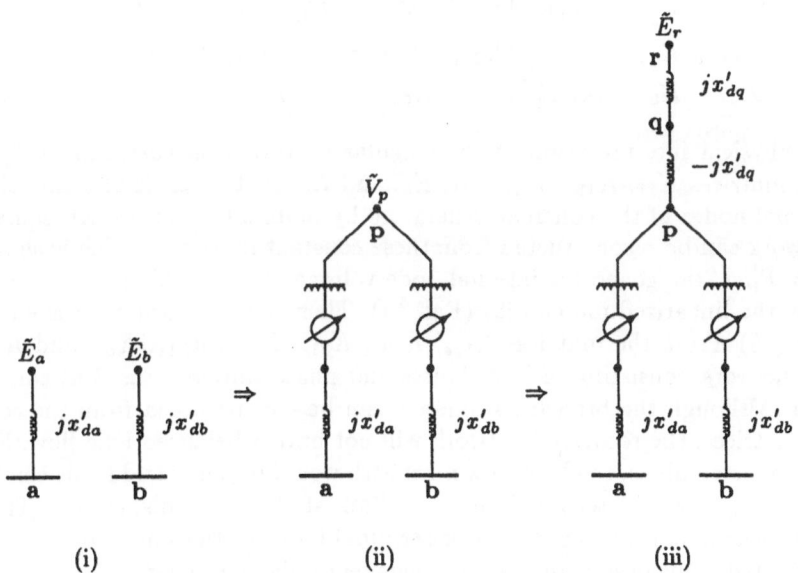

Fig. 2. Internal-Node Aggregation

The linking of the internal nodes creates an equivalent generator with multiple terminal buses, which is not accepted by conventional power system simulation programs. Taking $(x'_d)_{eq}$ from (32) as an equivalent transient reactance, the network is extended beyond Bus **p** by two buses with reactances of $-(x'_d)_{eq}$ and $(x'_d)_{eq}$ (Fig. 2-iii). The node **r** serves as the internal node of the equivalent generator and Bus **q** is the generator terminal bus. Bus **p** can be eliminated if desired. At this point the network aggregation is completed. If the aggregate generator is to be represented by an electromechanical model, then the inertia m_{eq} and the transient reactance $(x'_d)_{eq}$ of the equivalent generator can be computed from (32).

5.1.2 Impedance Compensated Aggregation

To improve on the internal-node aggregate network, we consider z to vary slowly with y in (26). As a first-order approximation, the *quasi-steady state* of z in (26) is obtained as

$$z_{\mathrm{qss}} = -K_{22}^{-1}(K_{21}y + K_{23}\Delta V_c + K_{24}\Delta V_e) \tag{34}$$

Eliminating z_{qss} from (26) results in the *impedance compensated aggregate* model

$$M_a \ddot{y}_{\mathrm{ic}} = K_{11s}y_{\mathrm{ic}} + K_{13s}\Delta V_c + K_{14s}\Delta V_e$$
$$0 = K_{31s}y_{\mathrm{ic}} + K_{5s}\Delta V_c + K_{6s}\Delta V_e \tag{35}$$

where y_{ic} approximates y and

$$K_{11s} = K_{11} - K_{12}K_{22}^{-1}K_{21}, \quad K_{13s} = K_{13} - K_{12}K_{22}^{-1}K_{23}$$
$$K_{14s} = K_{14} - K_{12}K_{22}^{-1}K_{24}, \quad K_{31s} = K_{31} - K_{32}K_{22}^{-1}K_{21}$$
$$K_{5s} = K_5 - K_{32}K_{22}^{-1}K_{23}, \quad K_{6s} = K_6 - K_{32}K_{22}^{-1}K_{24} \tag{36}$$

The physical interpretation of the singular perturbation correction (36) is that the matrices K_{11}, K_{13}, K_{14}, K_{31}, K_5, and K_6 are the result of connecting the internal nodes of the coherent generators by finite admittances. An equivalent network can be reconstructed from these constant matrices, which is shown in Fig. 3. First, the generator internal-node voltages E'_A and E'_B are computed to obtain the linearized model (35) (Fig. 3-i). Then the fast variable z is eliminated in (35). From the matrices K_{11s}, K_{13s}, K_{31s}, K_{14s}, K_{31s}, K_{5s}, and K_{6s}, a power network consisting of impedances and phase shifters (Fig. 3-ii) can be obtained. Although the branch parameters can be reconstructed from the connection matrices, the recovered network will not have a balanced load flow. For tightly connected areas, the load-flow mismatch would be small and loads can be added to the generation terminal buses to eliminate the load mismatches. Thus in the reconstruction, all the generator terminal buses in the same area will be interconnected. This interconnection is dependent only on the parameters within a coherent area.

In the slow coherency aggregation, the equivalent generator internal node is also linked to multiple terminal buses. Thus the approach used in the internal-node aggregation to build a single generator terminal bus can also be applied here, as shown in Fig. 3-iii.

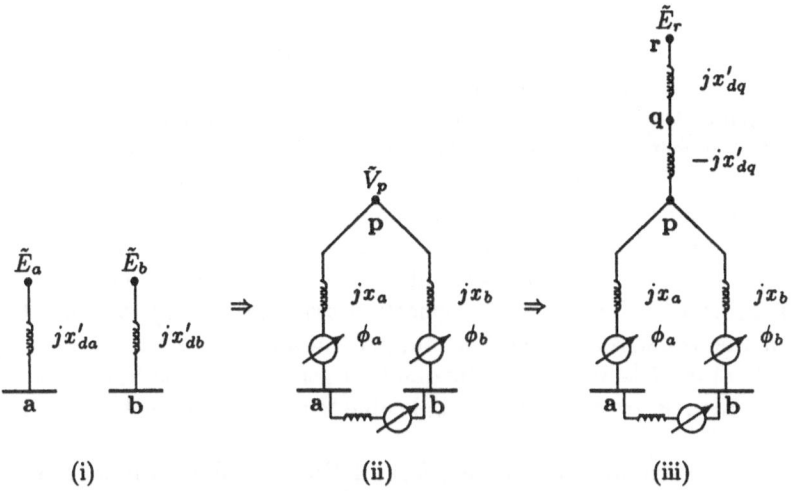

Fig. 3. Impedance Compensated Aggregation

The internal-node aggregation technique was applied to a 48-generator 140-bus system with 17 coherent areas and to a 2,000-bus system with 60 coherent areas. The inter-area modes of the internal-node aggregate models approximated those of the full models to within 5%. For a comparison, when the terminal-bus aggregation technique was used, the resulting reduced model had an error of about 10%.

The impedance compensated aggregation technique was applied to the same two systems. The frequencies of the inter-area modes of the aggregate models were also computed and found to approximate those of the full models with an error of only about 2.5%.

All three of the network aggregation techniques can be performed on a per-coherent-area basis; that is, the aggregation of a coherent area does not depend on the parameters of the other coherent areas [22]. This property allows the aggregation of the coherent areas to be performed in parallel. In addition, the aggregate network of a coherent area remains unchanged if any other coherent areas, for some specific analysis needs, are retained in detail. However, if the second- or higher-order terms are added to construct higher-order aggregates from the singularly perturbed model (26), the per-area aggregation property is lost since the correction terms will depend on the parameters in other coherent areas.

5.2 Generator and Control Device Model Aggregation

After the aggregate network of the reduced model (9) has been determined, the next step is to determine the dynamic equation (8) for the aggregate generator and its control devices. The main objective is to obtain the dynamic equation (8) that is consistent with the generator and control device models that are available in power system simulation programs.

Several schemes have been proposed to obtain the equation (8). In [8] all the generators of the full model of Area i are moved to the aggregate generator bus. Then the aggregate generator model is obtained from a weighted average of the generator inertias, currents, and impedances. The parameters of the control devices of the aggregate generator are obtained by approximating a weighted frequency response of the full system model with all the generators at the aggregate generator bus. Another scheme is to use singular perturbation techniques to obtain the aggregate generator models [24]. The decoupling of the variables is achieved using the weak connection assumption (13). An analytical aggregate model is derived for the coherent generators and their control devices having identical models and parameters and for zero system loading. However, the aggregate models are shown to be applicable to a system not satisfying these assumptions.

The new nonlinear model reduction formulation will allow a more rigorous approach to obtain the aggregate generator model and the control devices. Note that the full model (1) and (2) and the aggregate model (8) and (9) for Area i are both nonlinear input-output models. These models have the same inputs namely the current injections at the boundary buses, and the same outputs namely the voltages at the boundary buses. For disturbances that occur external to Area i and do not involving re-configuring the area, the current injections are the only forcing function of the area. Thus Area i can be disconnected from the rest of the system and a model reduction can be performed on Area i independent of the other areas.

For the generator and control device aggregation, we treat the reduction process as finding the dynamic equation (8) given the aggregate network (9) such that (8) and (9) are good approximations to the full model (1) and (2) as measured by the frequency response of the linearized input-output model or the time response of the output voltages subject to a particular set of input current injections. The input current injections may be the time response resulting from a set of specific disturbances. In this approach it is not necessary to first relocate all the generators in the coherent area to the aggregate generator bus as is required in [8]. Nor is it necessary to assume that the generators have identical parameters and the system loading is zero.

One advantage of the input-output model formulation is that it would also allow for the determination of the necessary detail of the aggregate generator model and its control devices. For example, if the sub-transient effects are not observed in the output voltages of an area, then a generator model including only the transient effect may be sufficient to represent the aggregate effect of the generators in the full-order model.

6   CONCLUSIONS

A new nonlinear model reduction formulation was proposed for the slow coherency and aggregation of large power system models. This formulation will allow the development of more rigorous algorithms to obtain the aggregate generator models and their control devices. Further investigations on the aggregate generator modeling will be reported later. Together with new coherency identification and network aggregation algorithms, accurate reduced-order models can be obtained for large power systems and used for more efficient dynamic security assessment.

ACKNOWLEDGMENT

Many colleagues have worked with me to develop the slow-coherency and aggregation approach. Among them I would like to acknowledge Pierre Acari, Said Ahmed-Zaid, John Allemong, Boza Avramovic, Ranjit Date, Brian Hurysz, Charles King, Ricardo Galarza, Hisham Othman, M. A. Pai, George Peponides, William Price, Peter Sauer, and Jim Winkelman. Many of us are indebted to Petar Kokotovic who had the foresight that singular perturbations would be useful for analyzing large power systems and guided us in this development.

This work is supported by NSF under grants ECS-9215076, by Electric Power Research Institute under project RP2447-02, Peter Hirsch, Project Manager, and by the Empire State Energy Research Corporation (ESEERCO) under project EP92-53, Charles King (New York Power Pool) and Ralph Wager, Project Managers. The ESEERCO members are the electric utilities of the State of New York: Central Hudson Gas & Electric Corporation, Consolidated Edison Company of New York, Inc., Long Island Lighting Company, New York Power Authority, New York State Electric & Gas Corporation, Niagara Mohawk Power Corporation, Orange and Rockland Utilities, Inc., and Rochester Gas and Electric Corporation.

References

1. J. H. Chow, editor, *Time-scale Modeling of Dynamic Networks with Applications to Power Systems,* Springer-Verlag, New York, 1982.
2. R. J. Piwko, H. Othman, O. A. Alvarez, and C. Y. Wu, "Eigenvalue and Frequency-domain Analysis of the Intermountain Power Project and WSCC Network," *IEEE Trans. Power Systems,* vol. PWRS-6, pp. 238-244, 1991.
3. Y. Mansour, "Application of Eigenanalysis to the Western North American Power System," in *Eigenanalysis and Frequency Domain Methods for System Dynamic Performance,* IEEE Publications 90TH0292-3-PWR, pp. 97-104, 1990.
4. L. Rouco and I. J. Pérez-Arriaga, "Multi-area Analysis of Small Signal Stability in Large Electric Power Systems by SMA", *IEEE Trans. Power Systems,* vol. 8, pp. 1257-1265, 1993.
5. J. H. Chow, R. A. Date, H. Othman, and W. W. Price, "Slow Coherency Aggregation of Large Power Systems," in *Eigenanalysis and Frequency Domain Methods for System Dynamic Performance,* IEEE Publications 90TH0292-3-PWR, pp. 50-60, 1990.

6. J. H. Chow, "New Algorithms for Slow Coherency Aggregation of Large Power Systems," *Proceedings of the Power System Workshop at IMA*, Springer-Verlag, 1994.

7. J. H. Chow, R. Galarza, P.Accari, and W. W. Price, "Inertial and Slow Coherency Aggregation Algorithms for Power System Model Reduction," Paper 94 SM 605-6 PWRS, presented at the IEEE Summer Power Meeting, San Francisco, 1994.

8. P. Kundur, G. J. Rogers, D. Y. Wong, J. Ottevangers, and L. Wang, *Dynamic Reduction*, EPRI Report TR-102234, April 1993.

9. J. H. Chow and K. W. Cheung, "A Toolbox for Power System Dynamics and Control Engineering Education and Research," *IEEE Trans. Power Systems*, vol. 7, pp. 1559-1564, 1992.

10. P. V. Kokotovic, H. K. Khalil and J. O'Reilly, *Singular Perturbation Methods in Control: Analysis and Design*, Academic Press, London, 1986.

11. G. Peponides, P. V. Kokotovic, and J. H. Chow, "Singular Perturbations and Time Scales in Nonlinear Models of Power Systems," *IEEE Trans. Circuits and Systems*, vol. CAS-29, pp. 758-767, 1982.

12. K. W. Cheung and J. H. Chow, "Stability Analysis of Singularly Perturbed Systems using Slow and Fast Manifolds," *Proceedings of the 1991 American Control Conference*, pp. 1685-1690.

13. J. H. Chow and P. V. Kokotovic, "Time-Scale Modeling of Sparse Dynamic Networks," *IEEE Trans. Automatic Control*, vol. AC-30, pp. 714-722, 1985.

14. J. Zaborszky, K.-W. Whang, G. M. Huang, L.-J. Chiang, and S.-Y. Lin, "A Clustered Dynamic Model for a Class of Linear Autonomous Systems using Simple Enumerative Sorting," *IEEE Trans. Circuits and Systems*, vol. CAS-29, pp. 747-758, 1982.

15. J. Lawler, R. A. Schlueter, P. Rusche, and D. L. Hackett, "Modal-coherent Equivalents Derived from an RMS Coherency Measure," *IEEE Trans. Power Apparatus and Systems*, vol. PAS-99, pp. 1415-1425, 1980.

16. R. Nath, S. S. Lamba, and K. S. P. Rao, "Coherency Based System Decomposition into Study and External Areas using Weak Coupling," *IEEE Trans. Power Apparatus and Systems*, vol. PAS-104, pp. 1443-1449, 1985.

17. B. Eliasson, *Damping of Power Oscillations in Large Power Systems*, Ph.D. Thesis, Lund Institute, Sweden, 1990.

18. P. Kundur, G. J. Rogers, D. Y. Wong, L. Wang, and M. G. Lauby, "A Comprehensive Computer Program Package for Small Signal Stability Analysis of Power Systems," *IEEE Trans. Power Systems*, vol. PWRS-5, pp. 1076-1083, 1990.

19. N. Martins, "Efficient Eigenvalue and Frequency Response Methods Applied to Power System Small-Signal Stability Studies," *IEEE Trans. Power Systems*, vol. 1, pp. 217-226, February 1986.

20. R. Podmore, "Development of Dynamic Equivalents for Transient Stability Studies," EPRI Report EL-456, 1977.

21. B. R. Copeland, *Reduced Order Dynamic Equivalents for Electric Power Systems: an Analysis of the Modal-coherent and Slow-coherent Equivalency Techniques*, M. S. Thesis, University of Tennessee, USA, 1986.

22. R. A. Date and J. H. Chow, "Aggregation Properties of Linearized Two-time-scale Power Networks," *IEEE Trans. Circuits and Systems*, vol. 38, pp. 720-730, 1991.

23. A. J. Germond and R. Podmore, "Dynamic Aggregation of Generating Unit Models," *IEEE Trans. Power Apparatus and Systems*, vol. PAS-97, pp. 1060-1069, 1978.

24. S. Ahmed-Zaid, P. W. Sauer, and J. R. Winkelman, "Higher Order Dynamic Equivalents for Power Systems," *Automatica*, vol. 22, pp. 489-494, 1986.

This article was processed using the LaTeX macro package with LLNCS style

Aggregation of Synchronous Generators and Induction Motors in Multimachine Power Systems

Said Ahmed-Zaid[1], Seog Sue Jang[1], Omer Awed-Badeeb[2] and Maamar Taleb[3]

[1] Clarkson University, Electrical and Computer Engineering Dept.
Potsdam, NY 13699, USA
[2] Sana'a University, Electrical and Computer Engineering Dept.
Sana'a, Yemen
[3] University of Bahrain, Electrical and Computer Engineering Dept.
Isa Town, Bahrain, P.O. Box 32038

Abstract. The aggregation theory of synchronous generators with small and large induction motors is investigated using a hybrid multimachine representation. A no-load linearization of a two-machine system composed of a synchronous generator and an induction motor illustrates the dynamical interaction of electromechanical and voltage variables in both machines. These aggregation concepts are extended to a realistic multimachine power system with three synchronous generators and three aggregate induction motor loads. Large percentages of induction machine loading are shown to be detrimental to the power system stability. The behavior of critical modes with increased induction motor loading is explained using a two-machine equivalent consisting of an aggregate synchronous generator connected to an aggregate induction motor load. These results are also verified by simulating the load bus voltages following a small change in the reference voltage setpoint of a synchronous generator.

1 Introduction

The present trend in power system simulation is toward increased modeling detail and longer simulation periods. Even though synchronous generators and their controls are still the primary components of interest, it has become increasingly important to include the parameters of large dynamic loads in certain studies where accurate simulation results are needed. Examples of such studies include induction motor protection setting, load shedding applications, system collapse and islanding, voltage stability, and long-term dynamics under abnormal conditions. In general, the dynamic performance of loads is particular important in studies which may involve frequency and voltage variations over a long period of time.

The accurate modeling of loads has remained a weak point in stability studies due to several factors including the large number of diverse load components; the changing load composition with time, weather, and temperature; the uncertain load mix and composition; and the uncertain load characteristics for studies involving large voltage and frequency variations [1-2]. Electric motor loads make up a large percentage of the total electric load in the United States. This rotating class of load is largely composed of single-phase and three-phase induction machines with a wide variety of ratings, electrical time constants, inertias, and shaft load characteristics. In particular, many industrial loads are composed of large synchronous and induction motors although there are notable exceptions such as electric furnaces, aluminum reduction plants and other electrolytic processes. The representation of these large motors in stability studies is particularly desirable if they make up an appreciable percentage of the load and if the inclusion of their parameters is critical in order to assess correct stability and power transfer limitations.

The structural or inner dynamics of three-phase induction motors depend largely on inertia and open-circuit time constants [3-7]. Smaller induction machines operate at relatively large slips and can adequately be represented by the well-known first-order slip model [5]. Large induction motors and generators are designed more efficiently with higher X/R ratios. They have higher open-circuit time constants and operate at relatively lower slips. Rotor flux transients play a dominant effect in these machines and can completely change the nature of the dynamic results obtained with the slip model [3,4,6,7]. The authors in [6] have shown that the dominant behavior in large machines is, in fact, characterized by a first-order voltage model around rated conditions, cautioning against the indiscriminate use of the slip model in simplified voltage studies. The first-order voltage model may also be inadequate near voltage collapse, in which case a third-order model for large induction machines is recommended [7].

In this paper, the authors investigate the different dynamic interaction of synchronous generators with large and small induction machines in a multimachine context. After developing a hybrid model, the structure of a simple two-machine system consisting of a synchronous generator connected to either a small or a large induction motor is examined at no load. An aggregation theory of electromechanical and voltage variables is used to explain the interaction of state variables in this model. These aggregation concepts are then extended to a realistic six-machine system composed of three synchronous generators and three induction motor loads. The interaction of synchronous generators with either small or large induction machines is investigated using aggregation theory, and eigenvalue and sensitivity analyses.

2 Multimachine Power System Model

In this section, the mathematical model of a hybrid multimachine power system composed of m synchronous generators and (n-m) induction motors is presented. Synchronous generators are represented by fourth-order two-axis models [8-9] and induction motors are represented by third-order models incorporating the effects of rotor electrical transients and shaft speed [6]. Such simple models are first used to illustrate the aggregation of synchronous and induction machine variables in unregulated power systems with no voltage regulators.

2.1 Synchronous Generator Equations

According to the nomenclature, a standard two-axis model of a synchronous generator neglecting transient saliency and written in its individual dq rotor reference frame is, for $i = 1, \ldots, m$,

$$T'_{doi} \frac{dE'_{qi}}{dt} = -E'_{qi} - (X_{di} - X'_{di})I_{di} + E_{fdi} \tag{1}$$

$$T'_{qoi} \frac{dE'_{di}}{dt} = -E'_{di} + (X_{qi} - X'_{qi})I_{qi} \tag{2}$$

$$\frac{d\delta_i}{dt} = \omega_s(\nu_i - 1) \tag{3}$$

$$M_i \frac{d\nu_i}{dt} = T_{mi} - (E'_{di}I_{di} + E'_{qi}I_{qi}) - D_i(\nu_i - 1) \tag{4}$$

2.2 Induction Motor Equations

Following the derivation in the appendix, a third-order induction machine model written in generator notation in a reference frame rotating with the resultant rotor flux is, for $i = m + 1, \ldots, n$,

$$T'_{doi} \frac{dE'_{qi}}{dt} = -E'_{qi} - (X_{di} - X'_{di})I_{di} \tag{5}$$

$$\frac{d\delta_i}{dt} = \omega_s(\nu_i - 1) - \frac{(X_{di} - X'_{di})}{T'_{doi} E'_{qi}} I_{qi} \tag{6}$$

$$M_i \frac{d\nu_i}{dt} = T_{mi} - E'_{qi}I_{qi} \tag{7}$$

The synchronous and induction machine equations can be recast into the following hybrid representation

$$T'_{doi} \frac{dE'_{qi}}{dt} = -E'_{qi} - (X_{di} - X'_{di})I_{di} + (1 - \alpha_i)E_{fdi} \tag{8}$$

$$T'_{qoi} \frac{dE'_{di}}{dt} = -E'_{di} + (X_{qi} - X'_{qi})I_{qi} \tag{9}$$

$$\frac{d\delta_i}{dt} = \omega_s(\nu_i - 1) - \alpha_i \frac{(X_{di} - X'_{di})}{T'_{doi} E'_{qi}} I_{qi} \tag{10}$$

$$M_i \frac{d\nu_i}{dt} = T_{mi} - (E'_{di}I_{di} + E'_{qi}I_{qi}) - (1 - \alpha_i)D_i(\nu_i - 1) \tag{11}$$

for $i = 1, \ldots, n$ and where $\alpha_i = 0$ for a synchronous machine and $\alpha_i = 1$ for an induction machine. Every induction machine is actually of the third order. For the sake of a unified representation, fictitious q-axis induction machine parameters are defined such that $X'_{di} = X_{qi} = X'_{qi}$ and E'_{di} is identically equal to zero at all times. This renders the second equation inactive for induction machines if the initial condition $E'_{di}(0)$ is chosen equal to zero during a simulation. If an eigenvalue analysis is performed, a superfluous eigenvalue is obtained which is equal to the negative of the reciprocal of the q-axis open-circuit time constant $T'_{qoi} = T'_{doi}$.

The network interface equations can be written in the following form, for $i = 1, \ldots, n$:

$$(I_{di} + jI_{qi}) = \sum_{j=1}^{n} (G_{ij} + jB_{ij})(E'_{dj} + jE'_{qj})e^{j\delta_{ji}} \tag{12}$$

where $\delta_{ji} = (\delta_j - \delta_i)$. The d- and q-axes components of these currents are, for $i = 1, \ldots, n$:

$$I_{di} = \sum_{j=1}^{n} [(G_{ij} \cos \delta_{ji} - B_{ij} \sin \delta_{ji})E'_{dj} - (G_{ij} \sin \delta_{ji} + B_{ij} \cos \delta_{ji})E'_{qj}] \tag{13}$$

$$I_{qi} = \sum_{j=1}^{n} [(G_{ij} \sin \delta_{ji} + B_{ij} \cos \delta_{ji})E'_{dj} + (G_{ij} \cos \delta_{ji} - B_{ij} \sin \delta_{ji})E'_{qj}] \tag{14}$$

3 Linearized Multimachine Model at No Load

A no-load eigenvalue analysis is now performed around a particular equilibrium point such that

$$T_{mi} = 0 \text{ pu}, \quad E_{fdi} = 1 \text{ pu}, \quad V_{ti} = 1 \text{ pu}$$

for $i = 1, \ldots, n$. All stator resistances are neglected as well as line charging for transmission lines and magnetizing inductances for transformers. All impedance loads are open circuited and synchronous machine damping coefficients are neglected. A shunt reactance with value $X_{ci} = -X_{di}$ provides local reactive power support for all induction machines ($i = m + 1, \ldots, n$). Essentially, this idealized situation assumes a flat voltage profile of one per-unit throughout the transmission network and all machines are synchronized even though no currents flow in the transmission lines. These operating conditions imply that

$$E'^o_{qi} = 1 \text{ pu}, \quad E'^o_{di} = 0 \text{ pu}, \quad \delta^o_i = 0 \text{ rad}$$
$$v^o_i = 1 \text{ pu}, \quad I^o_{di} = 0 \text{ pu}, \quad I^o_{qi} = 0 \text{ pu}$$

for all synchronous machines ($i = 1, \ldots, m$) and

$$E'^o_{qi} = \frac{X_{di} - X'_{di}}{X_{di}} \text{ pu}, \quad I^o_{di} = -\frac{1}{X_{di}} \text{ pu}$$
$$\delta^o_i = 0 \text{ rad}, \quad v^o_i = 1 \text{ pu}, \quad I^o_{qi} = 0 \text{ pu}$$

for all induction machines ($i = m + 1, \ldots, n$). The d- and q-axis current components of each machine are expressed as

$$I_{di} = \sum_{j=1}^{n} B_{ij}(-E'_{qj} \cos \delta_{ji} - E'_{dj} \sin \delta_{ji}) \tag{15}$$

$$I_{qi} = \sum_{j=1}^{n} B_{ij}(-E'_{qj} \sin \delta_{ji} + E'_{dj} \cos \delta_{ji}) \tag{16}$$

Linearizing these two relations around the above equilibrium point yields the following linearized current variables

$$\Delta I_d = -B\Delta E'_q \tag{17}$$

$$\Delta I_q = B\Delta E'_d - \bar{B}E\Delta\delta \tag{18}$$

where

$$\Delta I_d = \text{diag}(\Delta I_{di}), \quad \Delta I_q = \text{diag}(\Delta I_{qi})$$
$$\Delta E'_q = \text{diag}(\Delta E'_{qi}), \quad \Delta E'_d = \text{diag}(\Delta E'_{di})$$
$$\Delta\delta = \text{diag}(\Delta\delta_i), \quad E = \text{diag}(E^{'o}_{qi})$$

The matrix B is the susceptance matrix reduced to the n internal nodes behind transient reactances. The elements of the matrix $\bar{B}$ are related to those of B by the following expressions:

$$\bar{B}_{ij} = B_{ij} \tag{19}$$

$$\bar{B}_{ii} = -\sum_{j=1,\neq i}^{n} B_{ij}\frac{E^{'o}_{qj}}{E^{'o}_{qi}} \tag{20}$$

Upon substituting (17)-(18) into the linearized version of (8)-(11), two decoupled subsystems are obtained which will be referred to as the Pf and QV subsystems. These subsystems are described by the following matrix differential equations:

3.1 Pf Subsystem

$$\bar{T}'_{qo}\frac{d\Delta E'_d}{dt} = (-I_n + \bar{q}B)\Delta E'_d - \bar{q}\bar{B}E\Delta\delta \tag{21}$$

$$\frac{d\Delta\delta}{dt} = -\bar{\alpha}\bar{d}\bar{T}'^{-1}_{do}E^{-1}(B\Delta E'_d - \bar{B}E\Delta\delta) + \omega_s\Delta\nu \tag{22}$$

$$\bar{M}\frac{d\Delta\nu}{dt} = -E(B\Delta E'_d - \bar{B}E\Delta\delta) + \Delta T_m \tag{23}$$

3.2 QV Subsystem

$$\bar{T}'_{do}\frac{d\Delta E'_q}{dt} = (-I_n + \bar{d}B)\Delta E'_q + (I_n - \bar{\alpha})\Delta E_{fd} \tag{24}$$

where

$$\Delta\nu = \text{diag}(\Delta\nu_i), \quad \bar{T}'_{do} = \text{diag}(T'_{doi}), \quad \bar{T}'_{qo} = \text{diag}(T'_{qoi})$$
$$\bar{M} = \text{diag}(M_i), \quad \bar{\alpha} = \text{diag}(\alpha_i), \quad \bar{d} = \text{diag}(X_{di} - X'_{di})$$
$$\bar{q} = \text{diag}(X_{qi} - X'_{qi}), \quad I_n = n \times n \text{ identity matrix}$$

4 Aggregation of Synchronous and Induction Machines

Even though the dynamical structures of synchronous and induction machines are different, it is possible to describe a systemwide aggregation of electromechanical variables. This aggregation is useful in interpreting certain modes arising from the interaction of both types of machines. First, it is possible to transform (21)-(23) into an equivalent system resembling somewhat the structure of an n-machine system with only synchronous machines. To this effect, let us define the following linear transformation of angle

$$\Delta\theta = \Delta\delta - \bar{\alpha}\bar{d}\bar{T}_{do}^{\prime-1}\bar{M}E^{-2}\Delta\nu \tag{25}$$

The Pf subsystem transformed to the new variables is

$$\bar{T}_{qo}^{\prime}\frac{d\Delta E_d^{\prime}}{dt} = (-I_n + \bar{q}B)\Delta E_d^{\prime} - \bar{q}\bar{B}E(\Delta\theta + \bar{\alpha}\bar{D}\Delta\nu) \tag{26}$$

$$\frac{d\Delta\theta}{dt} = \omega_s\Delta\nu \tag{27}$$

$$\bar{M}\frac{d\Delta\nu}{dt} = -EB\Delta E_d^{\prime} + E\bar{B}E(\Delta\theta + \bar{\alpha}\bar{D}\Delta\nu) + \Delta T_m \tag{28}$$

where $\bar{D} = \bar{d}\bar{T}_{do}^{\prime-1}\bar{M}E^{-2}$. Due to the properties of the network matrix $K = E\bar{B}E$ [10] (that is, $K_{ij} = K_{ji}$ and $K_{ii} = -\sum_{j=1,\neq i}^{n} K_{ij}$), there are two zero eigenvalues in this subsystem which can be attributed to electromechanical center-of-inertia variables $\Delta\theta_c$ and $\Delta\nu_c$ defined by

$$\Delta\theta_c = \frac{\sum_{i=1}^{n} M_i\Delta\theta_i}{\sum_{i=1}^{n} M_i} \tag{29}$$

$$\Delta\nu_c = \frac{\sum_{i=1}^{n} M_i\Delta\nu_i}{\sum_{i=1}^{n} M_i} \tag{30}$$

For $i = m+1,\ldots,n$, the variables $\Delta E_{di}^{\prime}$ are identically zero and the two differential equations describing the center-of-inertia variables are

$$\frac{d\Delta\theta_c}{dt} = \omega_s\Delta\nu_c \tag{31}$$

$$M_c\frac{d\Delta\nu_c}{dt} = -\sum_{i=1}^{m} B_{io}\Delta E_{di}^{\prime} + \Delta T_{mc} \tag{32}$$

where $M_c = \sum_{i=1}^{n} M_i$, $\Delta T_{mc} = \sum_{i=1}^{n} \Delta T_{mi}$, and $B_{io} = \sum_{j=1}^{n} E_j B_{ij}$. If we assume that $B_{io} \cong 0$, then the electromechanical subsystem $(\Delta\theta_c, \Delta\nu_c)$ is near-aggregable with respect to (29)-(30) yielding the near-decoupled differential subsystem

$$\frac{d\Delta\theta_c}{dt} = \omega_s\Delta\nu_c \tag{33}$$

$$M_c\frac{d\Delta\nu_c}{dt} \cong \Delta T_{mc} \tag{34}$$

Assuming that $T'_{qoi} = T'_{qo}$ for all synchronous machines ($i = 1, \ldots, m$) and $B_{io} \cong 0$, the Pf is also near-aggregable with respect to the variable

$$\Delta E'_{dc} = \frac{\sum_{i=1}^{m} q_i^{-1} E_i \Delta E'_{di}}{\sum_{i=1}^{m} q_i^{-1} E_i} \tag{35}$$

This means that another real mode of the Pf subsystem can be approximately found from the following differential equation

$$T'_{qo} \frac{d\Delta E'_{dc}}{dt} \cong -\Delta E'_{dc} \tag{36}$$

Thus, one mode is almost equal to the negative of the reciprocal of each individual synchronous machine q-axis open-circuit time constant. Equations (33)-(34) and (36) are identical to the Pf equations of a single synchronous machine linearized around the no-load case [11].

Unfortunately, the widely-different time constants of synchronous and induction machines prevent a similar natural aggregation of the q-axis voltage variables from occurring in (24). Such an aggregation would approximately occur if all the d-axis open-circuit time constants were equal and if the matrix B is approximately a network matrix. Thus, the $\Delta E'_q$ variables may interact without a well-defined physical structure in a power system including induction machines.

5 Two-Machine Power System Example

The global aggregation concepts of the last section are first illustrated in a simple two-machine system composed of a synchronous generator connected to an induction motor through transformers and an inductive transmission line as shown in Figure 1. The series reactance X_{12} includes the reactance of the transmission line plus the series reactances of transformers at both ends of the transmission line. The induction motor is in parallel with a capacitor bank with reactance $X_c = -X_{d2}$. Thus, there is no current flowing in the transmission line and the capacitor provides local reactive power support for the induction machine.

A linearized analysis of the four cases displayed in Tables A.1 and A.2 in Appendix A is performed. This analysis is useful in revealing the different dynamic interaction of a large synchronous machine with different size induction machines. Case 1 uses a large aggregate of 500-hp induction machines and Case 2 uses an aggregate motor load 100 times smaller in size. Case 3 uses a large aggregate of 50-hp machines whereas Case 4 uses an aggregate motor load 100 times smaller. The eigenvalues of the two-machine system at no-load are shown in Table 1 for Case 1 through Case 4.

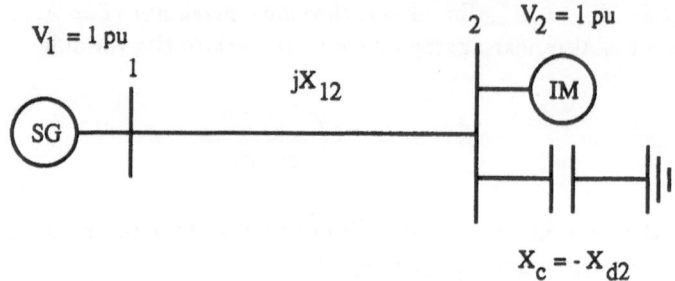

Fig. 1. Two-Bus Two-Machine Power System

Table 1   Eigenvalues of Two-Machine System

	Small I.M.		Large I.M.	
	Case 1	Case 2	Case 3	Case 4
E'_{q1}	-0.165	-0.167	-0.163	-0.166
E'_{q2}	-37.2	-94.9	-10.7	-25.1
E'_{d1}	-1.87	-1.85	-1.86	-1.85
θ_{12}	-20.1+j14.6	-112.	-6.62+j27.1	-12.8+j39.6
ν_{12}	-20.1-j14.6	-17.9	-6.62-j27.1	-12.8-j39.6
θ_c	0.0	0.0	0.0	0.0
ν_c	0.0	0.0	0.0	0.0

The no-load linearization of this simple system reveals the following dynamic interaction of state variables:

− Two zero modes are due to the center-of-area electromechanical variables.

− One real negative mode is approximately due to the synchronous q-axis damper variable.

− Two real negative modes are due to the interaction of q-axis voltage variables.

− One complex pair of eigenvalues (or two real negative eigenvalues for a small-size load with small induction machines) is due to the interaction of difference electromechanical variables.

This example shows that induction machines can exhibit intermachine oscillations which contribute to the system frequency and to electromechanical modes of oscillation. In system studies, this large rotating load inertia cannot be simply neglected or approximated by modifying generator inertia constants. In such cases, the damping of electromechanical modes of oscillation may not be captured correctly.

6 Six-Machine Power System Example

6.1 Machine and Power System Data

The above aggregation concepts are now illustrated on a more realistic power system composed of three synchronous generators and three aggregate induction motor loads as shown in Figure 2. The line data for this well-known system [8,9] is given in Table A.3 in Appendix A.

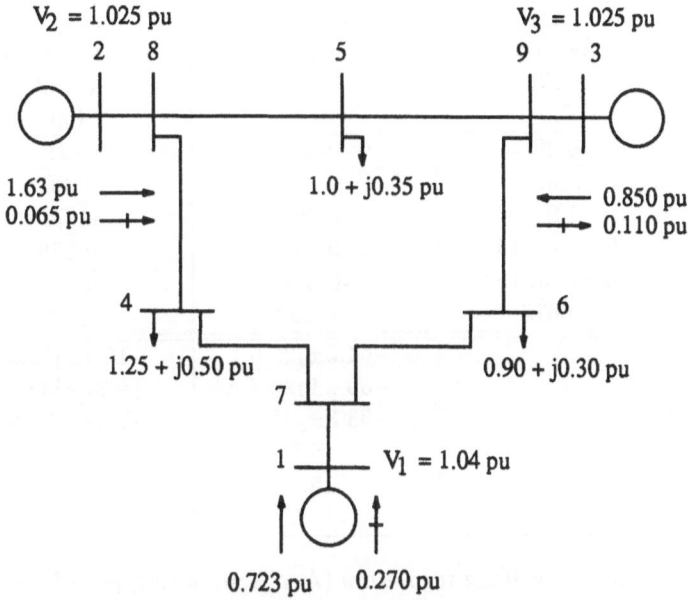

Fig. 2. Nine-Bus Six-Machine Power System

The machine data for the three synchronous generators ($i = 1, 2, 3$) and the three aggregate induction motor loads ($i = 4, 5, 6$) is given in Tables A.4 and A.5 in Appendix A. Every induction machine is in series with a transformer reactance $X_T = 0.05$ pu which has not been included in the machine data.

6.2 Unregulated Power System Analysis

An eigenvalue analysis of the linearized six-machine power system model reveals that there are three voltage modes due mainly to the three synchronous machine field winding flux linkage variables E'_{qi}, i=1,2,3. The three real parts of these three modes are monitored as we increase the percentage of induction motor loads with small and large machines. In Figures 3 and 4, the fastest voltage mode is the most sensitive to the percentage of induction machine loading and moves to the unstable region at a certain percentage of induction motor loads for both cases while the other two modes stay in the stable region. This critical mode moves slightly faster with large induction motor loads than with small induction motor loads. The eigenvalues of the six-machine system with 100% impedance loads or 100% small or large induction motor loads are shown in Table 2.

Table 2   System Eigenvalues with 100% Impedance Loads
or 100% Small or Large Induction Motor Loads

System Eigenvalues		
Impedance Loads	Small I.M. Loads	Large I.M. Loads
-1.11±j13.0	-1.16±j13.0	-1.14±j13.0
-1.39±j8.98	-1.39±j9.00	-1.37±j9.03
-0.172	-0.172	-0.172
0.0	0.0	0.0
-6.29	-6.24	-6.19
-4.62	-4.08	-4.06
-1.90	-2.58	-1.94
-0.45+j0.15	-0.347	-0.328
-0.45-j0.15	-0.277	-0.289
-0.0675	+0.229	+0.344
	-64.3±j22.6	-10.1±j33.5
	-60.2±j22.7	-9.77±j32.4
	-33.3±j23.1	-7.49±j26.5
	-15.5	-15.9
	-15.0	-15.0
	-11.5	-9.74

Without automatic voltage regulators (A.V.R.s), both types of loads are unstable for large percentages of either small or large induction motors. To determine the cause of the instability, a two-machine equivalent is extracted by physically aggregating the three synchronous generators into a synchronous machine equivalent and by aggregating the three induction motors into an induction machine equivalent using the aggregation theory developed in [10,11].

The following definitions are useful in building the transformation that displays the local and global variables once the number of equivalent machines with the same structure has been decided upon. Let U be an 6×2 partition matrix given by

$$U = \begin{bmatrix} 1 & 1 & 1 & 0 & 0 & 0 \\ 0 & 0 & 0 & 1 & 1 & 1 \end{bmatrix}^t \tag{37}$$

Let G be a 4×6 matrix given by

$$G = \begin{bmatrix} 1 & -1 & 0 & 0 & 0 & 0 \\ 1 & 0 & -1 & 0 & 0 & 0 \\ 0 & 0 & 0 & 1 & -1 & 0 \\ 0 & 0 & 0 & 1 & 0 & -1 \end{bmatrix} \tag{38}$$

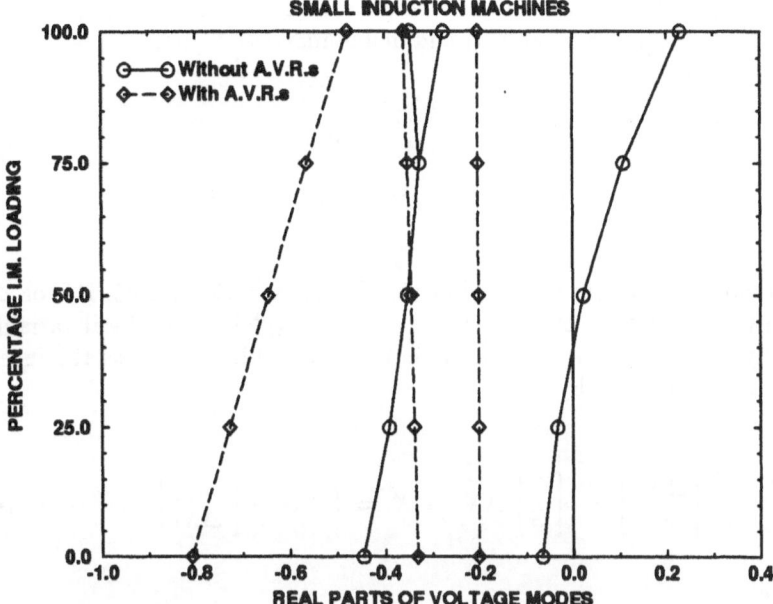

Fig. 3. Trajectories of Real Parts of Three Voltage Modes as Functions of the Percentage of Small Induction Machine Loading

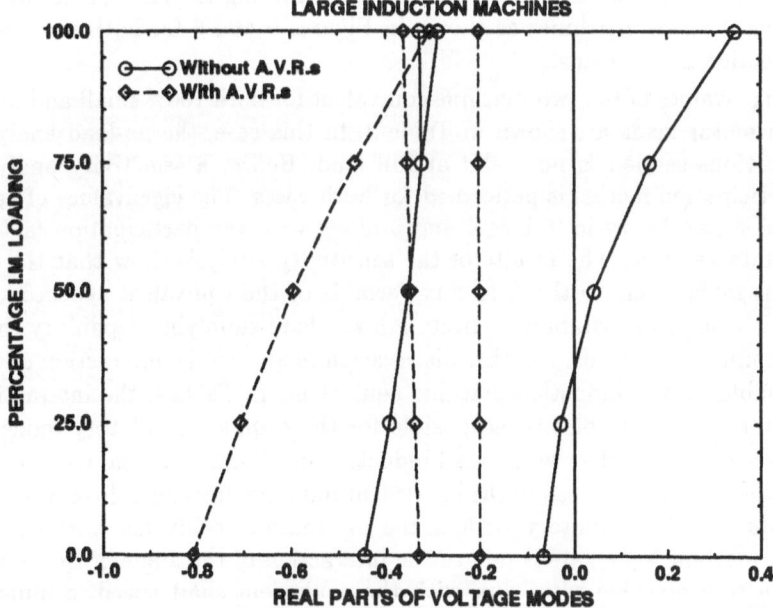

Fig. 4. Trajectories of Real Parts of Three Voltage Modes as Functions of the Percentage of Large Induction Machine Loading

Let T be a matrix of modified time constants ($T = \mathrm{diag}(T'_{qoi}/(X_{di} - X'_{di})$, $\mathrm{diag}(T'_{qoi}/(X_{qi} - X'_{qi})$, or $\mathrm{diag}(M_i)$) and define matrices C and G^+ by

$$C = (U^t T U)^{-1} U^t T \tag{39}$$
$$G^+ = T^{-1} G^t (G T^{-1} G^t)^{-1} \tag{40}$$

Let ΔX represent a vector of a kind of variable ($\Delta X = \Delta E'_q$, $\Delta E'_d$, $\Delta\delta$, or $\Delta\nu$) and let ΔX_a and ΔX_l be the associated aggregate (global) and local variables, respectively. The transformation to these latter variables as well as the inverse transformation are given by

$$\begin{bmatrix} \Delta X_a \\ \Delta X_l \end{bmatrix} = \begin{bmatrix} C \\ G \end{bmatrix} \Delta X, \quad \Delta X = \begin{bmatrix} U & G^+ \end{bmatrix} \begin{bmatrix} \Delta X_a \\ \Delta X_l \end{bmatrix} \tag{41}$$

provided the corresponding matrices of time constants are used each time. It is recalled that aggregate variables are averaged center-of-area variables and local variables are difference variables with respect to a chosen reference machine for each type of machine. The instability observed in the full-order model is also captured by the two-machine equivalent representing the global systemwide interaction of all six machines as shown in Figures 5 and 6 for both small and large induction motor loads.

The eigenvalues of the two-machine equivalent for both 100% small and large induction motor loads are shown in Table 3. In this case, the no-load analysis of the previous section is not valid at full load. Hence, a sensitivity analysis using participation factors is performed for both cases. The eigenvalues of each model are shown below in Tables 4 and 5 along with the participation factors of each state variable. The results of the sensitivity analysis show that the instability is mainly due to the field flux variable of the equivalent synchronous generator. Thus, this instability reflects a lower load-supplying capability than with pure impedance loads. Another observation relates to the interaction of the state variables in the induction machine equivalent. In Table 4, the interaction between rotor flux variables is responsible for the complex oscillatory mode in the two-machine equivalent with small induction machines. The real mode -11.6 is due to the rotor shaft speed of the equivalent induction machine. However, Table 5 shows that the oscillatory mode in the two-machine equivalent with a large induction machine aggregate is due to the interaction of the angle of the equivalent induction machine rotor flux with the equivalent shaft speed. Similarly, the real mode -9.60 is mainly due to the rotor flux magnitude in the induction machine equivalent. This latter mode interacts strongly with the q-axis damper variable of the global synchronous generator as observed in the 2×2 submatrix corresponding to $\lambda_{a3}, \lambda_{a4}$ in Table 3.

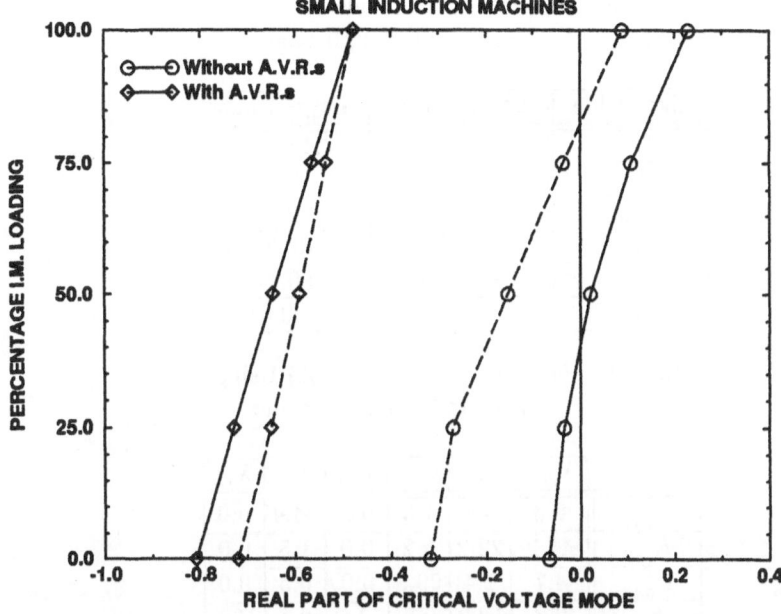

Fig. 5. Trajectories of Real Part of Critical Voltage Mode as Function of the Percentage of Small Induction Machine Loading. (Solid Line: Full-Order Model, Dashed Line: Aggregate Model.)

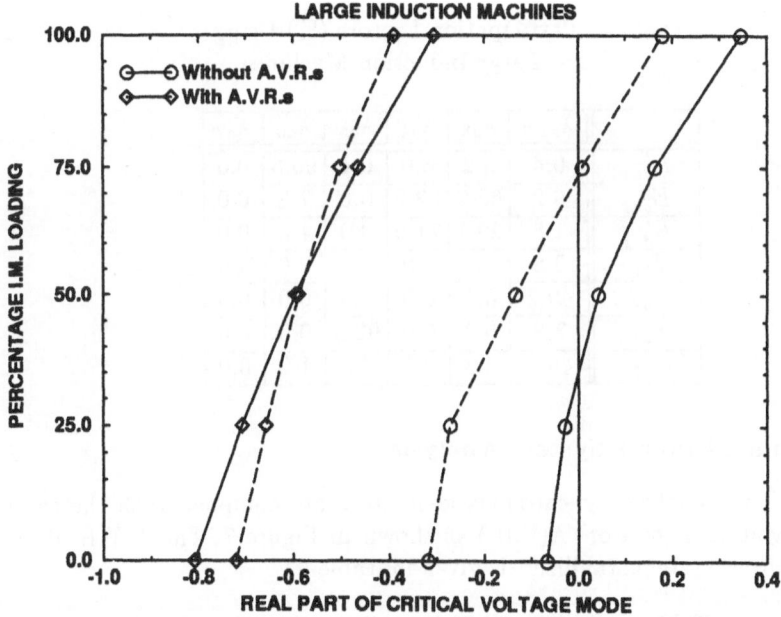

Fig. 6. Trajectories of Real Part of Critical Voltage Mode as Function of the Percentage of Large Induction Machine Loading. (Solid Line: Full-Order Model, Dashed Line: Aggregate Model.)

Table 3   Eigenvalues of Two-Machine
Equivalent at Full Load

Eigenvalue	Small I.M.s	Large I.M.s
$\lambda_{a1,2}$	-33.6±j22.9	-7.40±j26.6
λ_{a3}	-11.6	-9.60
λ_{a4}	-2.93	-2.20
λ_{a5}	-0.284	-0.294
λ_{a6}	+0.089	+0.179
λ_{a7}	0.0	0.0

Table 4   Participation Factors (%) Using
Small Induction Machines

	$\lambda_{a1,2}$	λ_{a3}	λ_{a4}	λ_{a5}	λ_{a6}	λ_{a7}
E'_{qa1}	0.3	3.7	1.5	0.3	94.4	0.0
E'_{qa2}	43.2	22.2	5.2	0.0	1.5	0.0
E'_{da1}	4.7	2.4	99.9	0.0	0.5	0.0
δ_{a1}	2.6	12.3	0.8	11.0	0.1	100.
δ_{a2}	66.3	44.1	0.8	11.0	0.1	0.0
ν_{a1}	2.6	12.6	0.9	88.8	0.0	0.0
ν_{a2}	20.9	95.8	5.7	10.9	3.5	0.0

Table 5   Participation Factors (%) Using
Large Induction Machines

	$\lambda_{a1,2}$	λ_{a3}	λ_{a4}	λ_{a5}	λ_{a6}	λ_{a7}
E'_{qa1}	0.4	3.2	6.0	0.1	90.3	0.0
E'_{qa2}	6.2	68.9	22.2	0.0	7.2	0.0
E'_{da1}	1.8	29.1	71.9	0.0	1.3	0.0
δ_{a1}	3.8	0.3	0.0	7.6	0.0	100.
δ_{a2}	49.3	5.1	0.0	7.6	0.0	0.0
ν_{a1}	3.8	0.3	0.0	92.3	0.0	0.0
ν_{a2}	45.9	3.3	0.1	7.6	1.2	0.0

6.3   Regulated Power System Analysis

In this section, the three synchronous generators are equipped with the same
automatic voltage regulator (A.V.R.) as shown in Figure 7. The A.V.R. data,
without a power system stabilizer, is given in Table 6.

Table 6   Automatic Voltage Regulator Data

T_A	T_E	T_F	K_A	K_E	K_F	A_{sat}	B_{sat}
0.06	0.5	1.0	25	-0.0445	0.16	0.1123	0.3043

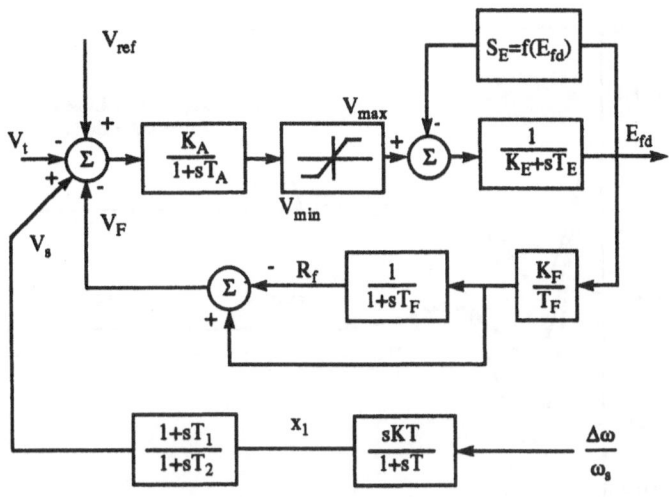

Fig. 7. Automatic Voltage Regulator

Table 7   System Eigenvalues with 100% Impedance Loads
or 100% Small or Large Induction Motor Loads

System Eigenvalues		
Impedance Loads	Small I.M. Loads	Large I.M. Loads
-1.09±j13.0	-1.14±j13.0	-1.13±j13.0
-1.37±j8.97	-1.37±j8.97	-1.35±j9.02
-0.319	-0.285	-0.297
0.0	0.0	0.0
-8.63±j8.18	-8.81±j8.46	-8.83±j8.46
-8.81±j8.49	-8.86±j8.58	-8.85±j8.21
-8.87±j8.59	-8.87±j8.06	-8.87±j8.58
-6.46	-6.41	-6.36
-4.77	-4.05	-4.14
-1.45	-2.35	-1.63
-0.81±j1.14	-0.48±j1.16	-0.31±j1.31
-0.33±j0.50	-0.36±j0.50	-0.36±j0.50
-0.20±j0.38	-0.20±j0.38	-0.20±j0.38
	-64.4±j22.6	-10.1±j33.5
	-60.2±j22.7	-9.77±j32.4
	-33.3±j23.1	-7.50±j26.5
	-15.5	-15.9
	-14.9	-15.0
	-11.1	-9.51

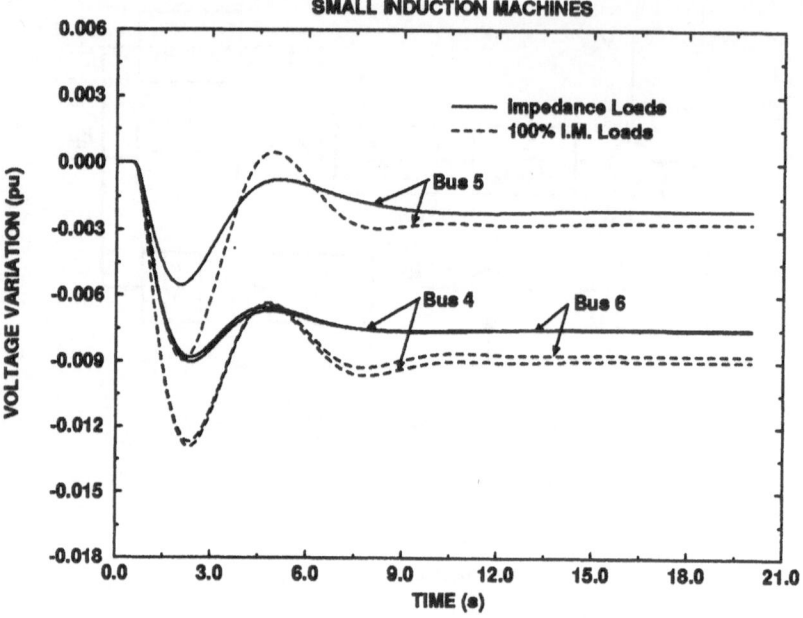

Fig. 8. Voltage Variations at Load Busses for a 1.5% Reduction in the Voltage Reference Setpoint of Synchronous Generator #1. (Solid Line: Impedance Loads, Dashed Line: 100% Small Induction Machine Loads.)

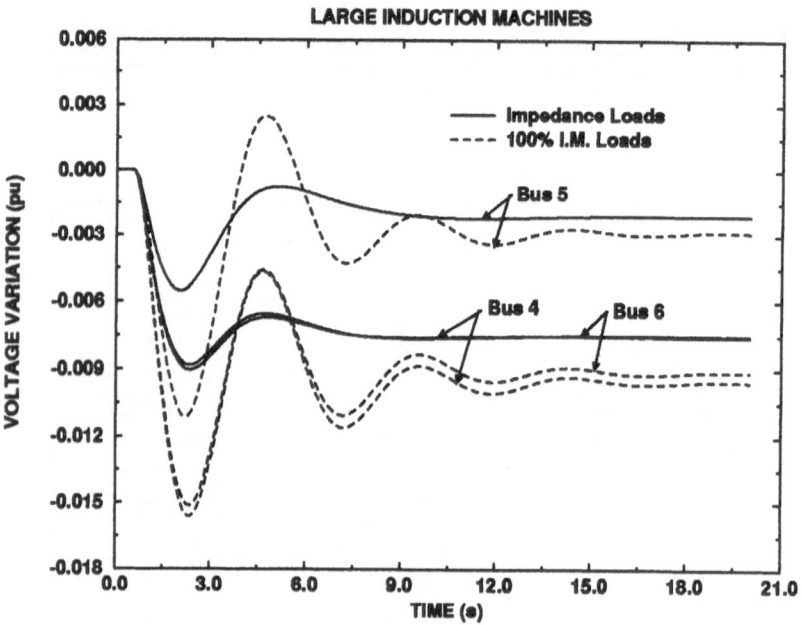

Fig. 9. Voltage Variations at Load Busses for a 1.5% Reduction in the Voltage Reference Setpoint of Synchronous Generator #1. (Solid Line: Impedance Loads, Dashed Line: 100% Large Induction Machine Loads.)

The eigenvalues of the six-machine system with 100% impedance loads or 100% small or large induction motor loads are shown in Table 7. In particular, the eigenvalues due to the synchronous generators and A.V.R. variables and those due to the induction motor variables have been separated. There are three system voltage modes which result from the interaction of the three synchronous generator field flux linkage variables E'_{qi}, i=1,2,3, and the three feedback compensator variables R_{fi}, i=1,2,3. These three complex voltage modes have also been grouped together in Table 7.

The real parts of the three voltage modes are again monitored as the percentage of induction motor loading is increased as shown in Figures 3 and 4. The regulated system is stable with large percentages of induction motor loads, but the critical voltage mode still possesses the tendency of moving towards the unstable region as in the unregulated case. The tendency of this critical mode to move to the right-half plane is well captured by a two-machine equivalent and is shown in Figures 5 and 6. To test the voltage sensitivity of the load buses with induction motor loads, the three load bus voltages are monitored for impedance loads and for 100% induction motor loads after a 1.5% drop in the A.V.R. reference voltage setpoint of synchronous generator #1. The results are shown in Figures 8 and 9. The voltages of the two load buses #4 and #5 adjacent to the disturbance are more affected than the relatively remote load bus #5. For the same voltage disturbance, the voltage drop at the load buses is higher if impedance loads are replaced by induction motor loads. As seen in Figure 8 and 9, when the loads are represented by large induction motors, the load-bus voltages oscillate more violently and drop more compared to small induction motors.

7 Conclusion

In this paper, the authors have investigated the dynamic interaction of synchronous generators with small and large induction machines in a multimachine power system. A no-load linearization illustrates the dynamical interaction of electromechanical and voltage variables in a two-machine power system consisting of a synchronous generator connected to an induction motor. These aggregation concepts are extended to a realistic six-machine system composed of three synchronous generators and three aggregate induction motor loads. In general, large percentages of induction machine loading are detrimental to the power system stability. The behavior of the critical mode with increased induction motor loading is explained using a two-machine equivalent consisting of an aggregate synchronous generator connected to an aggregate induction motor load. A sensitivity analysis shows that the dynamic interaction of synchronous generators with small induction motors is different from that with large motors. Induction motor loads are also more sensitive than impedance loads to voltage disturbances. These results are verified by simulating the load bus voltages following a small change in the reference voltage setpoint of a synchronous generator.

8 Aknowledgements

This research was supported in part by National Science Foundation Grant ECS-9058174 and in part by a scholarship from U.S. AID/AMIDEAST.

Nomenclature

E'_q	q-axis rotor voltage variable (pu)
E'_d	d-axis rotor voltage variable (pu)
I_d, I_q	d-, q-axis machine current (pu)
i_d, i_q	d-, q-axis system current (pu)
V_d, V_q	d-, q-axis machine voltage (pu)
v_d, v_q	d-, q-axis system voltage (pu)
X_d, X_q	d-, q-axis synchronous reactance (pu)
X'_d, X'_q	d-, q-axis transient reactance (pu)
T'_{do}, T'_{qo}	d-, q-axis open-circuit time constant (s)
δ	angle of q-axis w.r.t. system reference (rad)
ν	Normalized rotor shaft speed (pu)
ω_s	synchronous speed (rad/s)
H, M	Inertia constant (s) ($M = 2H$)
D	Damping coefficient (pu)
T_m	Input mechanical torque (pu)
T_L	Output load torque (pu)
$G_{ij} + jB_{ij}$	(i,j)th element of reduced admittance matrix (pu)

References

1. IEEE Task Force Report: Load Representation for Dynamic Performance Analysis. Paper 92 WM 126-3 PWRD, IEEE/PES 1992 Winter Meeting, New York, January 26-30, 1992.
2. Concordia, C. and Ihara, S.: Load Representation in Power System Stability Studies. IEEE Transactions on Power Apparatus and Systems **PAS-101** (1982) 969-977.
3. EPRI Report EL-2043: Application of Induction Generators in Power Systems, Final Report on Research Project 1945-1, Prepared by Power Technologies, Inc., October 1981.
4. EPRI Report EL-6126: Induction Motor Modeling in Stability Simulation, Final Report on Research Project 2473-2, Prepared by Minnesota Power, December 1988.
5. Brereton, D. S., Lewis, D. G. and Young, C. C. Young: Representation of Induction-Motor Loads During Power-System Stability Studies, AIEE Transactions **76** (1957) 451-461.
6. Ahmed-Zaid, S. and Taleb, M.: Structural Modeling of Small and Large Induction Machines Using Integral Manifolds, IEEE Transactions on Energy Conversion **6** (1991) 529-535.

7. Ahmed-Zaid, S., Taleb, M. and Price, W. W. Price: First-Order Induction Machine Models Near Voltage Collapse, 1991 International Conference on Bulk Power System Voltage Phenomena – Voltage Stability and Security, Deep Creek Lake, MD, August 4-7, 1991.

8. Anderson, P. M. and Fouad, A. A. Fouad: Power System Stability and Control. Ames, IA: Iowa University Press (1977).

9. Winkelman, J. R., Chow, J. H., Allemong, J. J. and Kokotovic, P. V. : Multi-Time-Scale Analysis of a Power System, Automatica **16** (1980) 35-43.

10. Chow, J. H. , Ed.: Time-Scale Modeling of Dynamic Networks with Applications to Power Systems. Lecture Notes in Control and Information Sciences, vol. 46. New York: Springer-Verlag (1982).

11. Ahmed-Zaid, S., Sauer, P. W. and Winkelman, J. R.: Higher Order Dynamic Equivalents for Power Systems, Automatica **22** (1986) 489-494.

Appendix A: Power System and Machine Data

Table A.1   Synchronous Machine Data in Two-Machine System

X_{12} (pu)	X_{d1} (pu)	X_{q1} (pu)	X'_{d1} (pu)	X'_{q1} (pu)	T'_{do1} (s)	T'_{qo1} (s)	H_1 (s)
0.2	0.8958	0.8645	0.1198	0.1198	6.00	0.540	6.40

Table A.2   Induction Machine Data in Two-Machine System

	Small I.M.s		Large I.M.s	
Parameter	Case 1	Case 2	Case 3	Case 4
X_{d2} (pu)	2.360	23.60	3.890	38.90
X_{q2} (pu)	0.1050	1.050	0.1680	1.680
X'_{d2} (pu)	0.1050	1.050	0.1680	1.680
X'_{q2} (pu)	0.1050	1.050	0.1680	1.680
T'_{do2} (s)	0.1557	0.1557	0.7834	0.7834
T'_{qo2} (s)	0.1557	0.1557	0.7834	0.7834
H_2 (s)	0.7916	0.07916	0.5269	0.05269

Table A.3   Line Data for Nine-Bus System

From	To	R (pu)	X (pu)	B/2 (pu)
1	7	0	0.0576	0
2	8	0	0.0625	0
3	9	0	0.0586	0
4	7	0.01	0.085	0.088
4	8	0.032	0.161	0.153
5	8	0.0085	0.072	0.0745
5	9	0.0119	0.1008	0.1045
6	9	0.039	0.170	0.179
6	7	0.017	0.092	0.079

Table A.4 Synchronous Machine Data in Six-Machine System

Machine	1	2	3
H (s)	11.15	6.4	3.01
D (pu)	9.6	2.5	1.0
X_d (pu)	0.765	0.8958	1.3125
X_q (pu)	0.714	0.8645	1.2578
X_d' (pu)	0.113	0.1198	0.1813
X_q' (pu)	0.113	0.1198	0.1813
T_{do}' (s)	4.8	6.0	5.89
T_{qo}' (s)	0.500	0.535	0.600

Table A.5 Induction Machine Data in Six-Machine System

Machine	Small I.M.s			Large I.M.s		
	4	5	6	4	5	6
X_d (pu)	1.89	2.36	2.62	3.11	3.89	4.32
X_d' (pu)	0.084	0.105	0.117	0.134	0.168	0.187
X_q (pu)	0.084	0.105	0.117	0.134	0.168	0.187
X_q' (pu)	0.084	0.105	0.117	0.134	0.168	0.187
T_{do}' (s)	0.156	0.156	0.156	0.783	0.783	0.783
T_{qo}' (s)	0.156	0.156	0.156	0.783	0.783	0.783
H (s)	0.990	0.792	0.712	0.659	0.530	0.474

Appendix B: Induction Motor Model

A simplified model of a single-cage induction machine is described by the following algebraic-differential equations written in a synchronously-rotating reference frame with stator transients neglected

$$(v_d + jv_q) = (R_s + jX_d')(i_d + ji_q) + j(e_q' - je_d')$$

$$T_{do}' \frac{de_q'}{dt} = -e_q' + (X_d - X_d')i_d + T_{do}'\omega_s(\nu - 1)e_d'$$

$$T_{do}' \frac{de_d'}{dt} = -e_d' - (X_d - X_d')i_q - T_{do}'\omega_s(\nu - 1)e_q'$$

$$M \frac{d\nu}{dt} = (e_d'i_d + e_q'i_q) - T_L$$

Using the following transformations,

$$E_q' = \sqrt{e_d'^2 + e_q'^2}, \quad \delta = \tan^{-1}\frac{-e_d'}{e_q'}, \quad T_m = -T_L$$

$$(I_d + jI_q) = -(i_d + ji_q)e^{-j\delta}$$

$$(V_d + jV_q) = (v_d + jv_q)e^{-j\delta}$$

a modified third-order induction machine model is, in a reference frame rotating now with the resultant rotor flux,

$$(V_d + jV_q) = -(R_s + jX'_d)(I_d + jI_q) + jE'_q$$

$$T'_{do}\frac{dE'_q}{dt} = -E'_q - (X_d - X'_d)I_d$$

$$\frac{d\delta}{dt} = \omega_s(\nu - 1) - \frac{(X_d - X'_d)}{T'_{do}E'_q}I_q$$

$$M\frac{d\nu}{dt} = T_m - E'_q I_q$$

This article was processed using the LaTeX macro package with LLNCS style

Lecture Notes in Control and Information Sciences

Edited by M. Thoma

1992–1995 Published Titles:

Vol. 180: Kall, P. (Ed.)
System Modelling and Optimization.
Proceedings of the 15th IFIP Conference,
Zurich, Switzerland, September 2-6, 1991
969 pp. 1992 [3-540-55577-3]

Vol. 181: Drane, C.R.
Positioning Systems - A Unified Approach
168 pp. 1992 [3-540-55850-0]

Vol. 182: Hagenauer, J. (Ed.)
Advanced Methods for Satellite and Deep
Space Communications. Proceedings of
an International Seminar Organized by
Deutsche Forschungsanstalt für Luft-und
Raumfahrt (DLR), Bonn, Germany,
September 1992
196 pp. 1992 [3-540-55851-9]

Vol. 183: Hosoe, S. (Ed.)
Robust Control. Proceesings of a Workshop
held in Tokyo, Japan, June 23-24, 1991
225 pp. 1992 [3-540-55961-2]

Vol. 184: Duncan, T.E.; Pasik-Duncan, B.
(Eds)
Stochastic Theory and Adaptive Control.
Proceedings of a Workshop held in
Lawrence, Kansas, September 26-28,
1991
500 pp. 1992 [3-540-55962-0]

Vol. 185: Curtain, R.F. (Ed.); Bensoussan,
A.; Lions, J.L.(Honorary Eds)
Analysis and Optimization of Systems:
State and Frequency Domain Approaches
for Infinite-Dimensional Systems.
Proceedings of the 10th International
Conference, Sophia-Antipolis, France, June
9-12, 1992.
648 pp. 1993 [3-540-56155-2]

Vol. 186: Sreenath, N.
Systems Representation of Global Climate
Change Models. Foundation for a Systems
Science Approach.
288 pp. 1993 [3-540-19824-5]

Vol. 187: Morecki, A.; Bianchi, G.;
Jaworeck, K. (Eds)
RoManSy 9: Proceedings of the Ninth
CISM-IFToMM Symposium on Theory and
Practice of Robots and Manipulators.
476 pp. 1993 [3-540-19834-2]

Vol. 188: Naidu, D. Subbaram
Aeroassisted Orbital Transfer: Guidance
and Control Strategies
192 pp. 1993 [3-540-19819-9]

Vol. 189: Ilchmann, A.
Non-Identifier-Based High-Gain Adaptive
Control
220 pp. 1993 [3-540-19845-8]

Vol. 190: Chatila, R.; Hirzinger, G. (Eds)
Experimental Robotics II: The 2nd
International Symposium, Toulouse,
France, June 25-27 1991
580 pp. 1993 [3-540-19851-2]

Vol. 191: Blondel, V.
Simultaneous Stabilization of Linear
Systems
212 pp. 1993 [3-540-19862-8]

Vol. 192: Smith, R.S.; Dahleh, M. (Eds)
The Modeling of Uncertainty in Control
Systems
412 pp. 1993 [3-540-19870-9]

Vol. 193: Zinober, A.S.I. (Ed.)
Variable Structure and Lyapunov Control
428 pp. 1993 [3-540-19869-5]

Vol. 194: Cao, Xi-Ren
Realization Probabilities: The Dynamics of
Queuing Systems
336 pp. 1993 [3-540-19872-5]

Vol. 195: Liu, D.; Michel, A.N.
Dynamical Systems with Saturation
Nonlinearities: Analysis and Design
212 pp. 1994 [3-540-19888-1]

Vol. 196: Battilotti, S.
Noninteracting Control with Stability for
Nonlinear Systems
196 pp. 1994 [3-540-19891-1]